Analytical Transmission Electron Microscopy

Jürgen Thomas • Thomas Gemming

Analytical Transmission Electron Microscopy

An Introduction for Operators

 Springer

Jürgen Thomas
Leibniz Institute for Solid State and
Materials Research (IFW) Dresden
Germany

Thomas Gemming
Leibniz Institute for Solid State and
Materials Research (IFW) Dresden
Germany

This book is the revised and slightly expanded translation of the German book edition by Jürgen Thomas and Thomas Gemming: "Analytische Transmissionselektronenmikroskopie— Eine Einführung für den Praktiker", Springer-Verlag, Wien 2013, ISBN 978-3-7091-1439-1.

ISBN 978-94-017-8600-3   ISBN 978-94-017-8601-0 (eBook)
DOI 10.1007/978-94-017-8601-0
Springer Dordrecht Heidelberg New York London

Library of Congress Control Number: 2014932544

© Springer Science+Business Media Dordrecht 2014
This work is subject to copyright. All rights are reserved by the Publisher, whether the whole or part of the material is concerned, specifically the rights of translation, reprinting, reuse of illustrations, recitation, broadcasting, reproduction on microfilms or in any other physical way, and transmission or information storage and retrieval, electronic adaptation, computer software, or by similar or dissimilar methodology now known or hereafter developed. Exempted from this legal reservation are brief excerpts in connection with reviews or scholarly analysis or material supplied specifically for the purpose of being entered and executed on a computer system, for exclusive use by the purchaser of the work. Duplication of this publication or parts thereof is permitted only under the provisions of the Copyright Law of the Publisher's location, in its current version, and permission for use must always be obtained from Springer. Permissions for use may be obtained through RightsLink at the Copyright Clearance Center. Violations are liable to prosecution under the respective Copyright Law.
The use of general descriptive names, registered names, trademarks, service marks, etc. in this publication does not imply, even in the absence of a specific statement, that such names are exempt from the relevant protective laws and regulations and therefore free for general use.
While the advice and information in this book are believed to be true and accurate at the date of publication, neither the authors nor the editors nor the publisher can accept any legal responsibility for any errors or omissions that may be made. The publisher makes no warranty, express or implied, with respect to the material contained herein.

Printed on acid-free paper

Springer is part of Springer Science+Business Media (www.springer.com)

Preface

What is the goal of a transmission electron microscope (TEM)? It is expensive and causes large running costs. The understanding of the results is sometimes difficult and they can be wrongly interpreted. It needs specialists and, connected with that, causes additional labour costs.

On the other hand, it is a microscope, i.e. its results are magnified pictures and everybody is able to understand pictures. Apparently, there are no problems understanding them. Why do we need an additional textbook about this topic?

Goethe: „Mikroskope und Fernrohre verwirren eigentlich den reinen Menschensinn." [1] (Microscopes and telescopes confuse the human mind.)

However, analytical transmission electron microscopy does not only include the microscopic imaging. Electron diffraction and compositional analysis by spectrometers for X-rays and energy losses of the electrons complement it. The analytical transmission electron microscope covers four challenging methods: electron microscopic imaging, electron diffraction, analysis of characteristic X-rays, and electron energy loss spectroscopy. There are specialists for each of these methods; nevertheless the operator at the microscope should hold an overview about the possibilities of analytical transmission electron microscopy. He should be able to handle them and should be familiar with the basics of the interpretation of the measured results.

It is the goal of this textbook to give such an overview. The idea to write this book was arising during the work in our electron microscopic laboratory of the Leibniz Institute for Solid State and Materials Research (IFW) Dresden. While teaching the students of materials science in lectures and practical courses about analytical transmission electron microscopy as well as while instructing graduate students, Ph.D. students, and technicians at the microscope we have attained experiences regarding frequently asked questions by the beginners and the didactical procedure to explain the function and the practical handling of a transmission electron microscope.

These experiences are integrated in this book. It addresses people who want or have to work at a transmission electron microscope and have not yet a further knowledge about this topic. The book should also be helpful for service engineers to ensure an overview of the basic knowledge about the microscopes maintained by them. Additionally, the book should be also a little bit amusing to receive interest in electron microscopy from non-experts, too.

The focus lies on explanations with the help of simple models and on hints for the practical electron microscopic work. The headlines of the chapters already indicate this matter. This is the difference to other introductions into electron microscopy. As far as practicable, we tried to avoid explanations based only on mathematical formalisms.

In this context we would like to give a comment on the model assumptions in physics: On the one hand, we speak about electrons as particles, on the other hand as waves. Or the position of the specimen within the electron microscope: Sometimes we draw the specimen outside of the objective lens, sometimes within the magnetic field of the lens. One or two of the readers will see a discrepancy here. However, it is not a discrepancy but a property of models in physics suitable to explain special features in nature. Dependent on the experimental setup we observe sometimes the particle and otherwise the wave character of the electrons. Or: To explain the multi-stage imaging of the electron microscope we draw the specimen outside of the magnetic field since the beam path within the lens does not play any role in this case. Only the beam path outside of the lens is important. On the other hand, when we discuss the imaging of magnetic samples the direct interaction between the sample and the magnetic lens field is essential and we have to use another model. In other words: The physical models used in this book are chosen to be as simple as possible to explain a specific outcome.

Despite the attempt of plausible explanations some correlations can be better understood with the help of mathematics. Formulas are necessary to obtain quantitative values. Especially the last chapter 10 considers this. There are some special basics explained in more detail, where it is applicable we point to such detailed explanations within the text. Here and there equations are also listed in the earlier chapters containing elements of the infinitesimal mathematics like differentials and integrals. Describing definitions and physical basics sometimes it is absolutely necessary. This should be no reason to stop reading the book even if the reader has problems with this kind of mathematics.

For specialists the chapter 10 cannot substitute textbooks about special topics of electron microscopy. We suggest some of such books in the hints for further reading at the end of this book.

Finally, we would like to thank: our academic teachers, friends and colleagues who introduced us into the topics of electron microscopy or had facilitated the work at modern instruments later on. Some names we would like to mention: Prof. Alfred Recknagel, Dr. Hans-Dietrich Bauer, and Prof. Klaus Wetzig in Dresden as well as Prof. Manfred Rühle, Prof. Frank Ernst, Dr. Günter Möbus, and Prof. Joachim Mayer working at the "Max-Planck-Institut für Metallforschung" in Stuttgart in the relevant time.

Electron microscopic investigations are only possible using suitably prepared thin samples. Many of the electron microscopic images could be only presented in this book because of the careful specimen preparation by Dipl.-Ing. (FH) Birgit Arnold and Dina Lohse.

In an early state of our book project we spoke with Prof. Josef Zweck from Regensburg and Prof. Klaus Wetzig from Dresden about our concept. They had encouraged us in our project and contributed by their advocacy essentially to the publication of the book by the Springer Verlag.

Stephen Soehnlen B.S., Wil Bruins, and Annelies Kersbergen were our negotiants of the Springer Verlag in Dordrecht. Without their benignity this book had not been published.

Nora Thomas M.A. was so kind to help us finding Goethe's citations with her special knowledge.

We are indebted to the persons mentioned here, as well as to those electron microscopists from Dresden who had been reading the German book manuscript or parts of it, as well as this English version later on, and gave helpful hints for its improvement but also to the colleagues, friends, and students who inspired us by questions and comments to think about facts and circumstances which seemed to be "completely clear".

Goethe: „Alles Gescheite ist schon mal gedacht worden, man muß nur versuchen, es noch einmal zu denken." [2] *(All the prudent things have been already thought. One just has to try to think about them once more.)*

Dresden October 2013

Jürgen Thomas
Thomas Gemming

References

1. von Goethe, J. W.: Wilhelm Meisters Wanderjahre, ed. Erich Trunz, Goethes Werke—Hamburger Ausgabe Bd. 8, Romane und Novellen III, 12. Aufl. München, II/Betrachtungen im Sinne der Wanderer (1989), p. 293
2. von Goethe, J. W.: ibidem, p. 283

Common remark: Biographic Data: Wikipedia—the free encyclopedia, http://de.wikipedia.org/wiki/Wikipedia

Contents

1 Why Such an Effort? 1
    1.1   The Problem with the Magnification 1
    1.2   The Limitation of Resolution 2
    1.3   Electron Waves 6
    1.4   The Role of Magnification 8

2 What Should we Know about Electron Optics and the Construction of an Electron Microscope? 11
    2.1   The Principle of Multistage Imaging 11
    2.2   Rotational-Symmetric Magnetic Fields as Electron Lenses 12
    2.3   Lens Aberrations 15
    2.4   Resolution Limit Considering the Spherical Aberration 19
    2.5   Electron Gun 20
    2.6   "Richtstrahlwert" (Brightness) 24
    2.7   We Construct an Electron Microscope 27
          2.7.1   Illumination System 27
          2.7.2   Imaging System 29
          2.7.3   Specimen Stage 30
          2.7.4   Acquiring the Images 32
          2.7.5   Vacuum System 34
          2.7.6   Miscellaneous 37
    References 39

3 We Prepare Electron-Transparent Samples 41
    3.1   What is the Challenge? 41
    3.2   "Classical" Methods 43
    3.3   Cutting, Grinding, and Ion Milling 47
    3.4   Focussed Ion Beam ("FIB") Techniques 51
    References 56

4 Let us Start with Practical Microscopy ... 57
- 4.1 What do We Peripherally Need? ... 58
- 4.2 We Put the Specimen into the Holder and Insert it into the Microscope ... 59
- 4.3 We Check the (alignment) State of the Microscope ... 61
- 4.4 Focussing the Image—Sharpness and Contrast ... 69
- 4.5 Contamination and Sample Damaging ... 70
- References ... 73

5 Let us Switch to Electron Diffraction ... 75
- 5.1 Why Diffraction Reflections? ... 75
- 5.2 Crystal Lattices and Lattice Planes ... 78
- 5.3 Selected Area and Convergent Beam Electron Diffraction ... 84
- 5.4 What Can We Learn from Selected Area Diffraction Patterns? ... 90
  - 5.4.1 Radii in Ring Diagrams ... 90
  - 5.4.2 Rules for Forbidden Reflections ... 93
  - 5.4.3 Intensities of the Diffraction Reflections ... 98
  - 5.4.4 Positions of Diffraction Reflections in Point Diagrams ... 99
  - 5.4.5 Indexing of Diffraction Reflections ... 103
- 5.5 Kikuchi- and HOLZ-lines ... 106
- 5.6 Amorphous Samples ... 110
- References ... 112

6 Why Do We See Any Contrast in the Images? ... 115
- 6.1 Elastic Scattering of Electrons Within the Sample ... 115
- 6.2 Mass Thickness and Diffraction Contrast ... 116
- 6.3 Brightfield and Darkfield Imaging ... 120
- 6.4 Bending Contours, Dislocations, and Semicoherent Particles ... 123
- 6.5 Thickness Contours, Stacking Faults, and Twins ... 128
- 6.6 Moiré Patterns ... 132
- 6.7 Magnetic Domains: Lorentz Microscopy ... 133
- References ... 136

7 We Increase the Magnification ... 137
- 7.1 Imaging of Atomic Columns in Crystals: Phase Contrast ... 137
- 7.2 Contrast Transfer by the Objective Lens ... 142
- 7.3 Wave-optical Interpretation of the Resolution Limit ... 145
- 7.4 Periodic Distribution of Brightness in Pictures: Fourier Analysis ... 147
- 7.5 Mass Thickness and Phase Contrast ... 150
- 7.6 Contrast of Amorphous Samples ... 152
- 7.7 Correction of Astigmatism ... 154
- 7.8 Measurement of the Resolution Limit ... 156
- 7.9 Correction of Spherical and Chromatic Aberration ... 158
- 7.10 Interpretation of High Resolution TEM Images ... 161
- References ... 162

Contents

8 Let Us Switch to Scanning Transmission Electron Microscopy 163
- 8.1 What Happens Electron-Optically? 163
- 8.2 Resolution or: What is the Smallest Diameter of the Electron Probe? 165
- 8.3 Contrast in the Scanning Transmission Electron Microscopic Image 170
- 8.4 Speciality: High Angle Annular Darkfield Detector 173
- References 174

9 Let us Use the Analytical Possibilities 177
- 9.1 Analytical Signals by Inelastic Interaction 177
  - 9.1.1 Emission of X-rays 178
  - 9.1.2 Electron Energy Losses 182
- 9.2 Energy Dispersive Spectroscopy of Characteristic X-rays ("EDXS") 185
  - 9.2.1 X-ray Spectrometers and Spectra 186
  - 9.2.2 Qualitative Interpretation of X-ray Spectra 190
  - 9.2.3 Quantifying X-ray Spectra 194
  - 9.2.4 Line Profiles and Elemental Mappings 202
- 9.3 Electron Energy Loss Spectroscopy ("EELS") 204
  - 9.3.1 Electron Energy Spectrometer 205
  - 9.3.2 Low-Loss and Core-Loss Regions of the Spectra 206
  - 9.3.3 Qualitative Elemental Analysis 209
  - 9.3.4 Background and Multiple Scattering: Requirements to the Sample 210
  - 9.3.5 Measurement of the Specimen Thickness 213
  - 9.3.6 Edge Fine Structure: Bonding Analysis 216
  - 9.3.7 Quantifying Energy Loss Spectra 219
- 9.4 Energy Filtered Imaging 221
- 9.5 Comparison Between EDXS and EELS 224
- References 225

10 Basics Explained in More Detail (with a Bit More Mathematics) 227
- 10.1 Diffraction at an Edge (Huygens' Principle) 227
- 10.2 Wave Function for Electrons 228
- 10.3 Electron Wavelength Relativistically Calculated 233
- 10.4 Electron Beam Paths in Rotational-Symmetric Magnetic Fields 234
- 10.5 Resolution Limit Considering Spherical Aberration 243
- 10.6 Schottky Effect 244
- 10.7 Electric Potential in Rotational-Symmetric Arrangements of Electrodes 247
- 10.8 Laue Equations and Reciprocal Lattice, Ewald Construction 249
- 10.9 Kinematical Model: Lattice Factor and Structure Factor 261
- 10.10 Debye Scattering 268
- 10.11 Electrons Within a Field of a Central Force 272

10.12	Mean Free Path for Elastic Scattering	277
10.13	Distances in Moiré Patterns	279
10.14	Contrast Transfer Function	282
10.15	Scherzer Focus	290
10.16	Delocalisation	294
10.17	Potential in Electrostatic Multipoles	297
10.18	Electron Probe and Aberrations	299
10.19	Classical Inelastic Collision	306
10.20	Efficiency of Energy Dispersive X-ray Detectors	307
10.21	Calculation of Cliff-Lorimer k-factors	313
10.22	Correction of Absorption for EDXS	318
10.23	Prisms for Electrons	320
10.24	Convolution of Functions	324
References		327

Summary and Outlook 329

Physical Constants 331

Hints for Further Reading 333

Index 335

Abbreviations

ΔE	energy spread
Δf	defocus
Δf_A	astigmatic difference of the focal lengths
Λ	mean free path
α	diffraction angle, aperture, angle between unit cell axes
β	illumination aperture, angle between unit cell axes, acceptance angle
γ	angle between unit cell axes
δ	resolution limit, distance between slits
δ_C	radius of the chromatic aberration disk
δ_D	radius of the diffraction disk, delocalisation
δ_S	radius of the spherical aberration disk
θ	Bragg angle
λ	wavelength, mean free path
(μ/ρ)	attenuation (or absorption) coefficient for X-rays
ν	frequency
Φ	(crystal) potential
ϕ	phase, phase shift
ρ	density
σ	visual angle, cross section of scattering
φ	angle, azimuth in polar and cylindrical coordinates
Ψ	magnetic potential
ψ	wave function
Ω	solid angle
ω	radial frequency, fluorescence yield
A	area
A_M	image ratio
a	exponent, fraction of the Kα X-ray peak, acceleration
$\mathbf{a_1, a_2, a_3}$	lattice vectors
$\mathbf{b_1, b_2, b_3}$	reciprocal lattice vectors
$\mathbf{B}$	magnetic induction
$\mathbf{b}$	Burgers vector
b	image distance

C	general constant
CBED	Convergent Beam Electron Diffraction
C_C	coefficient of chromatic aberration
C_S	coefficient of spherical aberration
CTF	Contrast Transfer Function
c	concentration
D	dispersion, damping function
D_{eff}	detector efficiency
d, d_{hkl}	lattice spacing, generally: distance
E	energy, electric field strength
E_0	primary electron energy
E_P	plasmon energy
EDXS	Energy Dispersive X-ray Spectroscopy
EELS	Electron Energy Loss Spectroscopy
ELNES	Energy Loss Near Edge Fine Structure
EXELFS	Extended Energy Loss Fine Structure
f	focal length
F	force
FIB	Focussed Ion Beam
F_{hkl}	structure factor
G	lattice factor
g	object distance
I	electric current, intensity
H	brightness
HAADF	High Angle Annular DarkField
hkl	Miller's indices
HOLZ	High Order Laue Zone
HRTEM	High Resolution Transmission Electron Microscopy
i	imaginary unit, integer number
j, j	current density, total momentum quantum number, integer number
k	wave number vector
k_{AB}	Cliff-Lorimer k-factor
L	camera length
l	orbital quantum number
M	magnification, integer number
M_r	atomic or molecular weight, respectively
m	magnetic quantum number
m, m_0	mass
N	integer number
n	refractive index, principal quantum number, integer number
NBED	Nano Beam Electron Diffraction
p	momentum, pressure
PED	Precessing Electron Diffraction
Q	ionisation cross section (probability), electric charge
q	space frequency

Abbreviations

R	richtstrahlwert (brightness)
r	radius (also in polar and cylindrical coordinates)
S, S	definite viewing range, refractive power, parameter of agreement
s	scattering vector, spin quantum number
s, s_{opt}	path, way, length, optical path
SAED	Selected Area Electron Diffraction
SDD	Silicon Drift Detector
STEM	Scanning Transmission Electron Microscopy
T	period of oscillation, absolute temperature
TEM	Transmission Electron Microscopy
T_W	windows transparency for X-rays
U	electric potential, voltage, background
U_0	acceleration voltage
U_{OV}	overvoltage ratio
$\mathbf{v}$	velocity
V_{UC}	volume of a unit cell
W	work, potential energy
W_A	work function
W_{Acc}	acceleration work
WDXS	Wavelength Dispersive X-ray Spectroscopy
W_P	potential energy
x	local coordinate
y	object size, local coordinate
y_D	delocalisation
y'	image size
Z	atomic number, general: integer number
z	local coordinate (also in cylindrical coordinates), optical axis

About the Authors

Jürgen Thomas (born in 1948) studied physics at the TU Dresden from 1966 to 1971. In 1970 he had the first contact with electron microscopy and received finally his diploma and doctoral degree on topics of electron microscopy and electron-solid-interactions under supervision of Prof. Alfred Recknagel in Dresden. Between 1978 and 1989 he was responsible for the development of technologies for electron-beam welding and vacuum drying in the industrial research. In 1990 he went back to the electron microscopy and joined the Leibniz Institute for Solid State and Materials Research (IFW) Dresden where he has been working in the laboratory for analytical transmission electron microscopy until today.

Thomas Gemming (born in 1969) studied physics at the University Karlsruhe from 1988 to 1994. He received his doctoral degree on high-resolution transmission electron microscopy in the group of Prof. Manfred Rühle at the Max-Planck-Institut für Metallforschung in Stuttgart in 1998. Afterwards he expanded his field of work to analytical transmission electron microscopy. In 2000 he moved to the Leibniz Institute for Solid State and Materials Research (IFW Dresden) where he is currently working as a department head for Micro- and Nanostructures. Additionally he is currently the executive secretary of the German Society for Electron Microscopy (DGE).

Chapter 1
Why Such an Effort?

Abstract Very tiny structures, so-called "nanostructures", play an increasingly important role in modern materials science: Cobalt-copper multilayer stacks with single layer thicknesses of only about 1 nm lead to unexpectedly strong changes of the electrical resistance when the magnetic field is changed ("giant magnetoresistance"). This facilitates reading information of computer hard disks. Nanoparticles on surfaces hinder the direct contact of water with the surface: The water drops drip off ("Lotus effect"). Nanoparticles improve the efficiency of catalysts by their large specific surface. The exciting question is: "How can we investigate and characterise such nanostructures?"

Let us start with an example. When we are waiting for a tram at a tram stop and the tram approaches but we cannot read the number of the tram line, we have to wait until the tram approaches us. The experience teaches: To see smaller details we have to shorten the viewing distance. Or in other words: The visual angle σ (Fig. 1.1) must be large enough.

1.1 The Problem with the Magnification

However, it is impossible to move arbitrarily close to the observed object. The physicists assume an optimal viewing distance of $S = 25$ cm (distance of most distinct vision). Using this assumption the visual angle σ is given by the height y of the object:

$$\tan \sigma = \frac{y}{S}$$

for small angles ($\sigma \ll 1$) this simplifies to

$$\sigma = \frac{y}{S}. \tag{1.1}$$

Fig. 1.1 Definition of the visual angle σ. (Drawn using CorelDraw Cliparts by the Corel Corporation)

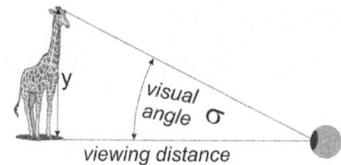

Fig. 1.2 Magnifying the visual angle by a loupe. (Drawn using CorelDraw Cliparts by the Corel Corporation)

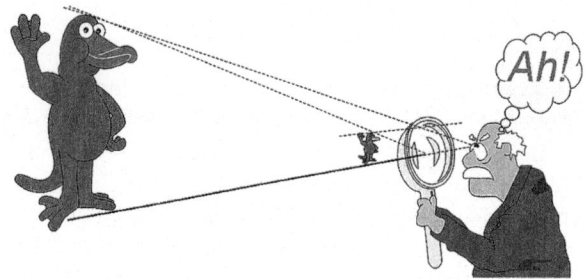

Following this, structures in the order of nanometres cause a visual angle of less than 10^{-5} mrad ≈ 0.0000006°. This would be the same as seeing a tram in a distance of about 100,000 km. We need an (optical) device which is suitable to increase the visual angle without approaching the object. The simplest variant is the loupe. Considering this, we can define the *magnification* of an optical device (Fig. 1.2):

$$\text{Magnification } M = \frac{\text{Visual angle with optical device}}{\text{Visual angle without optical device}} = \frac{\sigma'}{\sigma}. \quad (1.2)$$

The route to characterise nanostructures seems to be clear: We have to arrange a lot of loupes in series to get a sufficient visual angle and consider some optical laws to get a sharp image. Unfortunately, this conclusion is wrong. The reason will be described in the next paragraph "Limitation of resolution".

1.2 The Limitation of Resolution

Up to now we have not considered the nature of light: It can act particle-like as well as wave-like. Let us regard the light as a wave and imagine that the light wave runs up to a pinhole. How does the wave propagate behind the pinhole? A sharp limitation of both sides could be assumed (Fig. 1.3a).

However, this assumption is wrong. Looking at a small pinhole in a black sheet of paper we see only an unsharp edge of the pinhole. The diffraction of the wave around the obstacle is the reason. The physical basis is given by Huygens' principle[1]

---
[1] Christiaan Huygens, Dutch physicist, 1629–1695.

1.2 The Limitation of Resolution

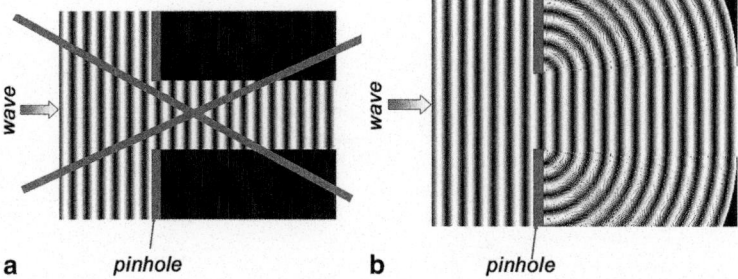

Fig. 1.3 Propagation of a wave behind a pinhole. **a**) Naive assumption. **b**) Reality (Huygens' principle). Sect. 10.1

of the wave propagation: *Each point of a wave front is the origin of an elementary wave. The overlay of the elementary waves yields the new wave front*. Considering this fact the assumption mentioned above has to be corrected as shown in Fig. 1.3b.

Furthermore, let us assume we had a second hole in the black paper *very* close to the first one. Because of the diffraction effect of light waves we could not separate the two holes, even if they do not touch each other. We cannot detect two holes.

Let us try to describe this situation mathematically, i.e. to find a formula to calculate the *resolution limit* of the light optical microscope. Ernst Abbe[2], the famous scientific co-worker of Carl Zeiss[3], developed this idea in about 1870.

In the geometric optics the imaging process requires that all the rays coming from one object point meet in one image point. This can be described by the thin lens equation

$$\frac{1}{f} = \frac{1}{g} + \frac{1}{b} \qquad (1.3)$$

which links the focal length f, the object distance g, and the image distance b. As consequence mnemonics arise like this: "Rays being parallel to the optical axis in the object space meet in the back-focal point." Parallel means that the object distance g is infinite and according to Eq. (1.3) the image distance b is equal to the focal length f.

How can we figure out the imaging process thinking about the wave character of light? Why is the image created just at a special distance from the lens? The explanation seems to be simple: The image is the result of the interference of all the waves scattered at the object and overlaid by the lens again. To get an overlay without phase shift by the lens the optical path s_{opt} must be equal for all these waves. The optical path is the product of the geometric path s and the refractive index n

---

[2] Ernst Abbe, German physicist and employer, 1840–1905.
[3] Carl Zeiss, German engineer and employer, 1816–1888.

Fig. 1.4 Wave paths for the imaging process

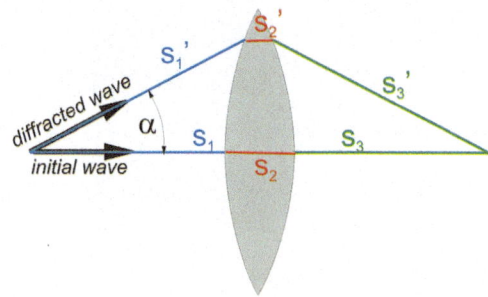

of the medium in which the light propagates. Generally, n is not constant along the complete path and we have to write:

$$s_{\text{opt}} = \int_0^{s_0} n(s) \cdot \mathrm{d}s. \tag{1.4}$$

The postulation for equal optical paths is another formulation of the lens equation. This is to be explained by the following sketch (Fig. 1.4): It seems to be clear that the geometric paths for both waves (characterised by their propagation directions) are different. However, the paths of the waves within the lens are also different: Due to the lens shape the initial wave on the optical axis has to run a longer distance within the lens than the diffracted wave. This is how the different geometric path lengths are balanced to have equal optical path lengths.

Let the refractive index of the surrounding area of the lens be 1 and of the lens material be n. For aberration-free imaging the following equation needs to be fulfilled:

$$s_1 + n \cdot s_2 + s_3 = s'_1 + n \cdot s'_2 + s'_3 \tag{1.5}$$

At the same time this equation determines a special shape of a functional lens.

As the image is the consequence of the overlay (interference) of waves at least two waves have to pass the lens: the initial and a diffracted wave. The next question is: How large is the diffraction angle α? To answer this question let us have a look at Fig. 1.5.

A wave propagates perpendicularly onto a double slit (distance between the two slits: δ). Elementary waves propagate behind the slits. These overlay with constructive interference only for (optical) path differences Δs of a whole-number multiple of the wavelength λ. From Fig. 1.5 one can deduce:

$$\Delta s = \delta \cdot \sin \alpha \tag{1.6}$$

and, with n as refractive index in the area behind the double slit (right side):

$$\Delta s = \delta \cdot n \cdot \sin \alpha, \tag{1.7}$$

1.2 The Limitation of Resolution

Fig. 1.5 Interference at a double slit

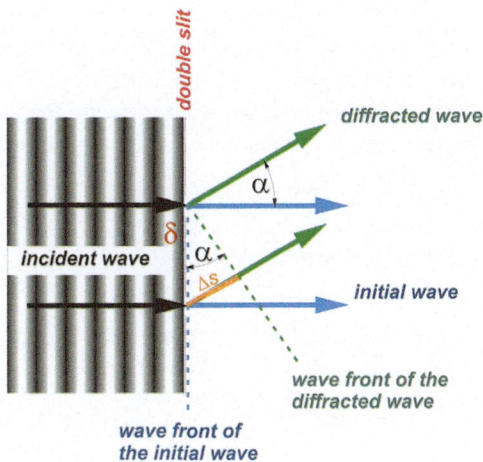

respectively. If the path difference is equal to the wavelength one yields

$$\lambda = \Delta s = \delta \cdot n \cdot \sin \alpha \quad \text{and} \quad \delta = \frac{\lambda}{n \cdot \sin \alpha}. \tag{1.8}$$

This equation describes the minimal resolvable distance δ between the slits for a given wavelength and *numeric aperture* $n \cdot \sin\alpha$. Or in general: It describes the minimal distance between two structure details which we can see separated in the image. Therefore Eq. (1.8) characterises the so-called resolution limit. A closer examination of the intensity distribution within the diffraction disks including the consideration of the *Rayleigh*[4]-*criterion* results in the resolution limit of a light optical microscope:

$$\delta = \delta_D = \frac{0.61 \cdot \lambda}{n \cdot \sin \alpha}. \tag{1.9}$$

We added the index D to indicate that the diffraction causes the limitation.

Let us now estimate the sizes of structure details which can be separated and investigated: The wavelength of the visible light amounts to about 0.5 μm, the refractive index of air, commonly the surrounding medium, is equal to 1 (using special immersion oils 1.4 can be reached) and $\sin\alpha$ is 1 at most. Using these values we get about 0.2 μm as resolution limit of the light optical microscope. Please keep in mind that this definition of resolution requires two objects to be observed as being separated.

---

[4] Lord John W. S. Rayleigh, English physicist, 1842–1919, Nobel prize in physics in 1904.

Rayleigh-criterion: Two point sources are considered to be just separated after imaging when the principal diffraction maximum of the diffraction disk from the first source coincides with the first minimum of the second one.

Therefore it is physically impossible to detect nanostructures using the approach described in the introduction. An increase of the magnification alone is not helpful in this case. We have to improve the resolution limit of the optical device. The route seems to be clear: If we want to retain the "classical" optical imaging[5] described above we have to substitute the light by something with shorter wavelength.

1.3 Electron Waves

Due to Louis de Broglie[6] it has been known since 1924 that particles with a rest mass greater than zero (e.g. electrons) have wave character, too. Their wavelength λ can be calculated using

$$\lambda = \frac{h}{p} \qquad (1.10)$$

(h: Planck's[7] constant, p: momentum).

On the other hand, it was known that electrons can be deflected by electric and magnetic fields. In 1926 Hans Busch[8] proposed the construction of lenses for electrons using such fields similarly to glass lenses for light. Therewith the electron optics was born and the possibility was given to improve the resolution limit of such an "Übermikroskop[9]" in comparison with the light optical microscope.

For a better understanding of this fact we have to calculate the wavelength of the electrons. Starting point is de Broglie's formula (1.10). The momentum of the electron is given by

$$p = m_0 \cdot v, \qquad (1.11)$$

i.e. by the product of rest mass m_0 and velocity v. The velocity can be calculated by the energy theorem, i.e. the work W_{Acc} to accelerate the electron is changed into kinetic energy of the electron. In physics the work is given by

---

[5] The resolution can be also improved by use of a very small spot for excitation as well as for acquiring of information instead of the illumination of the complete area that should be imaged ("non-classical" methods, e.g. scanning or near-field methods with light).

[6] Louis de Broglie, French physicist, 1892–1987, Nobel prize in physics in 1929.

[7] Max Planck, German physicist, 1858–1947, Nobel prize in physics in 1918, is the originator of quantum physics.

[8] Hans Busch, German physicist, 1884–1973, is the originator of electron optics.

[9] "Übermikroskop" was the name of the first commercial transmission electron microscope created by the company "Siemens" (introduction on the market 1939). The German electrical engineers Ernst Ruska (1906–1988, Nobel prize for physics in 1986), Bodo von Borries (1905–1955), and Max Knoll (1897–1969) were the leading developers. Before 1939 and later there were instruments in laboratories for development and testing of parts which influenced the development of the microscope as a complete instrument, e.g. in Manfred von Ardenne's (1907–1997) institute in Berlin-Lichterfelde.

1.3 Electron Waves

$$W_{Acc} = \int F_S \cdot ds \tag{1.12}$$

(F_S: force component in the direction of the path unit ds). The electron charge $-e$ (e: elementary electric charge) experiences in an electric field with the strength E the force

$$F_S = -e \cdot E. \tag{1.13}$$

Simplifying, we assume a homogeneous electric field between cathode and anode (distance d, accelerating voltage U_0). Considering the direction of the electric field from anode to cathode the strength of the field is given by

$$E = -\frac{U_0}{d}, \tag{1.14}$$

i.e.

$$W_{Acc} = \int_0^d e \cdot \frac{U_0}{d} \cdot ds = e \cdot U_0. \tag{1.15}$$

Therewith the energy theorem in its classical form is

$$\frac{m_0}{2} v^2 = e \cdot U_0 \tag{1.16}$$

with elementary electric charge e and rest mass m_0. Using Eqs. (1.10), (1.11), and (1.16) we get

$$\lambda = \frac{h}{\sqrt{2 \cdot e \cdot m_0 \cdot U_0}}. \tag{1.17}$$

Relativistic effects cannot be neglected at accelerating voltages greater than 50 kV. In this case Eq. (1.17) needs to be modified to

$$\lambda = \frac{h}{\sqrt{e \cdot U_0 \left(2 \cdot m_0 + \frac{e \cdot U_0}{c^2}\right)}} \tag{1.18}$$

with the additional constant c (velocity of light Sect. 10.3).

The wavelength of electrons accelerated by 300 kV yields 2.24 pm in classical calculation but 1.97 pm ≈ 0.002 nm in the relativistic variant (Fig. 1.6). Following the classical calculation the wavelength at an accelerating voltage of 300 kV has an error of 14 %! Nevertheless, we use the classical approximation for the simplified

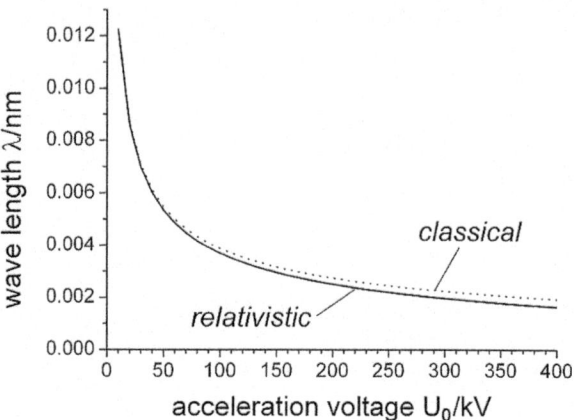

Fig. 1.6 Wavelength of the electrons in dependence on the accelerating voltage within the electron gun

explanations in this book as the relativistic description can easily plugged in using a correction term.

Taking Eq. (1.9) for the resolution limit of the light optical microscope we should reach a resolution limit of 1.6 pm (0.0016 nm) with the electron microscope. Currently in practice a resolution limit of about 0.1 nm is reachable. In Sect. 2.4 we will see one reason for this discrepancy.

1.4 The Role of Magnification

We have seen: The resolution limit is the essential feature of an optical instrument. However, which role does the magnification play?

The resolution limit is a distance regarding the sample (object). Typical for the transmission electron microscope is a value of about 0.1 nm. Of course, this length has to be magnified to detect it also by eyes, by a photo sheet or by the camera. The device for registration has got a resolution limit, too. Let us name it δ_{Reg}. For example, the human eye has a δ_{Reg} of about 0.1 mm.

"To see it by eyes" means that the resolution limit of the optical instrument must be magnified to 0.1 mm. Commonly, this *useful magnification* is given by

$$M_{useful} = \frac{\delta_{Reg}}{\delta}. \tag{1.18}$$

Higher magnifications than M_{useful} do not lead to an improvement of the recognisability of details. Contrariwise, they lead to unsharp images!

Using our eyes for registration we get a useful magnification of about 500 for the light optical microscope ($\delta \approx 0.2$ μm) and such one of about 1 million for the transmission electron microscope ($\delta \approx 0.1$ nm).

1.4 The Role of Magnification

Currently, in electron microscopy so-called "CCD cameras" (CCD: Charge Coupled Device—cf. Sect. 2.7.4) are often used to register the images. The digital images consist of pixels with a default size of 15–25 μm for the cameras which are currently used. The electron-optically useful magnification in the transmission electron microscope reduces to about 500,000 in the latter case. The magnification of the images displayed on the computer screen depends on the used pixel size of the computer display. Values between 0.2 and 0.3 mm are typical, i.e. they are larger than the resolution limit of the human eye.

Chapter 2
What Should we Know about Electron Optics and the Construction of an Electron Microscope?

Abstract Whenever the term "electron microscope" used in this book then the transmission electron microscope is meant and not the (conventional) scanning electron microscope. The transmission electron microscope is a microscope in its true sense, i.e. an instrument for imaging of a sample that is transmitted and optically imaged by rays. Nevertheless, it is possible to focus and to scan the electron beam on the specimen in most of these devices, too ("STEM"—see Chap. 8). But it sticks to the transmission of the beam through the sample. In principle the transmission electron microscope is constructed like a (transmission) light optical microscope. Accordingly, we will deal with some electron-optical fundamentals, lens aberrations, and important components of the electron microscope in this chapter.

2.1 The Principle of Multistage Imaging

The useful magnification of a light optical microscope (and even more of an electron microscope) is reached by multistage imaging. Figure 2.1 shows this for the two-stage imaging of a light optical microscope.

In principle, an electron microscope is just constructed like a light optical microscope. However, it is turned upside down in most cases, i.e. the condenser is the lens on top.

For a system of two (thin) lenses with the focal lengths f_1 and f_2 as well as the distance t between them the total focal length f is

$$\frac{1}{f} = \frac{1}{f_1} + \frac{1}{f_2} - \frac{t}{f_1 \cdot f_2}. \tag{2.1}$$

Within an electron microscope the image distance b is given by the fixed positions of the lenses and the screen or the camera. Changing the focal length allows to image object planes being in different distances onto the screen (cf. the lens Eq. (1.3)). Later, we will see which capabilities can be opened by this.

The total magnification M of multistage imaging with the single magnification steps $M_1, M_2, \ldots M_n$ is given by

$$M = M_1 \cdot M_2 \cdot \ldots \cdot M_n. \tag{2.2}$$

Fig. 2.1 Two-stage imaging at a light optical microscope. The light of the lamp (*arranged at the bottom*) penetrates a condenser lens and illuminates the specimen. The objective lens creates a magnified real intermediate image that is again magnified to the final image by the projective lens. The final image can be acquired by a photo plate, for example. Often, the projective is substituted by an eyepiece lens which allows seeing the magnified real intermediate image by eye with a loupe

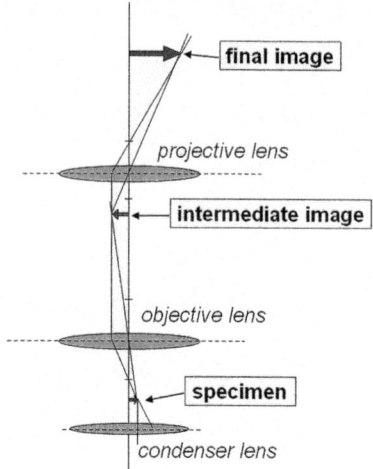

In normal speech the magnification is often put on one level with the image ratio, although, properly speaking, these are two different things: The magnification is equal to the quotient of the visual angles with and without the optical device (1.2); the image ratio A_M is equal to the quotient of the image size y' and the object size y. A simple geometric reflection with consideration of the algebraic signs shows that A_M goes together with the negative quotient of image and object distance b and g, respectively. Finally, b can be calculated from g and the focal length f by the lens Eq. (1.3) and it follows:

$$A_M = \frac{y'}{y} = -\frac{b}{g} = \frac{f}{f-g}. \tag{2.3}$$

Looking at the algebraic sign the difference between magnification and image ratio can be most notably seen: Contrary to the magnification the image ratio can be positive or negative. A negative image ratio means that the image turns upside down. Taking only the modulus of the image ratio the value is the same as the magnification.

2.2 Rotational-Symmetric Magnetic Fields as Electron Lenses

Let us think about the following question: Why do the rotational-symmetric magnetic fields commonly used in electron microscopes work as electron lenses, i.e. why do they influence the electrons in the same way like glass lenses the light? Fig. 2.2 shows a sketch of the construction of such an electron lens.

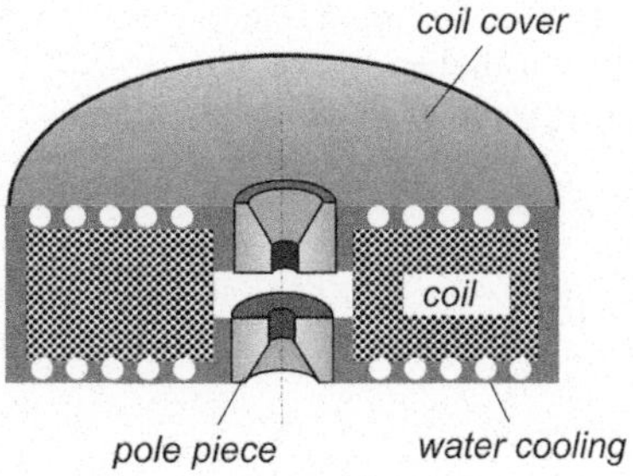

2.2 Rotational-Symmetric Magnetic Fields as Electron Lenses

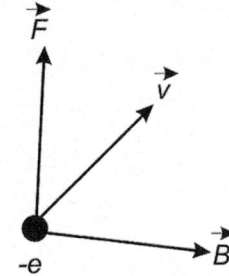

Fig. 2.2 Sketch of a magnetic electron lens. The coil current can reach up to 30 A. Water cooling causes the constant temperature within the lens. The pole piece shapes and concentrates the magnetic field

Fig. 2.3 Direction of the Lorentz force *F* concentrated on an electron with the charge –e moving with the velocity **v** within a magnetic field of induction **B**

For the desired lens effect all electrons which are not moving along the central axis ("optical axis") have to feel a power that moves them to this axis. We know that electrons moving with the velocity **v** within a magnetic field of strength (more precisely: of magnetic induction) **B** are influenced by the Lorentz force[1]

$$\mathbf{F} = -e \cdot \mathbf{v} \times \mathbf{B} \qquad (2.4)$$

with the elementary electric charge e. Additionally, it is considered that force **F**, velocity **v**, and magnetic induction **B** are vectors, i.e. they are denoted by modulus (numerical value) and direction. Commonly, such vectors are drawn by arrows with a length representing the modulus and an angle against a reference axis for representation of their direction. We are going to indicate a vector by a small arrow above the letter in the figures and by **bold letters** in equations and text.

How can we determine the direction of the Lorentz force following the "cross product" of Eq. (2.4)? In the mathematical sense **v**, **B**, and **F** create a "right-handed trihedron", i.e. the force **F** is perpendicular on the plane spanned by **v** and **B** (cf. Fig. 2.3).

Defining the direction we must not forget the negative sign in Eq. (2.4). For example, in practice one uses the "right-hand rule": *One spreads thumb, forefinger, and middle finger of the right hand so that they create a right angle against each other. Then one justifies the hand in such a way that the thumb points to the*

---

[1] Hendrik Antoon Lorentz, Dutch mathematician and physicist, 1853–1928, Nobel prize in physics in 1902

Fig. 2.4 Cylindrical coordinates in the region of the pole piece gap and field lines of the magnetic induction **B**

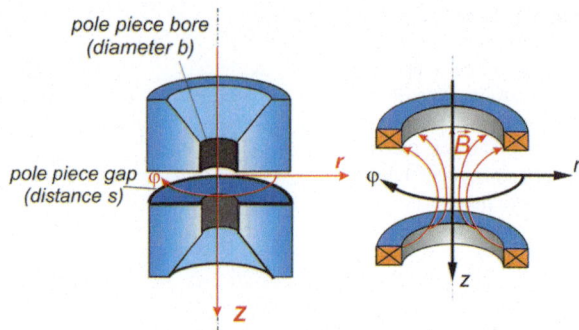

direction of the velocity and the forefinger to the direction of induction. In this case the middle finger points to the direction of the cross product (here: Lorentz force). Following this rule firstly we should get a force direction downwards in Fig. 2.3. Considering the negative sign in Eq. (2.4) the force direction reverses to upwards. This is shown in the figure.

Next, we have to think about the modulus F of the Lorentz force. For the cross product it is equal to the area of the plane spanned by the vectors **v** and **B**:

$$F = e \cdot v \cdot B \cdot \sin \sphericalangle (\mathbf{v}, \mathbf{B}). \tag{2.5}$$

We realise: There is no force on electrons having a parallel direction of velocity and magnetic field. In other words: A homogeneous magnetic field parallel to the optical axis does not influence electrons which run in the direction of the optical axis (these are "parallel rays" in geometric optics). For the lens effect we need a component of the magnetic field perpendicular to the movement direction of the electrons, i.e. we need an inhomogeneous magnetic field.

For the further discussion it is useful to introduce a coordinate system into the magnetic lens. Because of the rotational symmetry we select cylindrical coordinates r, φ und z (Fig. 2.4).

We realise that the field lines do not have any component in the azimuthal (φ) direction but only such ones in r- (B_r) and z-direction (B_z). The consequences for our parallel ray mentioned above should be explained with the help of Fig. 2.5. It can be seen from this figure that our electron that symbolised the parallel ray is really deflected to the optical axis (Sect. 10.4).

Thereby we recognise two essential properties of the rotational-symmetric magnetic electron lens:

1. Due to the azimuthal velocity component the electrons move on spiral paths with changing distances to the middle (optical) axis through the lens, i.e. the image is rotated against the object. In the following figures with beam paths the $r(z)$-plane is always drawn for easier understanding of the sketches.
2. The degree of the field inhomogeneity, i.e. the ratio B_r/B_z at each position within the lens field (shape of the field lines) essentially determines the refraction prop-

2.3 Lens Aberrations

Fig. 2.5 Movement of an electron within an inhomogeneous magnetic lens field. (**a**) The electron has only a component v_z of the velocity ("parallel ray"). Due to the radial component B_r of the magnetic induction a Lorentz force F_φ in azimuthal direction (i.e. into the drawing plane), acts on the electron and causes a velocity component v_φ in azimuthal direction. (**b**) Due to the new velocity component v_φ given now, the electron interacts with the axial component B_z of the magnetic induction and experiences a force directed towards the optical axis as expected for a lens

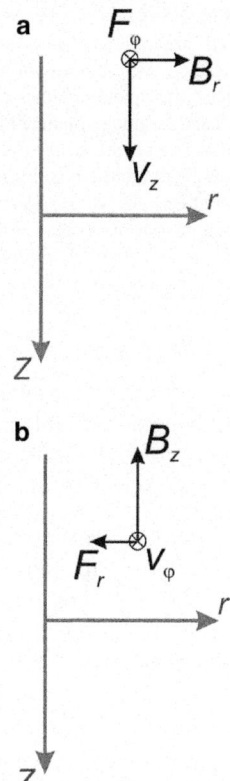

erties of the lens. Later on (Sect. 2.7.3) we will see that the specimen is positioned within the magnetic field of the objective lens. A specimen of magnetic material changes the field inhomogeneity in dependence on the exact position of the specimen. Therefore, we have to expect some additional difficulties when imaging such magnetic specimens.

2.3 Lens Aberrations

The next question is whether it is possible to set the "right" degree of the field inhomogeneity for lens refraction not only in principle but also for an aberration-free optical imaging. Or transferred to the light optics: What happens if we cannot fulfil the requirement of equal optical distances for all the imaginable waves between object and image point (cf. Sect. 1.2) since the shape of the glass lens is wrong?

In fact, this is a problem for rotational-symmetric electron lenses described here. With perpetuation of the rotational symmetry it is impossible to shape the field inhomogeneity as necessary for aberration-free imaging. This problem was identified

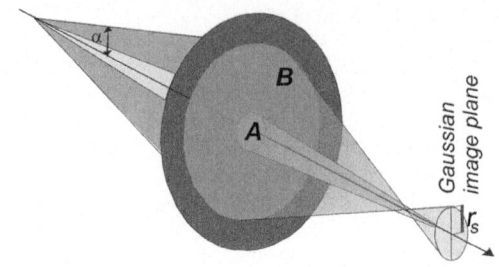

Fig. 2.6 Effect of the spherical aberration. The slightly opened ray bundle *A* creates the image point in the Gaussian image plane. On the other hand, the wider opened (angle α) bundle *B* is refracted stronger. As a consequence, within the Gaussian image plane a spherical aberration disk with radius r_s is created

by Otto Scherzer[2] already in 1936 [1]. As consequence the lens aberrations play a more important role in electron-optical devices than in light optical ones. Here, we want to deal with three important aberrations: the spherical aberration, the chromatic aberration, and the astigmatism.

Spherical aberration This aberration acts especially on rays with larger distances from the optical axis. The name already points to the reason for this aberration: The surface of the lens does not have the right "spherical" shape, it is not properly cut. Or, transferred to magnetic electron lenses, the field shape ("field inhomogeneity") does not meet the requirements for an aberration-free imaging (Sect. 10.4). The discrepancies between requirement and reality are the larger the weaker the rotational-symmetric magnetic field is. Commonly, the spherical aberration of lenses with larger focal lengths ("weak lenses") is larger than that of strong lenses.

The consequence of this "wrong" field shape is shown in Fig. 2.6.

The radius r_s of the spherical aberration disk depends on the opening angle α. Within the Gaussian[3] image plane r_s is given by

$$r_s = M \cdot C_s \cdot \alpha^3 \tag{2.6}$$

with C_s as spherical aberration coefficient of the lens and *M* as magnification at the imaging plane. The spherical aberration coefficient can be seen as a parameter of the lens quality. For rotational-symmetric magnetic lenses its value is approximately equal to the focal length.

Normally, aberration disks are referred to the object plane. There the radius is:

$$\delta_s = C_s \cdot \alpha^3. \tag{2.7}$$

---

[2] Otto Scherzer, German physicist and electron optician, 1909–1982
[3] Carl Friedrich Gauß, German mathematician and physicist, 1777–1855

2.3 Lens Aberrations

This is independent of the magnification and just relates to the spherical aberration. Because of the dependence of the size of the aberration disk on the cube of the opening angle of the beam the spherical aberration is called aberration of third order.

Chromatic aberration From Eq. (2.4) for the Lorentz force we can see that the force on the electron depends on the magnetic induction and on the velocity of the electron. On the other hand, we know in the meantime that the electron wavelength λ is determined by the electron energy, i.e. by the velocity (cf. Eqs. (1.15)—(1.17)). Different wavelengths but also, of course, different field strengths lead to different optical powers of the magnetic lens and lead to the chromatic aberration. Similar to the spherical aberration within the Gaussian image plane an aberration disk is created with radius δ_C (referred to the object plane) depending on the opening angle α, on the relative fluctuation $\Delta S/S$ of the optical lens power, and on the chromatic aberration coefficient C_C:

$$\delta_C = C_C \cdot \frac{\Delta S}{S} \cdot \alpha. \qquad (2.8)$$

There are two reasons for the fluctuation of the optical lens power: Changes of the velocity of the electrons (their wavelength, respectively) and changes of the magnetic lens field. For a plausible explanation we look at the squares of velocity and magnetic induction. The square of the velocity of the electrons is proportional to the electron energy (cf. Eq. (1.16)) and the magnetic induction is proportional to the coil current I (Ampere's[4] law). Because of the increasing optical lens power S by increase of the magnetic field and by smaller electron energy we can write:

$$S \propto \frac{I^2}{e \cdot U}. \qquad (2.9)$$

For small changes ΔS one yields:

$$\left|\frac{\Delta S}{S}\right| = 2 \cdot \left|\frac{\Delta I}{I}\right| + \left|\frac{\Delta U}{U}\right|. \qquad (2.10)$$

The objective lens is the lens essential for the quality of the imaging. Referring to the chromatic aberration we have to consider mainly the stability of the objective lens current. Fluctuations of the accelerating voltage, the energy spread of the electrons emitted by the cathode, and also inelastic interactions between electrons and specimen (i.e. interactions related to electron energy losses) can be the cause of fluctuations of the electron energy, too.

Let us additionally assume that the chromatic aberration should be irrelevant in comparison to the spherical aberration. That means:

$$\delta_C \ll \delta_S \text{ and } C_C \cdot \left(2 \cdot \left|\frac{\Delta I}{I}\right| + \left|\frac{\Delta U}{U}\right|\right) \cdot \alpha \ll C_S \cdot \alpha^3, \text{ respectively.} \qquad (2.11)$$

---

[4] André-Marie Ampére, French physicist, 1775–1836

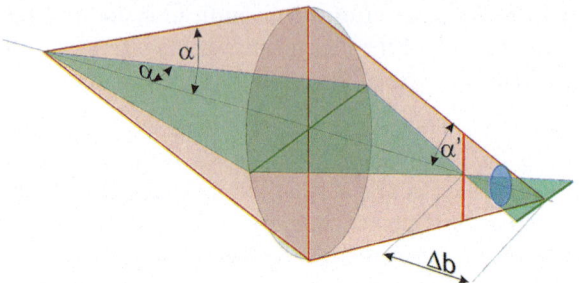

Fig. 2.7 Schematic illustration of a lens with two-fold axial astigmatism

The aberration coefficients C_S und C_C are of the same order, the aperture α amounts to about 10 mrad, therefore the postulation is:

$$2 \cdot \left|\frac{\Delta I}{I}\right| + \left|\frac{\Delta U}{U}\right| \ll 10^{-4}. \tag{2.12}$$

This implies a considerable demand for the stability of the power supply of an electron microscope. For the best state-of-the-art high resolution microscopes equipped with correctors of spherical and chromatic aberrations (cf. Sect. 7.9) a relative stability of $<10^{-8}$ has to be realised [2].

(Axial) astigmatism In a gedankenexperiment we take two planes from a bundle of electrons which are penetrating a lens (drawn green and red in Fig. 2.7). Astigmatism means that the focal lengths are different in these two (or more) planes.

Deviations from the rotational symmetry created by unroundness of the pole piece bore, by smallest inhomogeneities in the pole piece material, by dirty apertures but also by magnetic material within the lens field can be reasons for these differences. We want to confine ourselves to the two-fold astigmatism as sketched in Fig. 2.7.

The two planes are perpendicular to one another and have got the astigmatic difference of the focal lengths Δf_A connected with difference Δb of the image distances. Obviously, the radius of the (blue drawn) circle of the least confusion is given by:

$$r_{\min} = \frac{1}{2} \cdot \Delta b \cdot \alpha' = \frac{1}{2} \cdot \Delta b \cdot \frac{\alpha}{M} \tag{2.13}$$

with M as magnification. At high magnifications (object distance $g \approx$ focal length f) it follows for (small) differences of the image distances the relation (cf. Eqs. (10.266) and (10.267)):

$$\Delta b = \Delta f_A \cdot M^2. \tag{2.14}$$

Fig. 2.8 Radius of the spherical aberration disk (δ_S) and the aberration disk by diffraction (δ_D) in the object plane depending on the aperture α (parameter: $C_S=1.2$ mm, $U_0=300$ kV, i.e. $\lambda=1.97$ pm ≈ 0.002 nm)

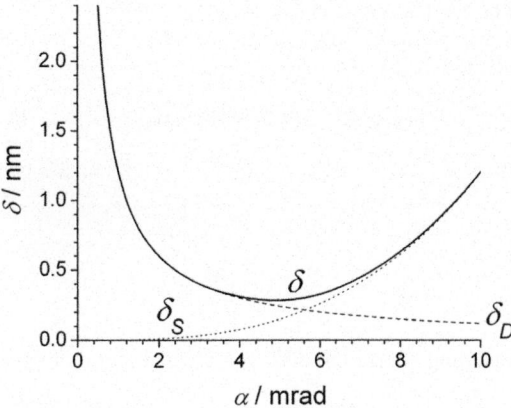

Usually, the size of the aberration disk is referred to the object side. Doing that, the radius has to be divided by the magnification M and we get:

$$\delta_A = \frac{r_{min}}{M} = \frac{1}{2} \cdot \Delta f_A \cdot \alpha. \tag{2.15}$$

2.4 Resolution Limit Considering the Spherical Aberration

At the end of Sect. 1.3 we promised to come back to the resolution limit. Let us remember: Using electron waves we expected a resolution limit of about 2 pm but in reality only about 0.1 nm (i.e. 100 pm) can be reached.

Meanwhile, we recognized a reason of this discrepancy: In contrast to glass lenses our rotational-symmetric magnetic lenses have a spherical aberration limiting the resolution additionally to the (wave specific) aberration by diffraction.

We already know how the radii δ_D und δ_S of both aberration disks in the object plane depend on the aperture α (Eqs. (1.9) and (2.7)). With consideration of very small apertures ($\sin\alpha \approx \alpha$) and of a refractive index $n=1$ (vacuum) these equations can be written as:

$$\delta_D = \frac{0{,}61 \cdot \lambda}{\alpha} \text{ and } \delta_S = C_S \cdot \alpha^3 \tag{2.16}$$

We see, there is an oppositional dependence of the aberration disk sizes on the aperture α (cf. Fig. 2.8). Therefore an optimal aperture α_{opt} exists for a minimal size δ of the

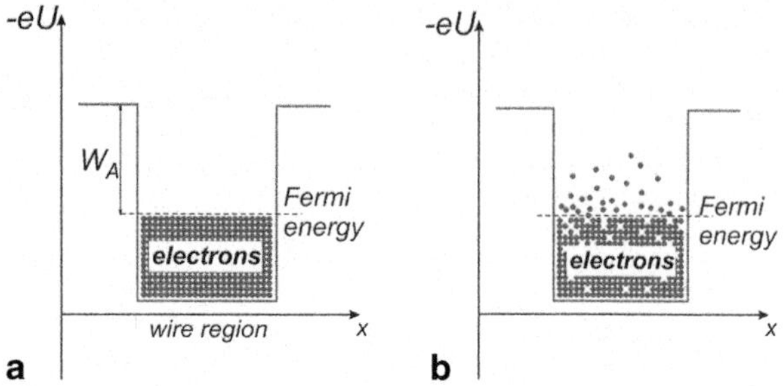

Fig. 2.9 Potential wall model for temperature $T=0$ K (**a**) and $T>0$ K (**b**)

resulting aberration disk. For an evaluation of δ we use the law of error propagation:

$$\delta = \sqrt{\delta_S^2 + \delta_D^2}. \tag{2.17}$$

From Fig. 2.8 we read for the minimum of δ: $\alpha_{opt} \approx 5$ mrad and $\delta_{min} \approx 0.3$ nm. The minimum can also be calculated following the equation

$$\delta_{min} = 0.9 \sqrt[4]{C_S \cdot \lambda^3} \tag{2.18}$$

(Sect. 10.5). This value is very close to the resolution limit which is practically reached. But unfortunately, it is a bit too large now. Therefore we need to revisit the resolution limit of a transmission electron microscope once more (Sect. 7.3).

2.5 Electron Gun

Now is the time to think about the possibilities to generate "free" electrons which interact with the specimen and create an optical imaging by lenses.

At first, we have to ask: "Why do the electrons normally run within a wire and do not step out?" The answer is given by the mean positive potential within the wire originated by the atomic nuclei. This potential keeps the electrons within the wire similar to water in a pot. Using this simple model we can easily understand how the electrons can escape the wire: by heating. Assuming we put the water filled pot onto a heat plate then we will see that the water starts to boil and finally the hot water bubbles over the wall of the pot.

In the same manner we can imagine the thermal emission of electrons. Instead of the pot we use a "potential wall model" (Fig. 2.9). The energy levels of the electrons are described by *quantum numbers*. Each level can be occupied by only one electron (Pauli's[5] principle). At the absolute zero of the temperature (0 K = −273.16 °C)

---

[5] Wolfgang Pauli, Austrian physicist, 1900–1958, Nobel prize in physics in 1945

2.5 Electron Gun

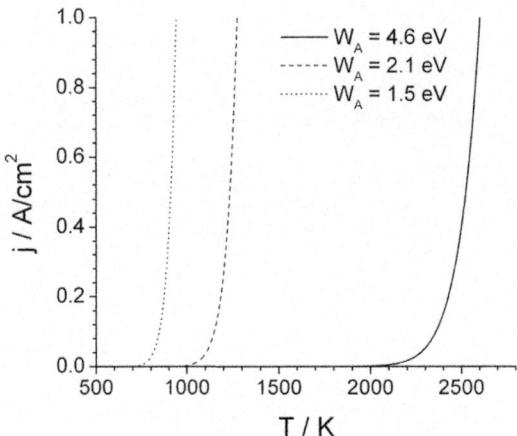

Fig. 2.10 Thermal emission according to Richardson's equation. Current density j in dependence on temperature T for three different work functions: 4.6 eV (tungsten), 2.1 eV (LaB$_6$) und 1.5 eV

the electrons fill all the energy levels which are minimal possible. The energy maximum reached in doing so is named Fermi[6] energy.

To escape the wire the electrons have to be lifted to the potential in the surrounding of the wire ("vacuum level"). The work necessary for this is called work function W_A.

For thermal emission this work is done by heating; as a consequence the wire is heated. It is quite evident that the necessary heat and, therefore, the wire temperature has to be higher at higher work function of the material. Additionally, the material must not melt. In practice, tungsten or lanthanum hexaborid (LaB$_6$) are used. In the simplest variant the cathode consists of a thin tungsten wire (diameter about 0.5 mm) bent like a hairpin ("tungsten hairpin cathode"). The work function of tungsten amounts to about 4.6 eV, the excitation of a sufficient thermal emission needs temperatures of more than 2500 °C. The emitted density of the electron current can be calculated by Richardson's[7] equation:

$$j = A \cdot T^2 \cdot e^{-\frac{W_A}{k \cdot T}} \tag{2.19}$$

(A: Richardson's constant, T: absolute temperature, k: Boltzmann's[8] constant).

Looking at Fig. 2.10 the consequence of a reduction of the work function can be seen: A current density of 1 A/cm², for example, requires a temperature of about 2700 K ≈ 2430 °C for the tungsten hairpin cathode, contrary to that an LaB$_6$ cathode only needs a temperature of about 1300 K ≈ 1030 °C. Beside the life time increase of the cathode the lower temperature has another additional effect. For understanding let us remember the pot with hot water: The water boils stronger at higher

---

[6] Enrico Fermi, Italian/American physicist, 1901–1954, Nobel prize in physics in 1938
[7] Owen Williams Richardson, English physicist, 1879–1959, Nobel prize in physics in 1928
[8] Ludwig Boltzmann, Austrian physicist, 1844–1906

Fig. 2.11 Reduction ΔW_A of the work function by the Schottky effect for two different electric field strengths ($E_1 = 30$ kV/cm, $E_2 = 4 \cdot E_1$)

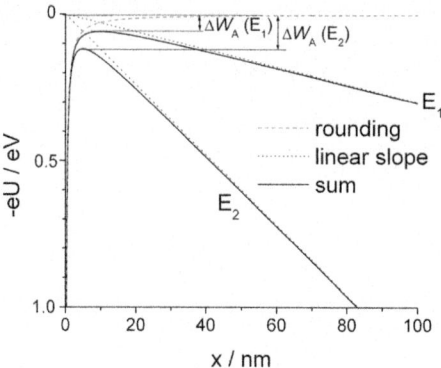

temperatures of the heating plate. Referring to the thermal emission of electrons, this implies that the energy spread of the emitted electrons becomes wider. The spread amounts to about 5 eV for a tungsten hairpin cathode and reduces to about 3 eV for an LaB_6 cathode.

What can we do to reduce the work function furthermore? Let us remember the Schottky[9] effect. It describes the reduction of the work function with a high electric field close to the cathode. To understand this we have to specify the potential wall model (cf. Fig. 2.11).

We assume that the electrons escaped from the wire are staying close to the wire surface for a short time and form a space-charge cloud at this position. In small distance from the wire the electrons feel also a force and therefore they possess a potential different from that of the vacuum level. That causes a "rounding" of the potential shape and, additionally, the high electric field originates a linear down slope of the potential outside of the wire (Sect. 10.6).

The sum of both parts leads to a decrease of the potential wall i.e. to a reduction of the work function proportional to the square root of the electric field strength. Additionally, the potential wall becomes thinner by the electric field and can be tunnelled by electrons (they have wave character, too). That increases the yield of free electrons furthermore. At extremely high electric field strengths this tunnelling effect can lead to a sufficient electron emission even at room temperature ("cold field emission cathode" otherwise: "Schottky field emission cathode"). Using a field emission cathode the energy spread of the emitted electrons amounts to less than 1 eV.

Beside the different velocities of the electrons escaping the cathode the Boersch[10] effect contributes to the energy spread, too [3, 4]. The repulsion of the electrons at positions with high current densities (e.g. the focussing of the electrons within the gun (cross over)) causes this effect. The Boersch effect can amount up to a few 0.1 eV and is already considered in our values of the energy width.

---

[9] Walter Schottky, German physicist, 1886–1976
[10] Hans Boersch, German physicist, 1909–1986

2.5 Electron Gun

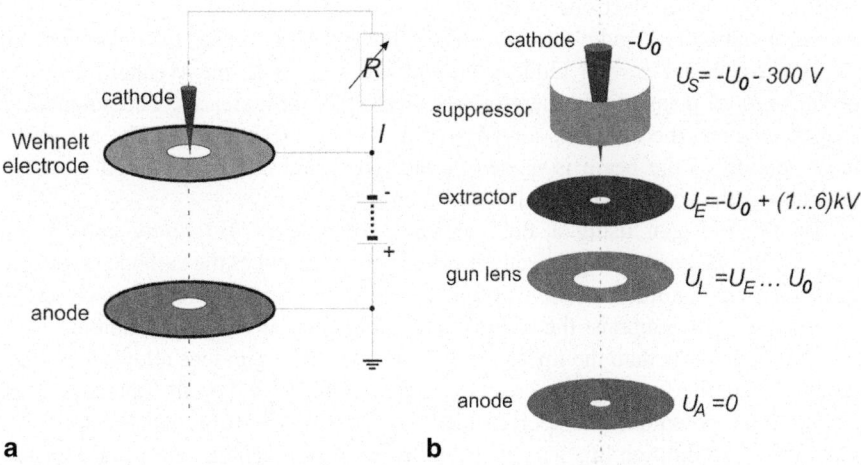

Fig. 2.12 Electrodes within the electron gun. (**a**) Triode system for hairpin and LaB$_6$ cathodes. (**b**) Electrode system within an electron gun with field emission cathode (labels of the electrodes and their potentials)

On a spherical surface the electric field strength E_r is inversely proportional to the spherical radius r:

$$E_r \sim \frac{1}{r}, \qquad (2.20)$$

i.e. for very small curvature radii of the cathode tip the electric field strength becomes very large. Therefore, field emission cathodes are extremely sharp (radius of curvature ca. 1 µm ... 2 µm) and their emission behaviour depends dramatically on the shape of the tip. A bombardment of the tip by residual gas ions would round it and has to be avoided. As consequence, the residual gas pressure in the cathode chamber must be extremely low, i.e. ultrahigh vacuum (pressure $< 10^{-8}$ Pa) is absolutely required.

To reach short electron wavelengths necessary for a high resolution of the electron microscope the electrons escaped from the cathode have to be accelerated. That happens within the electric field between cathode (tip) and anode (plate with a bore) (see also Eqs. (1.13) to (1.15)). Simultaneously, the electric field creates the electron beam. Additionally, it should be possible to vary the current coming from the anode. How this works depends on the type of the cathode.

For electron guns with tungsten hairpin or LaB$_6$ cathode a triode system is used that works similar to a triode electron tube (Fig. 2.12a). The heated cathode is at a negative high tension potential and emits the beam current I. That causes a voltage drop at the resistor R, i.e. the bottom of the resistor is more negative (up to a few 100 V) than the cathode itself. This potential is held by the Wehnelt[11] electrode

---

[11] Arthur Wehnelt, German physicist, 1871–1944

working as control electrode of the beam current *I*: The beam current is lower at a more negative electrode. Simultaneously, the circuit stabilises the beam current: A decreased beam current reduces the voltage drop at *R*, the Wehnelt electrode becomes more positive and therefore the beam current increases. The potential is shaped to focus the electrons already within the gun (Sect. 10.7). The narrowest cross section of the beam is named "cross over" and can be regarded electron-optically as the actual intrinsic electron source.

An electron gun using a field emission cathode is differently constructed (Fig. 2.12b). To create a high electric field strength in front of the cathode tip the extractor electrode holds a positive potential against the cathode. The suppressor cap is weakly negative against the cathode and avoids that any electrons coming from other cathode parts than the tip may enter the beam. The gun lens allows focussing the beam, i.e. the forming of a cross over that can be shifted on the optical axis by changing the potentials of the electrodes (U_E, U_L). It is possible that electrons are hindered to go through this lens at special ratios U_L/U_E. These settings are unsuitable to get a beam current and should be avoided.

2.6 "Richtstrahlwert" (Brightness)

Now, we want to ask: "Is there a parameter describing the quality of the electron gun relevant to its role within an electron microscope?" Yes, there are even several parameters. One of them is the energy width, the other two are related to the electron current (brightness of the beam). The more easily understandable parameter is the maximum number of electrons in the beam, the maximum beam intensity. The other parameter is the *richtstrahlwert (brightness)* R[12]. It is defined by the quotient of the electron current density *j* escaping the cathode and the solid angle Ω in that the electrons are emitted:

$$R = \frac{j}{\Omega}. \tag{2.21}$$

For rotational symmetry and small angles

$$\Omega = \pi \cdot \alpha^2 \tag{2.22}$$

is valid where α the opening angle of the beam is. Let us think about the influences on α now. We use a strongly simplified model of the electron gun only including cathode and anode (Fig. 2.13).

---

[12] introduced into electron optics by E. Ruska and B. von Borries in 1939. In English the name "brightness" is common practice but we believe that this name does not hit the nail on the head. Hence, we suggest using the German "Richtstrahlwert" in English, too.

2.6 "Richtstrahlwert" (Brightness)

Fig. 2.13 Sketch for calculation of angle α enclosed in the richtstrahlwert (**v** velocity of the electrons, $\mathbf{v}_T$ component by thermal emission, $\mathbf{v}_{Acc}$ component by the accelerating field)

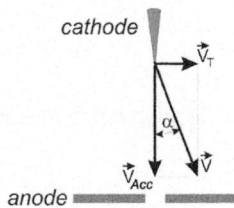

The largest angle α is reached by electrons which escape from the cathode perpendicularly to the central axis with the thermal energy

$$\frac{m}{2} \cdot v_T^2 = e \cdot U_T. \tag{2.23}$$

From Fig. 2.13 it follows for small angles ($\alpha \ll 1$) using the energy theorem (1.16):

$$\alpha = \frac{v_T}{v_{Acc}} = \sqrt{\frac{e \cdot U_T}{e \cdot U_o}} = \sqrt{\frac{U_T}{U_o}}. \tag{2.24}$$

For the thermal emission we calculate the current density using Richardson's Eq. (2.19) and get as an approximation of the richtstrahlwert

$$R = \frac{A \cdot T^2 \cdot e^{-\frac{W_A}{k \cdot T}} \cdot U_o}{\pi \cdot U_T}. \tag{2.25}$$

At an acceleration voltage $U_0 = 200$ kV the following richtstrahlwerts can be evaluated: for tungsten hairpin cathode ($W_A \approx 4.6$ eV, $T \approx 2700$ K, $e \cdot U_T \approx 5$ eV): $R \approx 3 \cdot 10^4$ A/cm² and for LaB$_6$ cathode ($W_A \approx 2.1$ eV, $T \approx 1800$ K, $e \cdot U_T \approx 3$ eV): $R \approx 10^7$ A/cm². At Schottky field emission cathodes a zirconium doped tungsten tip is used connected with a reduction of the work function to values of about 3 eV. Additionally, we have a further reduction of the work function by the Schottky effect. Using an extractor voltage of 3 kV and a tip radius of 1.5 μm the electric field strength in front of the cathode amounts to about $2 \cdot 10^4$ kV/cm. Due to this the work function decreases by ca. 1.7 eV. With $W_A \approx 1.3$ eV, $T \approx 1500$ K, $e \cdot U_T \approx 1$ eV we get only for the part of thermal emission a richtstrahlwert of about $7 \cdot 10^8$ A/cm². In reality the current density should be larger because of the tunnelling effect. For the Schottky cathode the richtstrahlwert is in the order of 10^9 A/cm².

For the cold field emission cathode the tip radius is smaller than that of the Schottky cathode to get a very high field strength and the energy spread amounts to only about 0.3 eV resulting in a richtstrahlwert greater than 10^{10} A/cm².

Let us ask for the practical consequences of a larger richtstrahlwert. What happens to it along the ray path for imaging? The beam current I_S should be formed by electrons escaped from a cathode area A_C and emitted into a solid angle

$$\Omega_C = \pi \cdot \alpha_C^2. \tag{2.26}$$

The richtstrahlwert referred to the cathode is given by

$$R_C = \frac{I_S}{A_C \cdot \pi \cdot \alpha_C^2}. \tag{2.27}$$

The condenser lens images the area A_C into an area A_B given by

$$A_B = A_C \cdot M^2 \tag{2.28}$$

(*M*: magnification). As consequence the (half) angle α_C of the beam opening is changed to

$$\alpha_B = \frac{\alpha_C}{M}. \tag{2.29}$$

If there are no apertures in the ray path the beam current I_S does not change. The richtstrahlwert R_B at A_B is given by

$$R_B = \frac{I_S}{A_B \cdot \pi \cdot \alpha_B^2} = \frac{I_S \cdot M^2}{A_C \cdot M^2 \cdot \pi \cdot \alpha_C^2} = R_C, \tag{2.30}$$

i.e. the richtstrahlwert does not change by the lenses[13]. Eq. (2.30) is also valid for the object plane. If we focus the electron beam to a preferably small spot with diameter d_S and beam current I_S then we yield

$$R = \frac{j}{\pi \cdot \alpha^2} = \frac{4 \cdot I_S}{\pi^2 \cdot d_S^2 \cdot \alpha^2}, \tag{2.31}$$

respectively

$$I_S = \frac{\pi^2 \cdot d_S^2 \cdot \alpha^2}{4} \cdot R, \tag{2.32}$$

i.e. the (along the ray path constant) beam current depends among others on the spot diameter d_S. The richtstrahlwert R indicates the parameter describing which maximal beam current is possible in a *small* electron probe. Later on, we will demonstrate additional consequences of a large richtstrahlwert for electron microscopy.

---

[13] According to Joseph Liouville's (French mathematician, 1809–1882) theorem the richtstrahlwert is a conserved physical property (like energy or momentum).

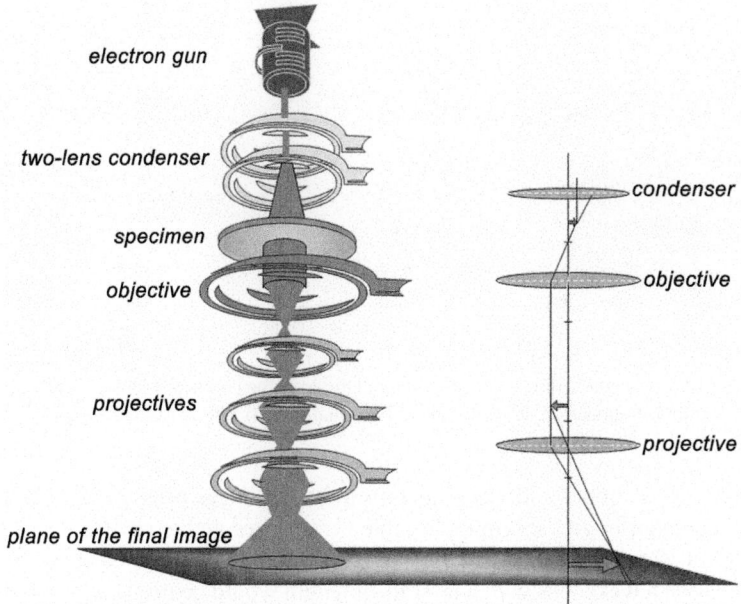

Fig. 2.14 Comparison between light optical microscope (*right*) and transmission electron microscope (*left*)

2.7 We Construct an Electron Microscope

After understanding the function of electron gun and electron lenses we are able to construct an electron microscope by use of our knowledge and some additions in a gedankenexperiment. The light optical microscope serves as an example but we rotate it in a way that the condenser will be the lens on the top (Fig. 2.14).

Of course, the lamp is substituted by an electron gun and the lenses are electric coils with pole pieces creating rotational-symmetric magnetic fields. Furthermore, the condenser and the projective are systems consisting of more than one lens. They will be described hereafter. Additionally, a special specimen stage is necessary to hold the sample with sufficient stability and to move it with high precision.

As humans do not have a sense organ for electrons we need special sensors to observe the image in the final image plane. Finally, we need a vacuum system to be able to work with free electrons within the microscopic column.

2.7.1 Illumination System

The illumination system includes an electron gun and a condenser system, i.e. "All things that are in the ray path before the specimen". The purpose of the illumination system is to set the electron current density on the sample to provide different

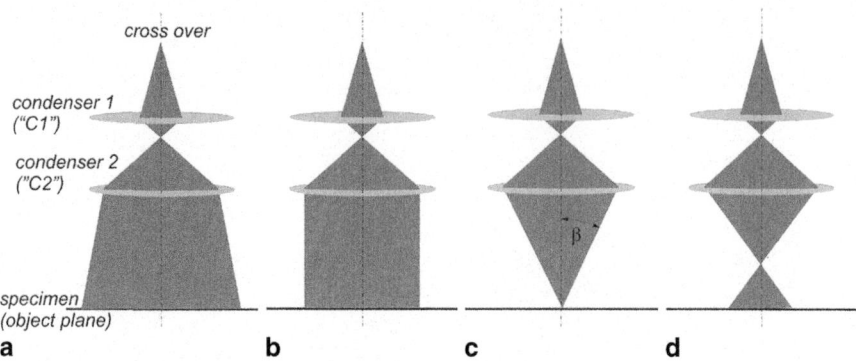

Fig. 2.15 Two-lens condenser with different excitations of the condenser lens 2: (**a**) Weak excitation. (**b**) Parallel illumination. (**c**) Convergent illumination. (**d**) Strong excitation

illumination conditions and, consequently, to set the image brightness. The distance between condenser and specimen is some centimetres and is rather large for electron-optical conditions. To image the cross over created by the electron gun onto the specimen a weak lens with a long focal length would be needed. Unfortunately, such lenses have a large spherical aberration and hence they are unsuitable for this purpose. The way out is given by the use of more than one condenser lens. The most common and easy to understand condenser system is the two-lens condenser (Fig. 2.15).

By shortening the focal length of condenser lens 1 the cross over will be more strongly demagnified, i.e. the spot on the sample will be smaller. Therefore this setting is named "spot size" on the control panel. But please do not forget the Sect. 2.6 about the richtstrahlwert: a smaller probe diameter means a smaller beam current, too.

By changing the focal length of the condenser lens 2 the current density at the specimen is set (therefore often labelled "intensity" or "brightness" on the control panel). The parallel and the convergent illumination are important special cases. For the latter one the illumination aperture β plays an important role (Fig. 2.15b and c).

Sometimes it is necessary to change this illumination aperture β at convergent illumination (cf. Chap. 5 Electron diffraction). This can be done by setting an aperture close to the condenser lens 2, the so-called "condenser aperture" (Fig. 2.16).

However, a larger number of electrons will be removed from the ray path by a smaller condenser 2 aperture: the beam current and also the richtstrahlwert become smaller. We used the two-lens condenser system to explain the illumination. It exists also other variants like three-lens condenser systems which have more flexibility (and complexity) and as Köhler[14]-type systems are optimised for parallel illumination [5].

---

[14] August Köhler, German optician, 1866–1948

Fig. 2.16 Setting of the illumination aperture β by a diaphragm close to the condenser lens 2 ("*C2-aperture*")

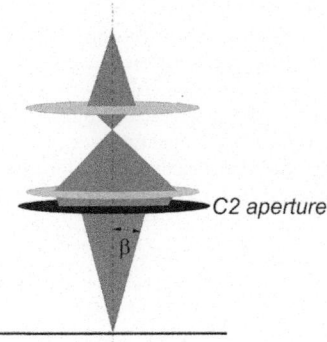

2.7.2 Imaging System

The imaging system produces an image of the electron-transmitted specimen and includes all lenses behind the specimen.

To get a multistage imaging an exact coordination of distances and focal lengths of the lenses creating the imaging system is absolutely necessary. One part of this coordination is the position of the specimen, i.e. the object plane. When the magnification at the microscope changes the focal lengths of the projective lenses change, the focal length of the objective lens remains constant. Due to the fixed final image plane different planes within the ray path are imaged by variation of the focal length f_{Proj} of the complete projective lens system (cf. lens Eq. (1.3)). For the imaging ray path this relates to the first real intermediate image plane.

The real intermediate image planes are *spatially selective planes*: The selection of an area by a small aperture in one of these planes is equal to the selection of an area on the specimen (object).

In this context another plane within the ray path has to be emphasised: The back-focal plane of the objective lens. The lens Eq. (1.3) shows that for all rays going parallel into the lens (object distance $g \to \infty$) the image distance b is equal to the focal length f, i.e. these rays meet in the back-focal plane (Fig. 2.17).

If these rays additionally run parallel to the optical axis ("parallel rays") they are joined in the back-focal point. If the rays are inclined relative to the optical axis the cross point moves away from the focal point on the optical axis but it remains approximately in the back-focal plane at small inclinations (Fig. 2.18).

When we put a small aperture in the back-focal plane and select a small area in this plane we choose a range of inclination angles of the incoming rays on the object-side of the lens. Therefore, the back-focal plane constitutes an *angle-selective plane* (diffraction plane, cf. Sect. 5.3).

Fig. 2.17 Ray path in a microscope with three-stage imaging (two projective lenses). Without consideration of the image rotation by magnetic lenses the final image is upside down in comparison to the specimen

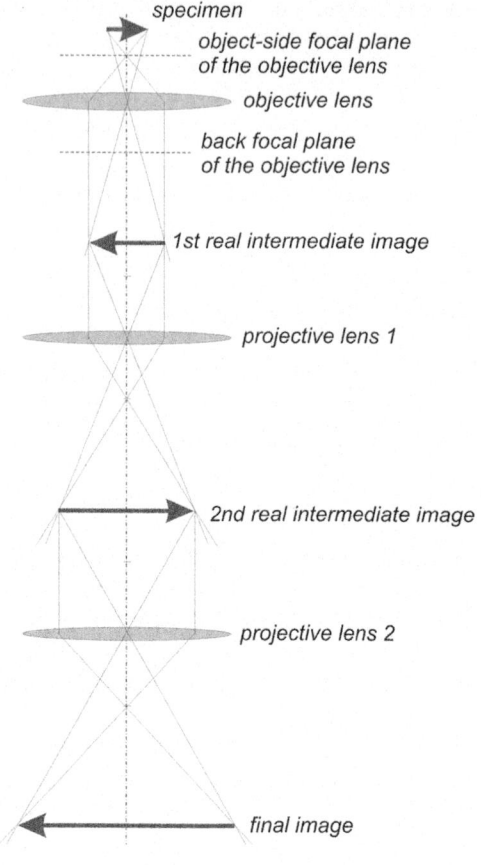

Fig. 2.18 Back-focal plane as angle-selective plane: A bundle of parallel rays inclined by an angle α relative to the optical axis meets in the back-focal plane in a distance $\alpha \cdot f$ from the optical axis ($\alpha \ll 1$)

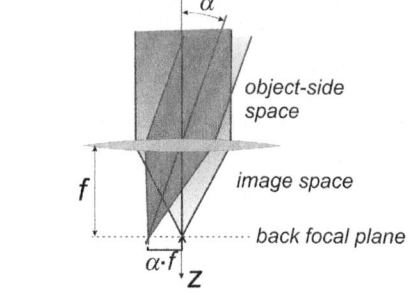

2.7.3 Specimen Stage

At a first glance it may seem to be surprising that we devote a special paragraph to the specimen stage of the electron microscope. But when we think about the useful magnification of the electron microscope in the order of one million the particu-

2.7 We Construct an Electron Microscope

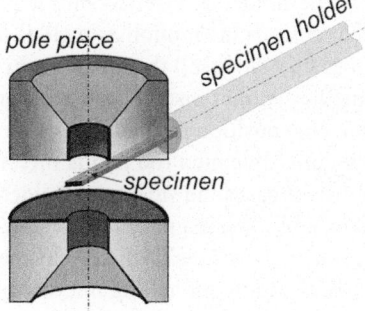

Fig. 2.19 Arrangement of the specimen within the pole piece of the objective lens (side-entry holder)

lar importance of the specimen stage can be understood better: When we record an image with a magnification of one million and the specimen moves by 0.1 nm during the recording time of e.g. 1 s the motion blur of the image reaches 0.1 mm corresponding to the resolution limit of the human eye. In other words: the image becomes unsharp!

Therewith an exceptional requirement for the stability of the specimen mounting is formulated: The motion of the specimen in a time of 1 s has to be less than one-tenth millionth millimetre. We know that the thermal expansion coefficient of metals is in the order of 10^{-5} K^{-1}. That means, a temperature fluctuation of only 1 K would change the length of an 1 mm long holder by 10 nm. The specimen mounting has to be exceptionally good in its mechanical stability and the fluctuations of the ambient temperature of the specimen should be less than 0.1 K with consideration of the thermal inertia.

Anything but that: To investigate special details of the specimen we need to be able to move the specimen accurately to a defined position which requires two perpendicular directions (x and y) in the specimen plane. As we know from the discussion about the imaging system (Sect. 2.1.2) there is an object plane that has to be approximately kept. Otherwise the coordination of the projective focal lengths does not properly work. That means, it must be possible to move the specimen along the objective axis (optical axis, z-direction) with high precision, too.

Finally (and we will understand this better after reading the Chaps. 5, 6, and 7), it is absolutely necessary for some investigations of crystalline materials to tilt the specimen around two axes for a suitable orientation between electron beam and crystal lattice.

The next point is to think about the position of the object plane (object distance to the objective) within the objective lens field. For the objective we use a lens with a short focal length to minimise the spherical aberration. Its focal length is about 1 mm ... 2 mm. Assuming that we want to get a magnification of about 100 by the objective lens then the distance between the object-side focal point and object plane amounts only to 1 % of the focal length (cf. Eq. (2.3)). That means the specimen is located within the pole piece of the objective close to the centre of the lens field (Fig. 2.19).

The holder is inserted into a device called "goniometer" or "specimen stage" that ensures the motion as well as the vacuum leakproofness. Beside the described possibility to insert the holder from the side into the pole piece ("side-entry holder") another variant inserts the specimen from the top or by a cartridge system into the imaging position within the pole piece ("top-entry holder"). The latter design avoids the long specimen rod and is therefore optimised for stability. Anyhow, in the majority of cases the side-entry holder is used and we confine ourselves to that one in this book. The main reason for side-entry stages is the opportunity to construct e.g. electrical wires inside the rod to enable in-situ experiments (heating, electrical bias, even scanning tunnelling microscopy, etc.).

2.7.4 Acquiring the Images

In microscopy one observes visually the final image plane. For this purpose the electrons have to be converted to visible light using an effect known as "cathodoluminescence". The electrons of a solid state atom are lifted on a higher energetic level by the incoming beam electrons. This "excited state" is not stable, the electrons in the higher levels go back to their ground state and the atom emits electromagnetic radiation (cf. Sect. 9.1.1). The wavelength of the emitted radiation depends on the energetic difference between the excited state and the ground state und therefore on the luminescent material. Visible light with a wavelength of about 500 nm (green light) is produced at an energetic difference of about 2.5 eV (e.g. cadmium sulphide, zinc selenide). The observation screen in the final image plane is a metal plate (usually aluminium) covered by a thin layer of the luminescent material. It is observed from the outside through a lead glass window.

In some cases (electron microscope completely covered by a housing to shield the microscope from noise and other outer influences or if the final image plane is very close to the floor) it is impossible to observe the screen directly. In these cases a television camera focussed on the screen is usual.

The next step is given by the documentation of the electron microscopic pictures. Until the 1990s this had mainly happened by photographs using special electron-sensitive emulsions. The display of the exposure time present even at modern electron microscopes is a remember of that time. An advantage of this method is the comparably large viewing field on the photographic plate. The disadvantage is the fact that the films or photographic plates have to be inserted into the microscope vacuum. Because of the water content of the emulsions a disturbance of the vacuum cannot be excluded even after vacuum drying before. And finally, the pictures are not immediately available. The development of the photos needs time, chemicals (developing bath, fixing bath) and last but not least a person experienced in photographic processing.

Currently, the pictures are often registered by a "CCD camera" (CCD: Charge-Coupled Device). It mainly consists of a transparent screen, a fibre optics, and a CCD array (Fig. 2.20).

2.7 We Construct an Electron Microscope

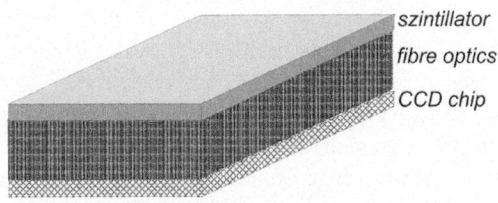

Fig. 2.20 Sketch of a CCD camera

The screen should have a high lateral resolution, hence an yttrium aluminium garnet ("YAG") crystal is used. It converts the signal of the incoming electrons into a light signal, which is transferred to the CCD elements by glass fibres. The thickness of the screen influences the yield of light (the thicker screen is more sensitive) as well as the spread of the light over adjacent CCD elements (the thicker screen spreads the information over a larger area). That needs a compromise for the electron energy (acceleration voltage) mainly used in the electron microscope.

The CCD element consists of silicon covered by a thin layer of silicon oxide and a (translucent) metal electrode ("MOS capacitor"). Impinging of light produces electron-hole pairs which charge the small capacitors. The quantity of this charge depends on the light intensity and on the exposure time (also known as "dwell time"). After the end of the dwell time the charges are readout and the CCD element is set back into its ground state. The camera is ready for the next image acquisition.

Due to the thermal motion of the charge carriers small charges are also produced without any incoming light. This "dark current" depends on the dwell time and is usually automatically measured before the image record starts. Later on, the picture will be corrected by the computer. The dark current can be reduced by Peltier[15]-cooling of the CCD array. An additional effect is the different efficiency of the individual CCD elements (pixels) which results in different output signals (counts) at an equal light intensity. To correct for that ("gain correction" or "flat field correction") a reference image is necessary that has to be recorded with uniform illumination of the complete camera area. Cameras with a larger viewing field are often put together of four single arrays ("quadrants"). The efficiency differences between them can be corrected in the described way, too.

If the number of electron-hole pairs becomes too large some of the charges flow to adjacent pixels ("blooming"). To avoid that the intensity of the electron beam or the dwell time has to be reduced. Be careful, a long time overload can permanently damage the CCD element.

In case of a weak intensity of the electron beam (image brightness) or intentionally short dwell times several pixels can be connected ("binning"). Binning 2 means, for example, that $2 \times 2 = 4$ pixels are connected. The sensitivity becomes four times higher, however at the expense of the lateral resolution by increase of the effective pixel size.

---

[15] Jean Peltier, French physicist, 1785–1845

Especially at very low intensities of the electron beam connected with long dwell times single pixels in the CCD camera image can appear extremely bright. X-ray quanta excited by the electrons cause this phenomenon. The higher energy of the X-rays (in comparison with light) leads to a very high yield of electron-hole pairs in the CCD element hit by an X-ray.

A disadvantage of the CCD camera, especially at overview pictures, is given by the comparably small viewing field. For a (usual) pixel size of 24 μm on the YAG screen and a so-called 1 K × 1 K camera (i.e. 1024 × 1024 pixels) the size of the viewing field only amounts to 24.6 mm × 24.6 mm. Cameras with larger pixel numbers have normally smaller pixel sizes connected with a somehow better resolution and the viewing field does not increase much, except for extremely large CCDs with e.g. 8 K × 8 K pixels. A software based solution to increase the field of view is a photomontage of single shifted frames.

CMOS-based direct detection cameras which have a high speed and sensitivity are a new development.

2.7.5 Vacuum System

There are several reasons why we need vacuum within the electron microscope:

- The electrons should not collide with air molecules inside the column.
- The heating filament of the cathode should not burn away.
- The tip of the cathode should not be rounded by ion bombardment.
- The electron gun should be voltage-proof, i.e. flashovers between cathode and anode should be avoided.
- The specimen should not be contaminated.

To fulfil these demands requirements for the maximal pressure within the microscope can be deduced. From the ideal gas law

$$p \cdot V = m \cdot R \cdot T \tag{2.33}$$

(p: gas pressure, V: gas volume, m: gas mass, R: gas constant, T: absolute temperature) follows the gas particle density

$$n = N_A \cdot \frac{m}{V} = \frac{N_A \cdot p}{R \cdot T} = \frac{p}{k \cdot T} \tag{2.34}$$

(N_A: Avogadro's[16] constant, k: Boltzmann's constant).

If the motion of the electrons should happen without any collisions with air molecules the mean free path between two collisions has to be much greater than the height h_M of the microscope (ca. 2 m). The mean free path is approximated by

---

[16] Amedeo Avogadro, Italian mathematician and physicist, 1776–1856

2.7 We Construct an Electron Microscope

$$\Lambda \approx \frac{1}{n \cdot \pi \cdot r_M^2} = \frac{k \cdot T}{p \cdot \pi \cdot r_M^2} \qquad (2.35)$$

with r_M as radius of a molecule. Because of $\Lambda \gg h_M$ it follows

$$p \ll \frac{k \cdot T}{h_M \cdot \pi \cdot r_M^2}. \qquad (2.36)$$

For oxygen und nitrogen r_M is ca. 150 pm is, i.e. at room temperature (293 K) the pressure within the microscopic column has to be considerably smaller than $3 \cdot 10^{-2}$ Pa. Assuming "considerably smaller" means "ten times smaller" a pressure $p < 10^{-3}$ Pa follows.

In this context a view on the gas particle density n is of interest. It amounts to about $3 \cdot 10^8$ molecules per mm^3 at a pressure of 10^{-3} Pa. The gas kinetics (Maxwell's[17] velocity distribution of the molecules) delivers a particle flow

$$\frac{dN}{dt} = \frac{n \cdot A}{4} \cdot \sqrt{\frac{8}{\pi} \cdot R \cdot T} = \frac{p \cdot A}{k \cdot T} \sqrt{\frac{R \cdot T}{2 \cdot \pi}} = p \cdot A \cdot \sqrt{\frac{N_A}{2 \cdot \pi \cdot k \cdot T}} \qquad (2.37)$$

from the half space onto a wall with area A. For air (1 mol = 29 g) and an area of 3 μm^2 (cathode tip with radius 1 μm) follows a particle flow of about 10^7 particles per second hitting the tip. Assuming one millionth of the particles had the energy to change the tip we had ten critical collisions per second. The importance of the very low pressure within the cathode chamber can be clearly seen. Following our estimation the number of critical collisions is reduced to less than 1 per hour at ultrahigh vacuum ($p < 10^{-8}$ Pa).

We learn that the vacuum requirements are different in the different parts of the electron microscope. In an electron gun equipped with field emission cathode the pressure should be less than 10^{-8} Pa, the other parts of the column need a pressure of less than 10^{-3} Pa. Small apertures between gun and column are used to stabilise this pressure difference. Please keep in mind that the evacuated part of the seemingly thick microscope column is only a liner tube of ca. 1 cm diameter.

We need pumps to generate and maintain these vacua. There are two different kinds of such pumps: ones which transport the gas and ones which save the gas. Transport pumps run with alternating volumes alternately sucking the gas from the chamber, compressing it and exhausting it to the environment (examples: rotary pumps, diaphragm pumps, roots pumps). Another variant of a transport pump induces a preferred direction to the Brownian[18] motion of the gas molecules. That happens by touching the gas molecules with the rotor of a turbomolecular pump

---

[17] James Clerk Maxwell, Scottish physicist, 1831–1879
[18] Robert Brown, Scottish botanist, 1773–1858

Fig. 2.21 Ion getter pump (housing opened using a saw, without permanent magnet). Interference colours of thin layers are visible in the centre

or by collisions with a directed beam of oil molecules (oil diffusion pump). These pumps cannot work against atmospheric pressure, they need a prevacuum generated by a transport pump as explained above.

Another type are storage pumps that save the pumped gas inside. The ion getter pump is the most usual variant in the high and ultrahigh vacuum range. In the simplest case it consists of a grounded stainless steel housing with a titanium grid inside (Fig. 2.21) being completely within a field of a permanent magnet. Between grid and housing a voltage of about 5 kV is applied. Below a pressure of some 10 Pa a glow-discharge burns kept up by collision ionisation. Titanium sputtered from the grid is deposited on the housing wall. The titanium layer chemically bonds oxygen and nitrogen and is permanently regenerated. Noble gases are only wafted by new layers after physisorption. Hence, the pumping speed of the ion getter pump regarding noble gases is rather poor.

The getter pump needs prevacuum only to start the glow-discharge. Later, during the "normal" work, a prevacuum pump is not necessary. The magnetic field extends the ion paths and increases the cross section for ionisation of the residual gas molecules.

The pump regulates itself: At higher pressure the glow-discharge and the pumping speed increase, at lower pressure they decrease. Therefore the pump current is an indication of the pressure, too. However, if the pump is running at a pressure within the prevacuum range for a longer time it gets very warm and the titanium layers release the bonded gas. The only help is to switch off the getter pump and to work with another pump until the getter pump has cooled down again.

An advantage of this pump is the absence of mechanically moving parts. This is helpful for the mechanical stability of the microscopic column.

Finally, the pressure has to be controlled, i.e. it must be measured. Commonly, two different measuring methods are used at an electron microscope working in different pressure ranges.

2.7 We Construct an Electron Microscope

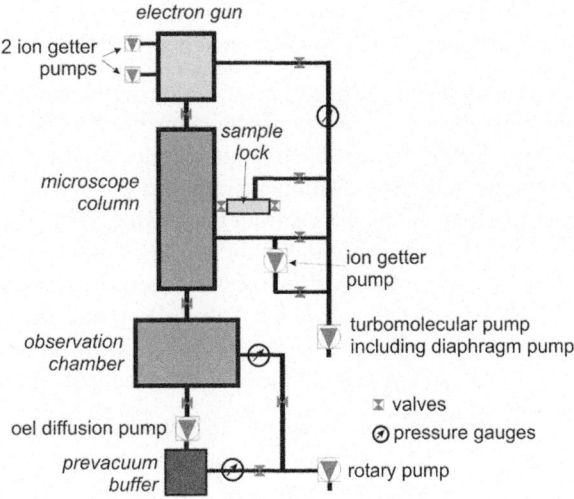

Fig. 2.22 Sketch of the vacuum system of an electron microscope equipped with Schottky field emission cathode

The first method bases on the heat conduction in gases which is caused by the motion of the gas molecules (convection). The gauge head consists of a current-carrying wire within a metal cylinder. The wire is heated during current flow, at the same time it is cooled by convection determined by the gas pressure. As the electric resistance of the wire depends on its temperature, the electric resistance is determined by the gas pressure. This gauge type is called "Pirani"[19] and works preferably in the pressure range between 1 Pa and 100 Pa.

The second method ("Penning"[20]) operates accordimg to the principle of the pressure dependent glow-discharge already explained in the description of the ion getter pump. However, as we want to measure and not to pump, the titanium grid is substituted by a single wire. This Penning gauge is used at pressures of less than 1 Pa and is often combined with a Pirani manometer.

Figure 2.22 shows the principle of a vacuum system of an electron microscope equipped with field emission gun. In case of an LaB_6 cathode the requirements for the vacuum within the electron gun are lower and, commonly, there are no additional pumps directly flanged at the gun. There is an outer tube connection to the pump at the microscope column with low flow resistance instead of the extra pumps.

2.7.6 Miscellaneous

In principle our electron microscope is complete now: We have got an illumination system, the specimen stage, the imaging system, the possibility to record the images

---

[19] Marcello Pirani, German physicist, 1880–1968
[20] Frans Michel Penning, Dutch physicist, 1894–1953

and we generate the required vacuum. Nevertheless, some important details are missing which should be mentioned here:

Beam alignment Despite of all accuracy during the production and the assembly of the parts the mechanical precision is insufficient for electron-optical requirements. Hence, it must be possible to align the electron beam, lenses, and apertures during operation. The first instruments allowed a mechanical tilting and moving of the electron gun and the lenses from the outside. This happens on cost of the mechanical stability and is unsuitable for modern microscopes. Instead of the mechanical shift the path of the electron beam is aligned to the geometric relations by a large number of deflection elements. The electron beam goes to the observation chamber within a pipe with a diameter of about 1 cm. The apertures are leakproofly inserted in this pipe and can be mechanically aligned from the outside.

Apertures The functioning of the condenser 2 aperture was already described (Sect. 2.7.1) Later, we will see that at least two additional apertures are needed (objective or contrast aperture and selected area aperture). These apertures must be aligned from the outside, too.

Shielding of ionising radiation Whenever electrons with high energy strike solids X-rays are emitted. This radiation is dangerous for the operator and its escape from the microscopic column has to be avoided. Therefore the column and the observation chamber are covered by lead and the observation window is made of lead glass.

Cold trap (liquid nitrogen cooling device) To minimise the contamination of the specimen the area around the sample is covered by copper plates which are cooled from the outside of the microscopic column by liquid nitrogen. Thus, the partial pressure of the hydrocarbons can be reduced (cf. Sect. 4.5). Water vapour and other "high-boiling" gases freeze out at the cooled plates and are released again after removing the liquid nitrogen reservoir. The getter pumps save a comparably large quantity of gas in this short time. This is prevented by start of a "cryo-cycle": After removing the liquid nitrogen reservoir the getter pumps responsible for the evacuation of the specimen chamber are switched off. During this time the cooled plates warm up and the specimen chamber is evacuated by another pump (in the vacuum system drawn in Fig. 2.22 this is the turbomolecular pump). After the plates are warm the strong gas yield is finished and the getter pumps can be switched on again.

Prevention of extraneous influences Low vibrations e.g. created by persons walking close to the microscope have to be kept away from the microscope column. For that purpose the column is mounted on air buffers. Additionally, at very high magnifications (loud) noises, air vortices, and alternating magnetic fields disturb. Even the operator can be the reason for troubles since he acts as heat source. Modern electron microscopes are often covered by a special soundproof housing whose walls consist of a metal with high permeability ("µ-metal"). The possibility of magnetic field compensation is often added. The operator sits outside of the cabin and remotly controls the microscope with the help of cameras and computers.

Power supply Operating an electron microscope needs a large number of supply voltages and currents. Special requirements for stability are given for the acceleration voltage (between cathode and anode), for the lens currents (cf. Sect. 2.3), and for the supply of deflection elements. Since the power supplies emit a lot of heat they should be placed in a separate room.

Air conditioner and water cooler with thermostat The heat produced by lens currents, electronics, operator etc. has to be reliably removed. Within the specimen chamber the temperature fluctuation should be less than 0.1 K (cf. Sect. 2.7.3) which is only reachable by direct water cooling of the lenses. The demands on the mechanical and temperature stability are very high, e.g. if there would be tiny air bubbles inside the cooling water pipes would produce inadmissible vibrations.

The microscopic laboratory has to be air conditioned (preferably to the temperature of the specimen holder inside the objective lens). The air conditioning system must avoid erratic air movements especially at the microscope column. Depending on the required performance of the microscope the above mentioned subjects may require a very careful planning and a substantial amount of money.

Computer Finally, a computer with high performance is absolutely necessary to control the microscope including accessories and measuring programs, monitoring important operations, allowing the data recording and, commonly, supporting the operator at the electron microscope.

References

1. Scherzer, O.: Über einige Fehler von Elektronenlinsen. Zeitschrift f. Physik. **101**, 593–603 (1936)
2. Haider, M., Müller, H., Uhlemann, S., Zach, J., Loebau, U., Hoeschen, R.: Prerequisites for a Cc/Cs-corrected ultrahigh-resolution TEM. Ultramicroscopy. **108**, 167–178 (2008)
3. Boersch, H.: Experimentelle Bestimmung der Energieverteilung in thermisch ausgelösten Elektronenstrahlen. Zeitschr. f. Physik. **139**, 115–146 (1954)
4. Bronsgeest, M.S., Barth, J.E., Schwind, G.A., Swanson, L.W., Kruit, P.: Extracting the Boersch effect contribution from experimental energy spread measurements for Schottky electron emitters. J. Vac. Sci. Technol. **B25**(6), 2049–2054 (2007)
5. Benner, G., Probst, W.: Köhler illumination in the TEM: Fundamentals and advantages. J. Microsc. **174**(3), 133–142 (1994)

Chapter 3
We Prepare Electron-Transparent Samples

Abstract During establishing of a new electron microscopic laboratory it has to be considered that proper thin specimens are absolutely necessary for transmission electron microscopy. As a matter of course the preparation of such samples ranges from simple to very difficult and demanding. This chapter gives an overview on suitable preparation methods for transmission electron microscopic specimens.

> The quality of the transmission electron microscopic results cannot be better than the quality of the electron microscopic specimen preparation!

3.1 What is the Challenge?

While electrons are transmitted through a solid they are interacting with the atomic nuclei, i.e. they experience the strong Coulomb[1] force (cf. Sect. 6.1). Therefore the electrons can be transmitted only through very thin foils. But, without transmission through the specimen our "transmission microscope" does not work. At electron energies between 60 and 300 keV as are usual in the transmission electron microscope the thickness of the specimen must be in the order of 100 nm (or lower). Later on, we will see more precisely that this amount announces only a coarse orientation. In practice the necessary specimen thickness depends on the specific question to be answered by electron microscopic methods. To image atomic columns we need extremely thin samples (thicknesses about 5 ... 20 nm), for measurements of electron energy losses thicknesses of 30–50 nm are optimal. In contrast, for conventional microstructure investigations, e.g. to image dislocations, the specimens should be thicker than 150 nm (cf. Chaps. 6, 7, and 9).

Let us illustrate the challenge by a simple example: We imagine that we want to investigate a ceramics by transmission electron microscopy. High temperature

---

[1] Charles Augustin de Coulomb, French physicist, 1736–1806.

superconductors[2] can be such ceramics for example. Porcelain is a more familiar kind of ceramics. In our little gedankenexperiment let us take a porcelain plate from the cupboard, cut out a small disc with 3 mm diameter from the plate ground and thin this disc in its centre to a thickness less than 50 nm. (By the way, a human hair is about 50,000 nm thick!) We can easily guess that this kind of preparation is time-consuming and not simple.

There are valid methods for this preparation, of course. In the following we want to specify some of them. We will confine ourselves to methods which are important in materials science and where our own experiences exist. For samples in life sciences and special cases we point to the literature (e.g. [1, 2], see also the Hints for further reading). Certainly, the electron microscopic sample preparation remains a challenge (almost) anyway.

And finally, if we prepare and investigate a suitable specimen new questions will appear:

- How far has the material been changed by the preparation?
- Is the tiny specimen prepared for transmission electron microscopy really representative for the complete material?

At first, let us answer the second question. Strictly speaking, we have to organise a sampling plan following the rules of mathematical statistics. The higher the number of samples the greater the statistical certainty of the result is. That, at least, is theory for the case. In the electron microscopic practice the time-consuming preparation often excludes this procedure demanded by the statistical rules. To defuse the question we assume that we have a box filled with 100 screws. Untypically, one of them has got a wrong thread. The probability to take just this untypical screw with one grab in the box is equal to 0.01. This is rather improbable but not impossible. This holds true for the task to determine typical structural features of a material. However, often the task is to determine structural reasons for a specific fault of a material. In this case the specimen has to be prepared at a specific pre-defined location.

The answer to the first question is more difficult. It extremely depends on the material and on the selected preparation method how far the sample has been already changed by the electron microscopic preparation. Possibly, the preparation is already a true scientific challenge, especially if there are no experiences with new kinds of materials. The description of the preparation procedures in this book is based on such experiences. As a rule, there are different methods for the preparation steps: either they are fast but strongly stress the material or they stress the material less but they are slow. It can be investigated by systematic variation of the methods how far the sample can be changed by the preparation.

Finally, we want to give a fundamental comment. "We need some transmission electron microscopic pictures of our sample" is a wish often raised by customers

---

[2] Superconductors conduct the electric current below a transition temperature without any electric resistance. At classical superconductors (metals) this temperature amounts to only some K. At high temperature superconductors it is considerably higher, e.g. larger than 77 K = $-196\,°C$ (temperature of liquid nitrogen).

of an electron microscopic laboratory. The electron microscopists favour a more precise question, e.g. what is the particle size? What is the layer thickness? Which morphology and which kind of phases exist? As a rule, it is purposeful to consider these questions already before the preparation. Or, in other words, if any transmission electron microscopic investigations are planned the necessary electron microscopic preparation should be included in the considerations as early as possible. The sample geometry can often be adjusted to the preparation requirements. Or a thin layer can be deposited onto a substrate that is better suited for later electron microscopic preparation.

3.2 "Classical" Methods

The challenge that very thin specimens are needed has been an integral part of transmission electron microscopy since its beginning. It does not surprise that the first scientifically relevant results could be obtained at biological samples. Microtome cuts were well-known in the 1930s. The main components hydrogen, nitrogen, and oxygen of biological samples weakly deflect the electrons because of their low atomic number; hence these specimens can be comparably thick. In 1939 Helmut Ruska[3] published first results and described cellulose lacquer as support film for electron microscopic specimens and the impregnation with metals as well as another targeted manipulation of single cell parts to change their density and/or their thickness to increase the contrast [3]. In that time the visualisation of viruses using an electron microscope played an important part. Hitherto the viruses seemed to be "invisible" because of the limited resolution of the light optical microscope.

Against this background we reach the first "classical" method of electron microscopic specimen preparation:

Bringing Small Particles Onto Support Films This method is suitable to measure shape and size distributions of small particles. As support films carbon or plastic films deposited onto a support grid of copper or another material are used (Fig. 3.1).

Instead of the grid other openings in a disc with a diameter of 3 mm are also usual: circular holes, slits, and arrangements of slits.

Silicon nitride windows are also available for supporting. Several windows about 100 µm times 100 µm in size are etched in a 200 ... 300 µm thick disc. The residual thickness of the windows is about 30 ... 100 nm.

It depends on the consistence of the particles how they are put onto the support film. If they are provided in the form of smoke a film covered support grid is fixed by tweezers at its edge and held in the smoke for a short time. Powders can be prepared by depositing a droplet of water or alcoholic suspension with the powder onto the support film using a thin glass rod. During this procedure the support grid must be fixed by a pair of tweezers. The droplet dries on the support film later. It

---

[3] Helmut Ruska, German physician, 1908–1973, Ernst Ruska's brother.

Fig. 3.1 Support grid for transmission electron microscopy (*thickness*: 10 … 30 µm, *mesh size*: 20 … 400 µm, *wire thickness*: 10 … 15 µm, *material*: copper, molybdenum, stainless steel, nickel, gold and others)

needs a little exercise; the droplet should not touch the tweezers and be removed by them. The beginner often produces specimens with a particle density being too high. A small tweezers tip of powder for a test tube of liquid is enough. If the powder tends to conglomeration a treatment of the suspension in an ultrasonic bath before depositing the droplet is useful.

If the particles are electron transparent additional analytical transmission electron microscopic investigations at such specimens are possible: determination of phases by electron diffraction, high resolution imaging, and analyses of the chemical composition.

In some circumstances larger particles can be pulverised using mortar and pestle to achieve electron transparency.

Gluing on Support Grids A preparation method proved to be successful for felts of nanoscaled fibres (e.g. carbon nanotubes) is the gluing of small felt pieces on a support grid. By doing this the support grid is held by tweezers and brushed with a tiny glue film by careful use of a glue stick. Subsequently, this support grid is pulled over the felt. Hereto the rule is: the less material the better. The advantage of this method over the deposition on a support film is the free-standing of the fibres (i.e. vacuum is in their surrounding) and the support film will not impair the contrast of the electron microscopic image. The disadvantage is that the single fibres can easily move and there is a risk that a motion blur arises during the recording of a micrograph. Often a compromise has to be found for electron microscopic investigations in such a way that fibres are used which are supported on both ends by more heavy felt.

Bringing Thin Films on Support Grids For samples that are free-standing thin films it is possible to put these films directly on a support grid. Normally, the films swim as small pieces (ca. 3×3 mm) on distilled water. The support grid is gripped by tweezers, dipped next to a film piece into the water, moved under the film piece, and hauled out together with the film. A little practice is absolutely necessary to find out the right speed of the lift-off procedure. The film piece should stay on the grid and not rinse off from it. The subsequent drying should be on clean and non-absorbent paper. A high absorbency could remove the film piece from the grid.

The next question is how we can get such free-standing films. One of the possibilities is the deposition of a thin (i.e. some few 10 nm thick) film on a single crystal of sodium chloride. The film is scratched by a scalpel or a razor blade in little squares (about 3×3 mm in size). The crystal is caught by tweezers and carefully

3.2 "Classical" Methods

shifted in distilled water with the film on the top. The water penetrates between film and sodium chloride and separates the film pieces from the crystal. After washing in clean distilled water the pieces are fished up as mentioned above. In case of crystalline samples it is possible to analyse the grain morphology, the phases, intermixtures and segregations as well as properties of grain boundaries. The use of heat-resistant support grids (e.g. such ones of molybdenum) and suitable specimen heating holders allow a heat treatment of the films within the microscope during the investigation. This enables the observation of grain growth and phase transformations in situ.

This kind of preparation allows the investigation of growth processes on substrates, too [4]. For this growth steps (i.e. crystallographic defects at the surface) are decorated by evaporation of a gold layer thin enough to avoid a continuous gold film onto the (heated) surface. The gold atoms move to steps or other energetically preferred positions on the crystal surface ("decoration"). Subsequently, a thin (10 … 20 nm) carbon film is evaporated; the gold-carbon film is removed from the crystal in distilled water as mentioned above. The special distribution of the gold particles in the carbon film resembles the arrangement of the growth steps.

The preparation of extraction replica can be performed similarly. The goal is the investigation of segregations on the surface of solids. To reveal the segregations in a first step the surface has to be selectively etched. Secondly, a thin carbon film is evaporated and reinforced by lacquer. Thirdly, the carbon-lacquer film has to be carefully stripped and brought onto a support grid. The lacquer must be chemically removed before starting the electron microscopic investigation. Alternatively, instead of a reinforcing by lacquer we can use a suitable chemical that dissolves the solid surface but not the carbon with segregations. In this case the film can be removed similar to the handling of films on sodium chloride crystals in water. We get a carbon film with embedded segregations which can be investigated concerning morphology, phases, and chemical composition. The choice of suitable substances and conditions for etching is very important for success and is often the most time-consuming part of the procedure.

Until the 1970s the preparation of replicas of solid surfaces was a widespread preparation method. The analysis of the topography on the sub-micrometre scale was the goal of this procedure. For doing that a matrix has to be impressed onto the surface that is to be investigated. For example, a small (ca. 0.5 mm thick and about 1 cm^2 in size) piece of acrylic glass can be used. The acrylic glass can be dissolved on its surface by a droplet of chloroform and subsequently it is pressed onto the sample surface. The acrylic glass resembles the topography of the surface and cures after evaporation of the chloroform. After stripping of the matrix its surface is obliquely coated with a thin (ca. 10 nm) film of a heavy metal (e.g. chromium) and subsequently with carbon from different directions (Fig. 3.2).

Due to the oblique evaporation the topography of the sample surface is reflected by the "light-shadow distribution" within the chromium layer. Because of the excellent mobility of carbon atoms the carbon layer covers the embedded chromium particles and generates a stable film.

Fig. 3.2 Preparation of a replica with acrylic glass matrix (evaporation step)

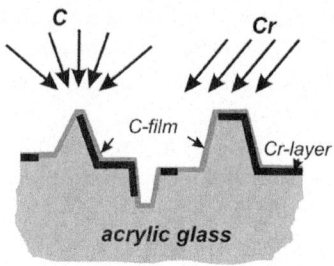

Fig. 3.3 Principle of the electrolytic thinning

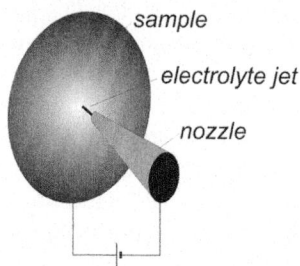

The matrix of acrylic glass is partitioned into small squares (ca. 3 × 3 mm) which are put on support grids. Finally, the acrylic glass has to be dissolved within an atmosphere of chloroform (within an extractor hood, of course).

The relevance of this method to prepare replicas of solid surfaces has drastically decreased with the availability of scanning electron microscopes. The surface topography can also be investigated by means of scanning electron microscopy on the sub-micrometre scale and the effort for preparation is much lower.

Electrolytic Thinning For electron microscopic investigations of the crystallographic morphology (crystallographic defects) of metals the electrolytic polishing has been established since the 1950s. It is used in case of electrically conducting, homogeneous samples.

The sample (disc with a diameter of 3 mm, ca. 0.5 mm thick) connected as anode is hit by an electrolyte jet sprayed from a nozzle connected as cathode (Fig. 3.3).

The end of the thinning process is reached after a little hole arose close to the centre of the sample. The sample thickness at the edge of this hole increases in a wedge-shaped manner and one can find in this area thicknesses well suitable for electron microscopic investigations. The different thicknesses of the wedge allow the use of the optimal thickness for answering special electron microscopic questions (cf. Sect. 3.1).

The selection of a well-suitable electrolyte and its correct temperature are the main problems of the electrolytic thinning. The current-voltage characteristic has to be measured to discover the range of the polishing region. Otherwise, the sample is etched and rough surfaces are generated, i.e. strongly fluctuating sample thicknesses. Before starting the electrolytic preparation a literature research is especially

Fig. 3.4 Geometry of thinned TEM specimens. The outer diameter must be equal (3.0±0.1) mm. Otherwise it is possible that the specimen is incompatible with the holder and cannot inserted

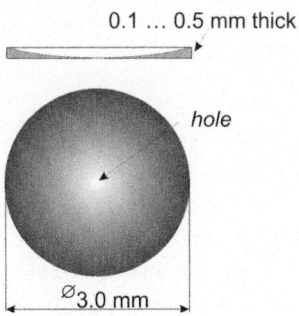

important to use other author's experiences with the material (see also: Hints for further reading, Electron microscopic preparation).

Ultramicrotome Thin sections of organic samples as usual in light optical microscopy are also suitable for electron microscopic investigations [5]. However, they must not be thicker than ca. 0.1 μm; hence they are called ultramicrotome sections. Usually the sample is embedded in epoxy before starting the ultra thin cut. Sometimes such sections are also produced from hard inorganic materials. The speciality of the ultramicrotome diamond knives is the comparably large cutting angle of more than 35°. The cut is similar to the planing in joinery but everything is much smaller than there. The sections are only some 10 μm in size. After the cut they drop into a water bath and are fished up on film covered support grids. An often experienced drawback of this method is a compression or coiling of the cut pieces.

Cleavage In special cases (e.g. silicon) it is possible to crack thin plates along crystallographic planes inclined to the plate surface. In this way one can get small wedges with wedge angles of about 20° and atomically smooth fracture planes [6]. At the tip such wedges are thin enough to electron-microscopically investigate them. If this method can be applied successfully very clean specimens can be obtained.

3.3 Cutting, Grinding, and Ion Milling

The currently most usual method for the electron microscopic preparation of materials science specimens can be described by cutting, grinding, and ion milling. This method fulfils the requirements to answer most electron microscopic questions and facilitates the preparation of inhomogeneous samples, e.g. layered stacks. The goal is a specimen geometry as already reached by the electrolytic thinning: a disc with a diameter of 3 mm, its thickness decreases from the outside towards the centre and close to the centre a little hole with a diameter of less than 1 μm exists (Fig. 3.4). The specimen preparation procedure described now is usually meant when using the term "conventional preparation".

Fig. 3.5 Principle of the core hole drilling

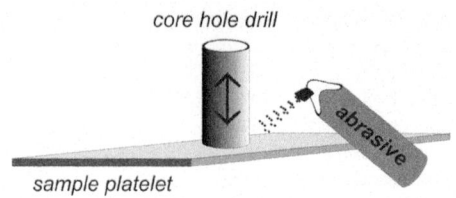

There are two different kinds of preparation. For the investigation of bulk materials or thin surface layers the sample is prepared in the "plan-view" manner, i.e. the thin specimen is taken parallel to the sample surface. This allows the analysis of grain morphology, phases, chemical composition, and the determination of the chemical bonding; at amorphous samples also the analysis of short-range orders. The other possibility is the cross-section preparation taking the specimen perpendicular to the sample surface. This is desirable for the analysis of layered stacks (single layer thicknesses, interfaces) but also for the determination of gradients (depth profiles) of grain structure, phases, and chemical composition in single layers.

Plan-View Specimens Firstly, we need a small disc with a diameter of 3 mm. Its thickness should be between 0.3 and 0.5 mm. Square pieces with a diagonal of 3 mm are possible, too. Dependent on the material this disc can be carved by punching, core hole drilling, cracking or sawing. Ductile materials can be punched, platelets of silicon and suchlike are carved by an ultrasonic core hole drill. This tool is made of a thin steel tube with an inner diameter of 3 mm and a hardened cutting edge, which is punched onto the sample with high (ultrasonic) frequency (Fig. 3.5).

Cutting edge and sample surface are coated by an abrasive suspension. The result of this procedure is a circle-shaped disc with a diameter of 3 mm. We can see that a plate-like geometry is advantageous. In other case the first step is sawing out a platelet from the sample bulk by diamond cutting wheel or wire saw. The diamond saw is a little circular saw; its saw blade is coated with splinters of diamond. The wire saw reminds us of a hacksaw with thin (ca. 0.5 mm thick) tungsten wire that is coated by an abrasive suspension and swings at the cutting position. Similar to the ribbon of an old typewriter the tungsten wire is continuously repositioned.

Normally, two cuts are necessary: At first, a reference plane is sawn and after adjusting by an (integrated) micrometre screw the second cut is placed in a small distance (e.g. 0.5 mm) from the reference plane. Of course, the thickness of the saw blade or the wire diameter must be considered to obtain the pre-defined platelet thickness.

Differences between both sawing methods exist on the one hand in the degree of stress for the sample and on the other hand in the time needed for sawing. The diamond wheel sawing is faster but stresses the sample more than the wire sawing. As mentioned above the general rule is: *The faster procedure is the more gruelling for the sample, i.e. changes of the sample by the preparation have to be expected especially in surface-near regions.*

3.3 Cutting, Grinding, and Ion Milling

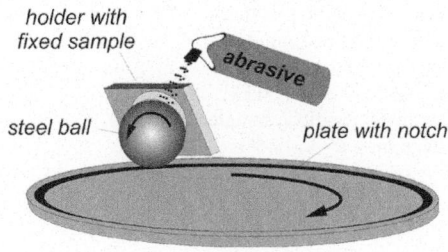

Fig. 3.6 Principle of the dimpling by a ball

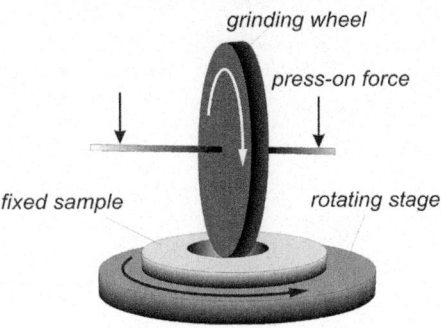

Fig. 3.7 Principle of the dimpling by a grinding wheel

If the disc is not thin enough it has to be thinned by parallel grinding using machines known from metallographic laboratories and sand papers or abrasives with increasing fineness.

The next step is concavely grinding the disc (dimpling). This is the grinding of a spherical calotte into the thin sample disc getting the sample geometry as shown in Fig. 3.4. There are two possibilities for this step, too. The first device we can imagine as a kind of record player with a plate of steel and a notch at its outside in which a steel ball runs. This steel ball is bedabbled by an abrasive (e.g. diamond paste of different granulation) and grinds the obliquely fixed sample (Fig. 3.6).

The second variant works with a grinding wheel and a rotating sample (Fig. 3.7). Also in this case a calotte-shaped indentation is generated. The advantage over the ball method is that the press-on force can be set as an additional parameter to the granulation of the abrasive yielding in an acceleration of the dimpling process. But here is also valid: A faster processing is more gruelling.

Instead of dimpling the sample can be also grinded in a wedge-shaped manner using a special device ("Tripod" [7]) to reach very small wedge angles (only a few degrees).

Unfortunately, the surface-near region of the specimen can be changed by the mechanical grinding up to a depth of some micrometres for finest granulation of the abrasive. This deformed layer has to be removed. Hence, the specimen is in the most cases not mechanically thinned to the final thickness. The last 10 ... 20 μm are removed by sputtering with argon ions ("ion milling" –Fig. 3.8) using an argon ion source with ion energies between some 100 eV and 5 keV.

Fig. 3.8 Last step of the preparation: Ion milling, i.e. sputtering by Ar⁺ ions (ion energy: 200 eV … 5 keV, inclination angle to sample stage surface: 5 … 15°)

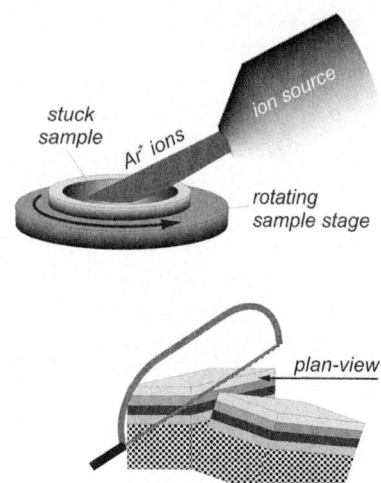

Fig. 3.9 Cross-section preparation: By sawing of a layer stack a new plan-view is created

Mechanical thinning to the final thickness is only possible in special cases and requires a tremendous deal of experience.

It has to be considered that the ion sputtered material is deposited in the surrounding of the sample ("redeposit") and that the sample stage of the milling machine may also be part of the sputter target. The stage should consist of a material with a slow sputter rate (tantalum, graphite) or at least be covered by them. At inhomogeneous samples a preferential sputtering can often be observed. In some cases these regions may get completely lost during preparation and the remaining parts are not representative anymore. This can be counteracted by rotation of the sample. Since the sputter rate also depends on the incidence angle of the ions (perpendicular incidence—low sputter rate) the surface roughness increases at uniform ion bombardment. The sample rotation counteracts this, too. To smooth the surface and to remove any dirty on the sample surface both sides of the specimen should be ion milled.

To investigate thin layers on substrate dimpling and ion milling are only done from the substrate side. The layer side is "gently" cleaned by a short ion milling with low ion energy on this side as last step of the procedure.

Cross-Section Specimens Let us imagine now, we want to look at a cross-section of a layer stack. This requires that the stack is sawed as shown in Fig. 3.9. The result is a new plan-view perpendicular to the sample surface. In principle, the procedure is now as described above.

In practice, such cross-sections are somewhat differently prepared. In fact, we have to ensure that details of interest, e.g. interfaces between different layers, are really electron-transparent.

Let us assume we had a thin (e.g. 50 nm thick) layer on a substrate with a thickness of 0.5 mm. From this two 1 mm wide stripes are separated (scratching and breaking or sawing) which are glued together face-to-face (i.e. layer to layer)

3.4 Focussed Ion Beam ("FIB") Techniques

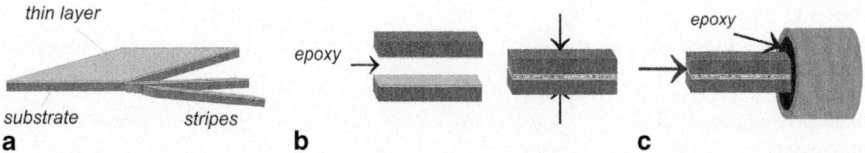

Fig. 3.10 Cross-section preparation. **a** Separation of two stripes. **b** Gluing face-to-face. **c** Gluing into a tubule of ceramics or metal (outer diameter 3 mm, wall thickness ca. 0.5 mm)

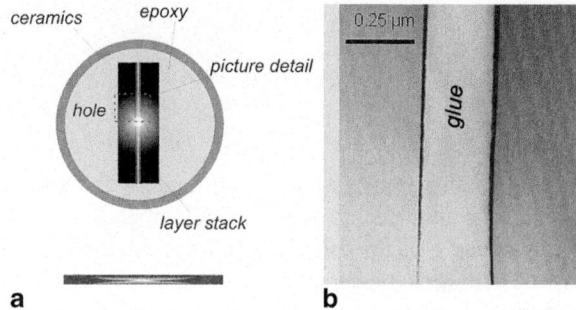

Fig. 3.11 Result of a face-to-face cross-section preparation. **a** Schematic illustration. **b** Transmission electron microscopic image

(Fig. 3.10). Highly fluid epoxy and a device to compress both stripes are recommended to get a gap as small as possible.

The complete face-to-face layer system is glued into a thin tubule (often Al_2O_3 ceramics). After hardening of the epoxy a thin disc (thickness: 0.3 ... 0.5 mm) is sliced from the tubule by two cuts (flat reference plane and "true" cut) which is handled by grinding, dimpling, and ion milling as described above to prepare plan-view specimens.

The result is a transmission electron microscopic specimen with the possibility to observe the interface between substrate and layer on four positions which are electron-transparent (Fig. 3.11).

3.4 Focussed Ion Beam ("FIB") Techniques

To use focussed ion beam techniques firstly a major investment is required: We need a new instrument, a "scanning ion microscope". It works like a scanning electron microscope but with an ion probe instead of the electron probe. An ion optics using electrostatic and/or magnetic lenses originates a very small ion probe with a diameter of only few nanometres that scans the sample surface point by point. During this procedure secondary electrons are emitted from the sample. They are used for the regulation of the brightness of the correlated picture points ("pixels") on a monitor. The image is generated in series, i.e. pixel by pixel. It is not a "true" optical imaging as given in the transmission electron microscope.

Fig. 3.12 Protection bar on the surface of the FIB lamella which will be cut at this position later on (scanning electron microscopic micrograph)

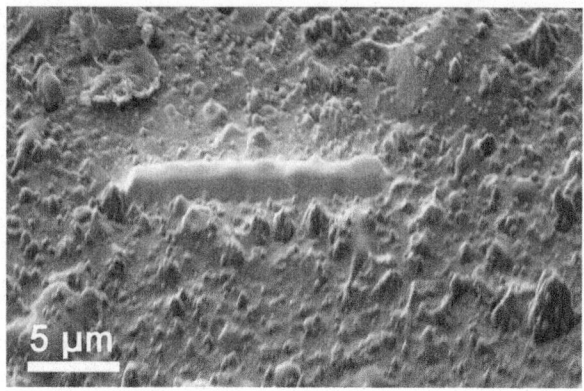

It is advantageous to use an instrument with two columns. The first one is taken from a scanning electron microscope and generates a fine-focussed electron beam for scanning electron microscopic imaging, the second one works with a fine-focussed gallium ion beam and works like the scanning ion microscope as mentioned above. Gallium becomes liquid even at a temperature of 30 °C and it is comparably easy to construct a gallium ion source.

The ion probe works like a milling machine equipped with an end mill with a diameter of only a few nanometres and the sample processing is done under scanning electron microscopic observation using the first column. The use of electrons for imaging avoids that the sample is already milled during imaging.

To perform the milling in a reasonable time (1–2 h for one lamella) standardly a comparably high ion energy of 30 keV is used. This involves two adverse effects: an amorphisation of the surface and an implantation of gallium ions into the sample. The implantation depends on the incidence angle and energy of the ions. Especially, the ion bombardment perpendicular to the sample surface is critical. To protect the top edge of the cut lamella a small protection bar is deposited on the sample surface before starting the milling procedure (Fig. 3.12).

The protection layer is deposited by a hydrocarbon gas containing a heavy metal (platinum, tungsten) that is led onto the sample surface by a local gas injection system and the area that should be protected is scanned by the focussed electron or ion beam. An energy input is connected with this leading to a network of hydrocarbons. At the position of the electron or ion bombardment a solid protection layer is generated containing hydrocarbons and heavy metals (cf. Sect. 4.5 Contamination).

Now, let us look for the benefits of our "nano mill tool" in the preparation of thin cross-section lamellas for transmission electron microscopy.

H-Bar Method The H-bar method starts like the "classical" cross-section preparation: At first, a narrow stripe has to be separated from the sample and glued onto a copper (or another material) half ring (Fig. 3.13). Contrary to the classical preparation this stripe must be very thin, preferably well smaller than 100 µm. This stripe is milled by the FIB machine in a way that an H-shaped form is created. A bridge remains as lamella that is thin enough for electron-transparency.

3.4 Focussed Ion Beam ("FIB") Techniques

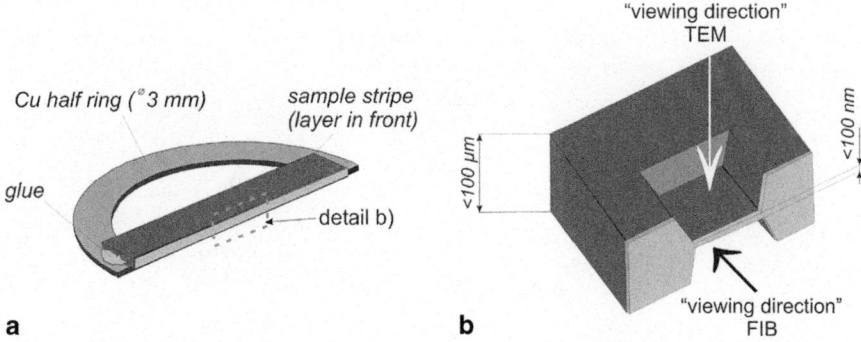

Fig. 3.13 Illustration of the FIB H-bar-method. **a** Glued sample stripe. **b** Geometry of the milled lamella

The stable fixation of the TEM specimen (glued stripe) is an advantage of this method. It is possible to recut the lamella by FIB if the observation in the transmission electron microscope shows that the lamella is not sufficiently thin. The disadvantage is the small lamella window connected with comparably high borders at the edges of this window. Especially during investigations of the crystal structure by diffraction methods this edge can lead to an improper limitation of the tilt angles. The relatively big volume of material close to the thin lamella can also be a critical circumstance. Especially for X-ray spectrometry (EDXS—cf. Sect. 9.2) this can lead to artefacts (shading of low-energetic X-rays). Later on, during the description of the electron microscopic practice, we will refer again to such problems. Nevertheless, this preparation method is reported to be the quickest if the procedure is optimised.

Lift Out and Placing the Lamella onto a Support Film To avoid the disadvantages mentioned above the lamella can be completely cut from the bulk sample. It does not require the separation of a narrow stripe before. We use the sample as received and deposit the protection bar onto the surface. The further procedure can be seen in the sketches of Fig. 3.14.

After laying down the lamella adheres tightly on the support film by adhesive forces. Sometimes there are problems of adhesion with lamellas having mechanical strains and are therefore bent.

In Fig. 3.14f we see that the thickness of the lamella is not uniform. It contains two thicker "bars" for stabilising. Let us remember: The thickness of the lamella is less than 0.1 µm (i.e. approximately 500 times thinner than a human hair). Although its size is only about 10×10 µm it has a tendency to bend, i.e. it must be mechanically stabilised by the bars. Additionally, we see that the support film is not contiguous; it rather looks like a spider web. This is also part of the concept. The support film increases the effective specimen thickness. The electron microscopic investigation should take place at a lamella area positioned over a hole of the support film. To find such an area one needs a bit of luck as it is often necessary during experimental work.

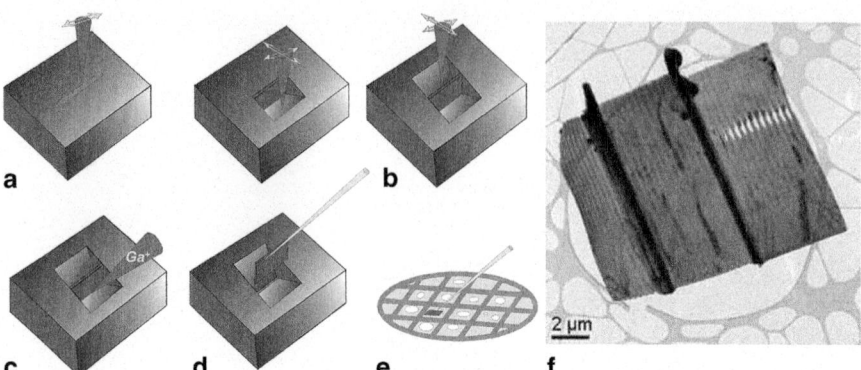

Fig. 3.14. Cutting of an FIB lamella and laying down onto a support grid covered by a film. **a** Deposition of the protection bar. **b** Milling of the lamella. **c** Cutting the lamella. **d** Separating the lamella by a glass needle. **e** Laying down of the lamella onto a support film (film with holes). **f** Transmission electron microscopic micrograph of a lamella on support film

Separating the lamella from the sample does not need luck alone but mainly experience. The procedure takes place using a glass needle fixed at a micromanipulator under light optical microscopic observation. Remembering the resolution limit of this microscope of about 0.2 µm we know that it can be difficult to see lamella and needle without any experience.

Unfortunately, this method has a serious disadvantage, too. It is impossible to recut the lamella if it is too thick for the electron microscopic investigation. In this case the complete procedure has to be started again.

Lift Out and Welding of the Lamella on a Support Grid The drawback of the above mentioned procedure was the irreversible deposition of the lamella onto the covered support grid. To avoid this the lamella is welded on a girder under scanning electron microscopic control. This sounds incredible but it really works. The girder can then be conveniently handled like a "normal TEM specimen".

Up to the separation the procedure is the same as described in Fig. 3.14. However, the lamella remains about 1 µm thick. With the micromanipulator that is positioned within the FIB chamber and equipped by a sharp needle the lamella can be handled. The tip of the needle is welded to the lamella. After cutting at its edges the lamella is carefully broken out, led to a special girder, held on its side and attached by welding. Then the needle is cut free and retracted. The welding is rather a sticking on procedure. For doing so, the same mechanism is used like for the deposition of the protection bar. The hydrocarbon gas containing heavy metal ions is brought to the welding position by the gas injection system. This area is repeatedly scanned by the electron or ion beam. The hydrocarbon network creates a fixed connection between girder and lamella. In this state the final thinning by FIB milling can be performed.

Often it is helpful to leave back some thicker bars for mechanical stabilisation (Fig. 3.15). The described procedural method allows a reinsertion of the support grid with lamella into the FIB machine for further thinning in case that should be necessary.

3.4 Focussed Ion Beam ("FIB") Techniques

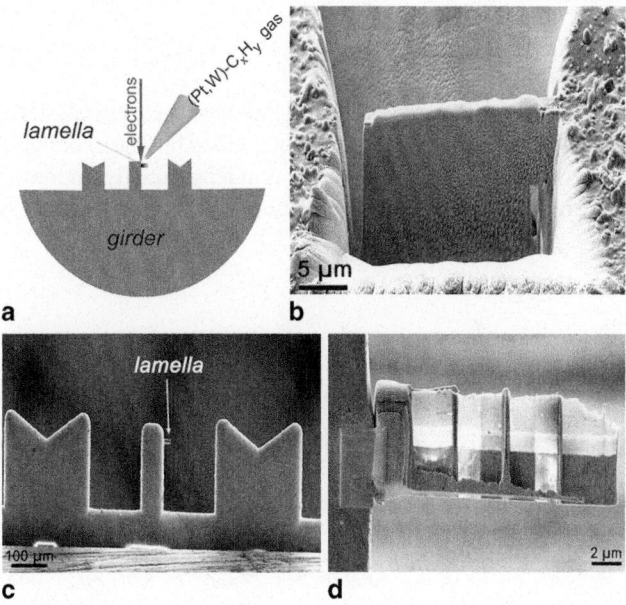

Fig. 3.15 Welding of a FIB lamella. **a** Sketch. **b–d** Scanning electron microscopic micrographs: **b** Stand-alone lamella in the sample. **c** Girder with welded lamella. **d** Welded lamella after final milling

Improvement of the Quality of the FIB Lamellas The quality of FIB lamellas immediately after preparation in the way described above is good enough to answer many electron microscopic questions: analysis of grain sizes, phase determination, investigations of the real lattice structure, also many analytical experiments, i.e. determination of the chemical composition with high lateral resolution. In other cases the amorphous layers which are created by the high energy ions trouble or the lamellas are simply too thick in order to image the atomic columns in high resolution TEM or for electron energy loss spectroscopic investigations of the fine structure of the edges (see Sect. 9.3.6).

These demanding cases need a reduction of the energy of gallium ions during the final milling. Additionally, the lamella should be inclined by a few degrees from its perpendicular position to compensate for the beam convergence. Commonly, it is possible to work with ion energies down to a few keV in FIB machines. At a further reduction discharges and the higher failure proneness of the slower ions with respect to outer influences impact the quality of the ion probe. This leads to larger cross sections and instabilities of the ion probe and, therefore, to the impossibility of a targeted thinning.

Alternatively, the welded lamellas can be finally thinned using another instrument e.g. working with argon ions of energy between 200 eV and 1 keV. It has to be considered that the girder is never hit by the ions, i.e. in this instrument the focussing of the argon ion beam to a probe with about 1 μm diameter should be possible. Otherwise, the specimen can be contaminated by redeposit coming from the compact girder.

The specimen preparation is inextricably linked to the transmission electron microscopy. It can need a formidable instrumental effort but the experience of the specialists for electron microscopic preparation is at least equally important. A detailed and comprehensive overview about preparation techniques is given e.g. in [8, 9].

References

1. Lang, G.: Histotechnik—Praxislehrbuch für die biomedizinische Analytik. Springer-Verlag, Wien (2006)
2. Allen, T.D. (ed.): Introduction to Electron Microscopy for Biologists. Academic Press, Elsevier (2008)
3. Ruska, H., v. Borries, B., Ruska, E.: Die Bedeutung der Übermikroskopie für die Virusforschung. Arch. ges. Virusforsch. **1**, 155–169 (1939). doi:10.1007/BF01243399
4. Bethge, H.: Oberflächenstrukturen und Kristallbaufehler im elektronenmikroskopischen Bild, untersucht am NaCl. Physica status solidi (b) **2**, 3–27 and 775–820 (1962)
5. Galetzka, W., Gnägi, H., Godehardt, R., Lebek, W., Michler, G.H., Vastenhout, B.: Ultramikrotomie in der Materialforschung. Hanser, München (2004)
6. McCaffrey, J.P.: Small-angle cleavage of semiconductors for transmission electron microscopy. Ultramicroscopy **38**, 149–157 (1991)
7. Benedict, J., Anderson, R., Klepeis, S.J.: Recent developments in the use of the tripod polisher for TEM specimen preparation, Specimen preparation for transmission electron microscopy of materials-III. MRS Symp. Proc. **254**, 121–140 (1992). doi:10.1557/PROC-254-121
8. Ayache, J., Beaunier, L., Boumendil, J., Ehret, G., Laub, D.: Sample Preparation Handbook for Transmission Electron Microscopy—Methodology. Springer, New York (2010)
9. Ayache, J., Beaunier, L., Boumendil, J., Ehret, G., Laub, D.: Sample Preparation Handbook for Transmission Electron Microscopy—Techniques. Springer, New York (2010)

Chapter 4
Let us Start with Practical Microscopy

Abstract In the Chaps. 1, 2 and 3 we roughly described what happens electron-optically within the microscope, we have learnt how the samples have to be prepared and now, we are sitting in thought in front of a transmission electron microscope. It is time to make us familiar with its appearance. Of course, the appearance depends on the type of the microscope; it depends on the manufacturing company but also on the production series. On the other hand, it is similar to a car: There are many similarities and we want to confine ourselves to these features. After all we do not get the driver licence only for a special type of car. At the beginning of the microscope session the alignment state of the microscope has to be controlled and corrected if necessary. Finally, we want to hint at possible changes of the specimen during the irradiation by electrons.

Figure 4.1 shows a sketch of a transmission electron microscope. The most important parts are labelled so that we will know later on what we speak about.

The control of the alignment state is the first working step during a transmission electron microscopic session. The best images can be obtained using paraxial rays. Within the multistage imaging system the single lenses must be very well aligned one to each other. One question is: What does "very well" mean? Let us use the comparison with a light optical microscope. There the mechanical adjusting during the production in the manufacturing company is sufficient. Let us remember the useful magnification: at a light optical microscope it amounts to 500 … 1,000, but at a transmission electron microscope to about 1 million. Roughly spoken, the alignment of the electron microscope must be 1,000 times better than that of a light optical microscope.

Generally, this accuracy cannot be reached only by carefulness during manufacturing and assembling. The lenses and the apertures have to be aligned while the electron microscope is working. Until the 1970s the alignment had been done (at least partly) by mechanical shift of the lens bodies. The improvement of the resolution limit leads to increasing importance of the mechanical stability of the microscopic column. The possibilities to shift the components as mentioned above inhibit a tight combination of the column parts and impair the mechanical stability. Therefore at modern electron microscopes not the lenses are aligned against each other but the electron beam is "bent" in a way that it optimally passes the lens fields. At non-rotational-symmetric devices (e.g. stigmator—cf. Sects. 7.7 and 10.17) the lens fields can be electrically or magnetically shifted and adjusted to the ray path. Only the apertures are mechanically aligned furthermore. Hence each aperture has two

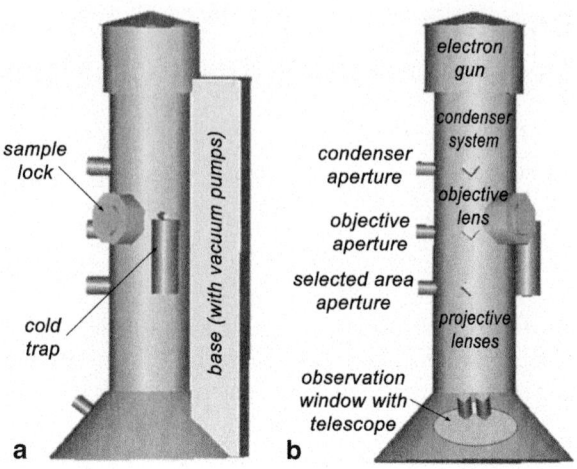

Fig. 4.1 Sketch of a transmission electron microscope (TEM). **a** View from the right. **b** Front view

perpendicularly arranged mechanical drives which allow its fine alignment. One of the drives additionally possesses a coarse shift to insert apertures with different sizes into the ray path. Currently, these aperture drives are often motor-controlled and the gearing mechanisms reproduce the aperture positions with high accuracy. These positions can be saved in a computer memory and recalled if necessary facilitating the operator's work.

Before starting the alignment we have to manage some other things.

4.1 What do We Peripherally Need?

Certainly, in the first place we have to mention the electron-transparent specimen. From Chap. 3 we know that beside of preparator's experience a considerable instrumental effort can be necessary for its production. The selection of the preparation instruments depends on the tasks of the electron microscopic investigation which must be known before establishing a new electron microscopic laboratory or after changing the research topics. Commonly, in materials science grinding machines as known from metallography are needed. Additionally, we need tweezers, scalpels, measuring instruments, light optical microscopes, hotplates, saws, and dimplers as well as machines for ion milling. Possibly, the purchase of a device for electrolytic thinning or an ultramicrotome makes sense. For all these instruments consumable material as abrasive papers, grinding compounds and aerosols, epoxy and other glues, gases (e.g. argon) etc. are needed. All these things have to be considered during the financial planning.

The specimen has to be carefully inserted into the specimen holder. That can be done using tweezers or vacuum tweezers; the latter is a thin hollow needle connected with a little diaphragm vacuum pump. It allows to suck the small sample disc and to deposit it at the right position on the specimen holder. Often the small specimen has to be inserted in a special orientation and its fixture has to be controlled. That should be done by use of a light optical stereo microscope. A reflected-light microscope with a magnification between 5 and 100 times and a working distance of at least 3 cm can be used for this purpose.

Later on (cf. Sect. 4.5), we will explain that the contamination of the sample surface during electron irradiation by hydrocarbons can influence the quality of the electron microscopic measurements. Hence it is useful to clean the specimen and its holder immediately before inserting it into the microscope by a plasmacleaner. In this process the carbonic coatings on the sample surface burn-up within a low-energy plasma, i.e. an inert gas containing oxygen is necessary. Often argon including about 20% oxygen is recommended but normal air is suitable, too.

For the same purpose, i.e. the reduction of the hydrocarbon contamination, the specimen within the transmission electron microscope is covered by cooled copper plates. The cooling of the plates is done via a copper rod that leakproofly protrudes into the microscopic column and is cooled by liquid nitrogen on the outside ("cold trap"). This cooling is especially helpful for quick specimen exchanges. Hence, we need equipment to handle liquid nitrogen. If the electron microscope is equipped with an energy dispersive X-ray detector some systems also need liquid nitrogen for cooling of the detector crystal (cf. Sect. 9.2.1).

Turbomolecular pumps are vented after closing the valves between pump and recipient and switching off the pump. This happens using clean nitrogen gas (99.999%) without any humidity. This gas is also required to vent the microscopic column if necessary. An appropriate gas bottle must be purchased. Since we speak about gases: the tank for generating the high tension and the box containing the electronics at the electron gun chamber are filled by sulphur hexafluoride (SF_6) up to a pressure of some bars to avoid flashovers. The appropriate gas bottle has to be kept in stock, too.

4.2 We Put the Specimen into the Holder and Insert it into the Microscope

Now, we want to really start and insert the specimen into the sample holder. At first, let us familiarise with a typical sample holder used for a transmission electron microscope. We confine ourselves to the side-entry holder because it is the most common in practice (cf. Sect. 2.7.3 and Fig. 4.2).

The specimen is inserted and fixed at the designated position (Fig. 4.2, right side) under light optical microscopic control. Dependent on the holder type the fixation is done in a variety of ways: for example by a spring-loaded flap or by a threaded ring or a circlip, respectively. Special sample holders often include special devices to fix the specimen.

The sample lock consists of a tube closed by a valve at its end on the side of the microscopic column. The holder is inserted into this tube, by doing so the pin has to

Fig. 4.2 Sample holder of a transmission electron microscope (side-entry holder)

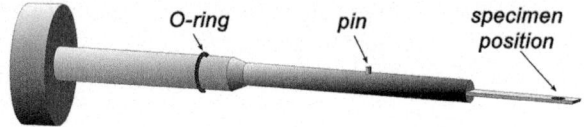

be slid into a dedicated groove within the tube. Now, the O-ring of the holder seals the outer end of the tube and a closed volume is generated between inner valve and O-ring which has to be evacuated by a vacuum pump. By rotating the holder at its handhold the pin opens the inner valve and the sample holder slides into its working position. It is not necessary to shift the holder by physical strength; the atmospheric pressure causes the motion. Quite the contrary, it is necessary to slow down the holder movement.

There are a lot of commercial sample holders for transmission electron microscopes: single and double tilt holders, rotation tilt holders, heating and cooling holders, holders for tensile tests, holders with Faraday cups, and holders with a tip opposite the specimen which can be piezoelectrically moved (tunnelling microscope within the transmission electron microscope). Additionally, customer's modifications for special investigations, e.g. in magnetic fields, exist.

We want to confine ourselves to the basic sample holders needed for the electron microscopic investigations described in this book. In other words: before selecting the sample holder it must be clarified which properties need to be measured using which methods. For grain sizes and phases in polycrystalline material, nanoparticles, and short-range orders in amorphous samples a single tilt holder is entirely sufficient. Diffraction and high resolution investigations on single crystals and epitaxially grown layers need a double tilt holder to select suitable crystal orientations. As a rule, in this holder the slot for the specimen is a small basket whose inner diameter is only sparsely greater than 3.0 mm. If the sample disc is greater it cannot be used and the effort for its preparation was unsuccessful. Analyses by energy dispersive X-ray spectroscopy ("EDXS") need a "low-background holder" with a sample slot of beryllium (cf. Sect. 9.2.2). It is worth to think about these problems before inserting the specimen into the holder. Each insertion and removal of the specimen is dangerous; the specimen can fracture or can be lost in another way.

At all handlings one must take care to cleanness of holder and specimen. The part of the holder beginning at the O-ring should not be touched by naked fingers; to touch the specimen with a naked hand is already impossible because of its littleness. If possible, the inserted specimen should be cleaned within a plasmacleaner together with the holder. But be careful, please, the plasma cleaning is mainly a plasma assisted oxidation of carbon in an atmosphere containing oxygen. Hence, specimens with carbon as support film or materials with affinity to oxidation (e.g. copper) should not be treated within the plasmacleaner.

It is time to take the cold trap (liquid nitrogen cooling device) into operation, i.e. we fill liquid nitrogen into the Dewar[1] bottle.

---

[1] Sir James Dewar, Scottish physicist, 1842–1923.

4.3 We Check the (alignment) State of the Microscope

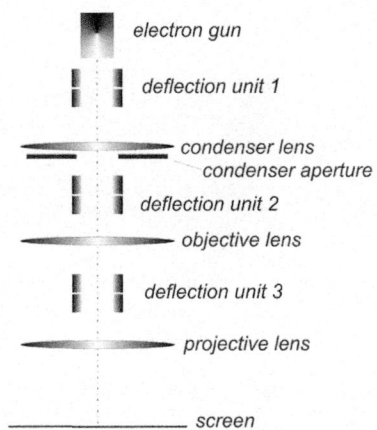

Fig. 4.3 Model of the electron microscope for understanding of the alignment steps. The deflection units are drawn in the x-z plane. Additional, identical units are arranged in the y-z plane perpendicular to the x-z plane to deflect the electron beam into all x-y directions

Finally, we want to comment on the handling of ferromagnetic samples: From Chap. 2 we know that the specimen is positioned within the strong magnetic field of the objective lens, i.e. there is a strong magnetic force which acts on the sample. An insufficiently fixed specimen (e.g. only by a spring-loaded flap) can be pulled out of the holder and touch the pole piece. At best, only the astigmatism increases, in the worst case the specimen hinders the electron beam and the microscopic column has to be demounted to remove the specimen. Magnetic samples should be tightly fixed within the holder. During the insertion of such samples into the microscope the objective lens current should be switched off or at least strongly reduced (switching into the "low magnification" range).

4.3 We Check the (alignment) State of the Microscope

The specimen is inserted, a look at the pressure indicator shows that the vacuum within the microscopic column is good enough ($p < 10^{-3}$ Pa) after a waiting time of some minutes: The time has come to open the valve between electron gun and column. Because of the excellent stability of modern transmission electron microscopes a complete and time-consuming alignment is needed only after changes and/or demounting of the microscopic column. But it is useful to check some features of the alignment before starting the electron microscopic investigation.

To understand the principles of the alignment we use a simplified model of the electron microscope only consisting of the electron gun, one condenser lens, the objective lens, and one projective lens. Between these parts deflection units are arranged which allow shifting and tilting the electron beam into two perpendicular directions (Fig. 4.3). Our model contains electrostatic deflection units for easier understanding of the sketches. In practice, magnetic systems are often used but this has no influence on the principle of operation. Additionally, our model includes an

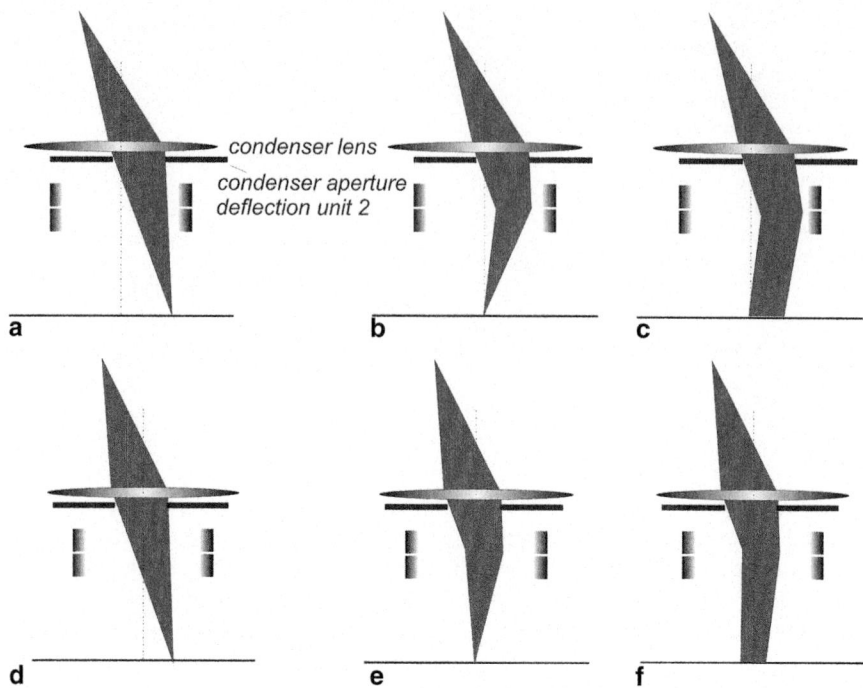

Fig. 4.4 For explanation of the alignment of the condenser aperture, ray paths. **a–c** Decentred aperture. **d–f** Centred aperture. Additional explanations are in the text

aperture: the adjustable condenser aperture in the plane of the condenser lens which is often called "condenser-2 (C2) aperture". There are five steps to check the alignment of the microscope.

1. Alignment of the Condenser Aperture ("C2 aperture"): The condenser C2 aperture influences the illumination of the viewing field and has to be firstly aligned. The alignment procedure is explained by the sketches of the Figs. 4.4 and 4.5. In the sketches only those parts of our model are drawn which are important for the explanation.

Let us start with an inclined illumination of the condenser lens (this should be corrected later on). At a moderate magnification (< 10,000) the focal length of the condenser lens ("intensity") is set in a way that we see the spot on the screen as small as possible (Figs. 4.4a and d). This spot is shifted by the deflection unit 2 ("beam shift") into the centre of the screen (Figs. 4.4b and e). By doing so, a possible misalignment of the condenser aperture does not play any role. The change of the condenser focal length leads to an increase of the spot on the screen and we observe the influence of the condenser aperture alignment: With misaligned aperture the centre of the increasing spot shifts (Figs. 4.4c and 4.5a, respectively); with correctly centred aperture the spot opens concentrically (Figs. 4.4f and 4.5b, respectively).

4.3 We Check the (alignment) State of the Microscope

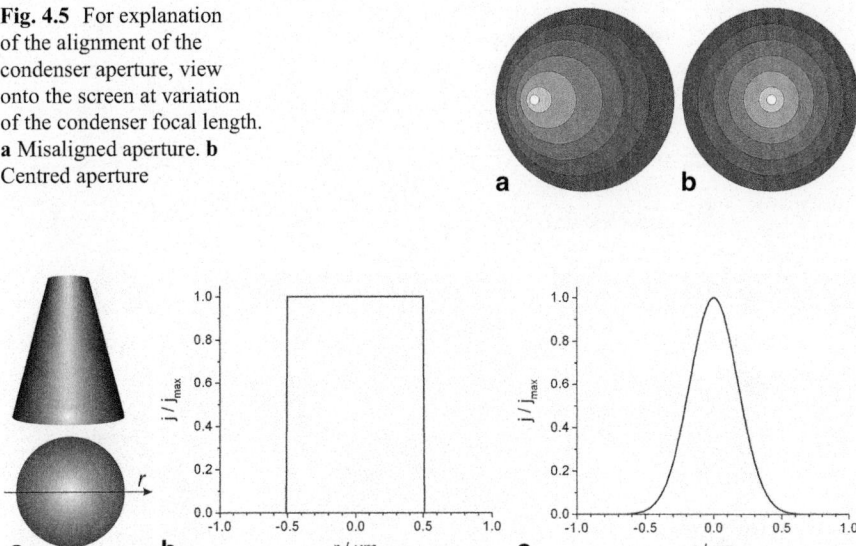

Fig. 4.5 For explanation of the alignment of the condenser aperture, view onto the screen at variation of the condenser focal length. **a** Misaligned aperture. **b** Centred aperture

Fig. 4.6 Current density in the beam cross section. **a** Side view and plane view. **b** Box-shaped profile (incorrect). **c** Gaussian profile (sufficient approximation for an aligned beam)

At "playing" with the condenser focal length, starting with a focussed spot, we see that the illuminated area opens up independently on the rotational direction of the intensity knob. Readers who are surprised by this fact should read the Sect. 2.7.1 (illumination system) once more.

The condenser aperture has to be shifted mechanically until the concentric case drawn in Fig. 4.5b is reached. If the spot is not circular but elliptical and the direction of the long axis of the ellipse is changed by about 90° after transition through the "smallest spot position" a two-fold condenser astigmatism is given. This can be corrected by the condenser stigmator.

2. Alignment of the Electron Gun: The goal of this alignment step is to reach a preferably bright and uniformly illuminated image. We want to think about ray path and appearance on the screen using the simplified model in this case, too. Doing so, the intensity profile within the beam cross section has to be considered. It is not box-shaped but it has its maximum in the centre and slopes slowly from the centre to the edge. Mathematically it can be sufficiently described by a Gaussian function (Fig. 4.6).

Using this knowledge we can specify the demand for a "uniformly illuminated image": The beam profile should be symmetrical to the centre as drawn in Fig. 4.6a. With the help of Fig. 4.7 we want to think about the possibility to correct an inclined illumination due to a slight misalignment of the electron gun by use of the deflection units.

In Fig. 4.7 the heart of the ray bundle with the highest intensity (maximum of the current density) is drawn brightly. We start with a misaligned state, i.e. the electron

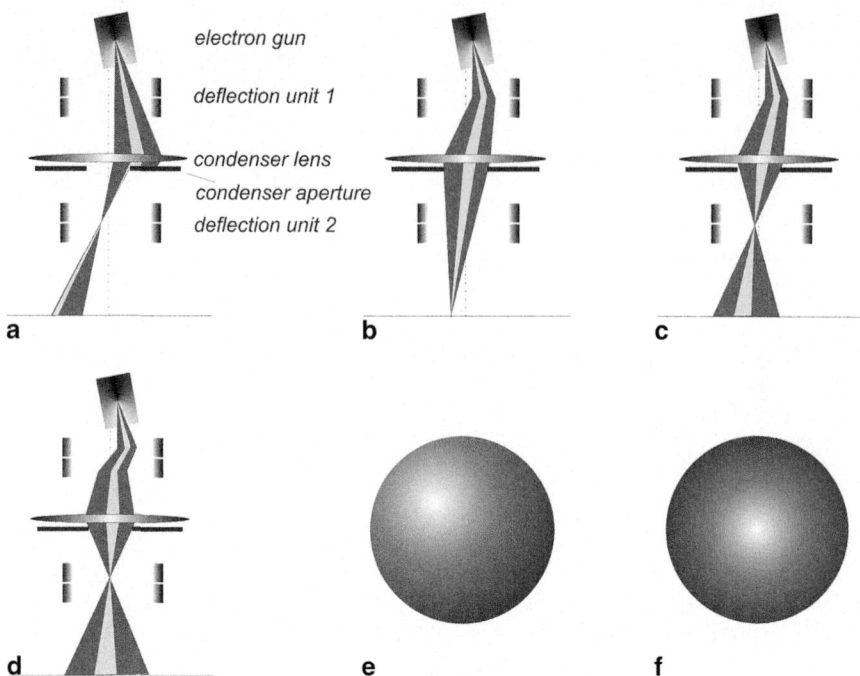

Fig. 4.7 Alignment of the electron gun. **a** Ray path in a misaligned state. **b** After beam tilting with focussed beam. **c** After beam tilting with broad beam. **d** After beam tilting and shifting with broad beam. View on the screen at (**e**) insufficient beam shifting and (**f**) aligned state

gun is tilted and displaced (Fig. 4.7a). The electron beam partly hits the condenser aperture; the intensity (i.e. the current) on the screen is reduced. At first, at low magnification (ca. 5,000) we focus the beam by the condenser ("intensity") so that we see a small bright spot on the screen. If necessary, this spot can be shifted to the centre of the screen by the deflection unit 2 ("beam shift"). Hereafter the beam is tilted using deflection unit 1 ("gun tilt") until the spot reaches its brightness maximum. If the spot moves away from the screen centre it can be corrected by "beam shift" anytime.

It is not so simple to visually control the brightness of the spot. Hence we use the displayed exposure time. This display is a reference to the time when the electron microscopic micrographs were recorded on films or photo plates. To set the correct exposure time the electron current reaching the screen is measured as basis for the determination of this time after calibration regarding the used photo material. Although this photographic technique is rarely used in electron microscopes today, the measurement of the current and the display of the exposure time survived. The higher the current the shorter is the exposure time. We use this fact to control the brightness within the spot. Therefore we change "gun tilt" until the displayed exposure time reaches its minimum (Fig. 4.7b).

Now, we change the focal length of the condenser lens ("intensity") and increase the magnification until the picture sketched in Fig. 4.7e can be seen on the screen: Within a circular broad disk a bright small spot can be observed. At sufficiently pre-centred field emission cathodes that should be possible without any problems. At LaB_6 cathodes the heating current must eventually be reduced; the outer edge of this figure is not necessarily a circle. By additional shifting of the beam using deflection unit 1 (it has two stages—"gun shift") the bright spot moves into the centre of the broad spot (cf. Figs. 4.7d and f). At LaB_6 cathodes a high symmetry of the figure on the screen is the goal. This is the criterion for a uniform illumination of the image.

As already mentioned for the alignment of the condenser aperture, possibly occurring condenser astigmatism has to be corrected in between by the condenser stigmator. A larger gun misalignment must possibly be corrected in more than one iterative step.

3. Setting of the Eucentric Height: For optical imaging the image is exactly focussed when the lens equation

$$\frac{1}{f} = \frac{1}{g} + \frac{1}{b} \qquad (4.1)$$

(*f*: focal length, *g*: object distance, *b*: image distance) is fulfilled. Obviously, there are three possibilities reaching this: the object distance is adjusted to focal length and image distance (example: light optical microscope), the image distance is adjusted to focal length and object distance (photo-apparatus) or the focal length is adjusted to object and image distance (eye). Concerning the electron microscope the last method seems to be advantageous because of the easy, sensitive, and precise setting of the focal length of magnetic lenses by changing of the lens current and in this way, the focussing does not need any mechanical displacement.

Otherwise, we know from Sect. 2.1 that the image ratio also changes if the focal length is changed. This has to be avoided for an electron microscope used as a measuring machine of smallest lengths. Furthermore, in an imaging system consisting of more than one lens the focal lengths have to be adjusted one to each other. A strong change of one of these focal lengths influences this adjustment and the other ones have to be readjusted.

To avoid these problems the focussing within the transmission electron microscope is done in two steps: A coarse adjustment is done by a change of the object distance as for the light optical microscope; the fine focussing is done by change of the focal length of the objective lens.

Let us think about criteria for the quality of the coarse adjustment. The obvious way is doing it in the same kind as for the light optical microscope: The sharpening of the image happens by shifting of the specimen stage in *z* direction (optical axis). It is sufficiently sensitive for the coarse adjustment. But, contrary to the light optical microscope at the electron microscope the objective focal length is changeable and before shifting of the sample stage we have to set a reference focal length. The operator presses a button at the control panel named e.g. "eucentric focus". A preset lens current is set and the reference focal length is given.

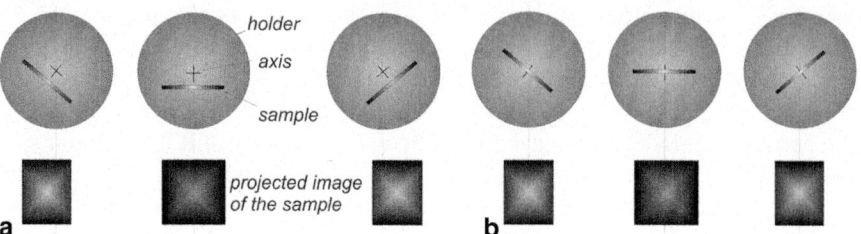

Fig. 4.8 For explanation of the setting of the eucentric height. **a** Sample *below* the goniometer axis. **b** Sample in the height of the goniometer axis ("eucentric height")

The next question is: How is it possible to control this reference? Or in other words: Is there another criterion checking for the "right" height of the specimen stage independent on the correct focussing? Let us remember: the sample holder is inserted into a goniometer and can be rotated around its axis. The goniometer axis is suitable to be the reference height independent on the focal length of the objective lens. Figure 4.8 sketches what we have to expect in cases of deviation or agreement between sample height and goniometer axis.

On the screen of the microscope we see a (magnified) projected image of the sample. Rotating the goniometer together with the sample leads to a distortion of the image. If the sample is located below the goniometer axis the sample details visible in the image are shifted sideways systematically dependent on the rotation direction (Fig. 4.8a). At the same height of sample and goniometer axis we only see the distortion; a shifting does not happen (Fig. 4.8b). Therewith we found a criterion for the setting of the eucentric height independently on the focal length. Because of the "insensitivity" of the mechanical shifting the magnification should not be too large (< 10,000). At modern instruments the periodic rotating is done by a computerised motor ("wobbling").

4. Setting of the Pivot Points for the Beam Tilt: There are contrast phenomena in transmission electron microscopy which depend on the orientation between sample and electron beam. This is to be explained in more detail in Chap. 6. To analyse such kind of contrasts the orientation has to be changed, i.e. specimen or beam must be tilted. As already mentioned, mechanical movements are critical in the electron microscope at high magnifications. An electric (and magnetic, respectively) beam tilting is a preferable alternative which can be realised using the deflection units 2 and 3 (Fig. 4.3). Certainly, the same sample area must be illuminated during the beam tilt, i.e. its pivot point has to be exactly in the object plane. The object plane is not drawn in Fig. 4.3 but it is important for the further understanding and is added in Fig. 4.9. It can be seen from this figure that the beam can be tilted in the desired manner by use of both stages of the deflection unit 2. The pivot point is shifted along the *z* axis (optical axis) into the object plane by setting of the "right" ratio of the deflection angles of both stages. We see that the focus point of the bundle below the objective lens is shifted away from the optical axis. The deflection unit 3 causes that the ray bundle optimally enters the projective lens even with the tilted beam.

4.3 We Check the (alignment) State of the Microscope

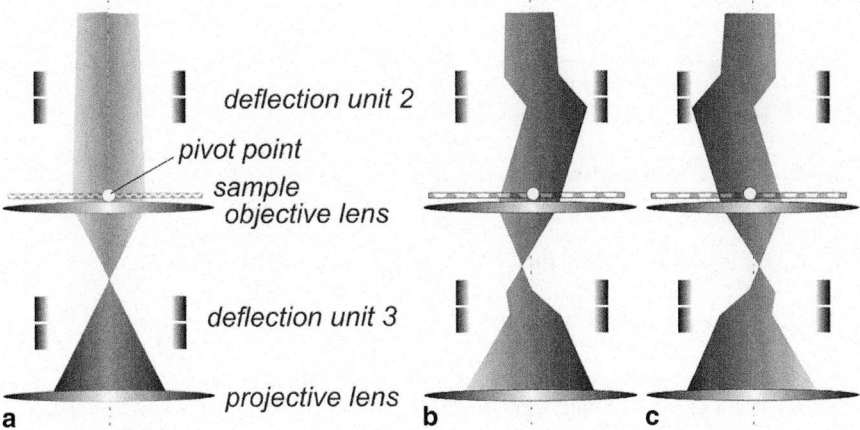

Fig. 4.9 Ray paths at beam tilt. **a** Untilted. **b** Tilt to the *right*. **c** Tilt to the *left*

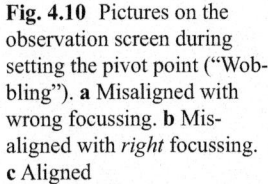
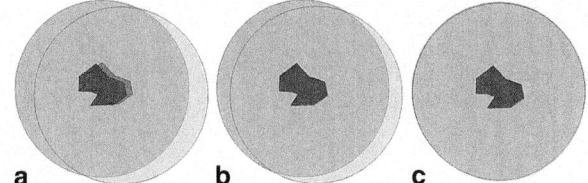

Fig. 4.10 Pictures on the observation screen during setting the pivot point ("Wobbling"). **a** Misaligned with wrong focussing. **b** Misaligned with *right* focussing. **c** Aligned

It is the goal of this alignment step to shift the pivot point exactly into the sample plane. The procedure in practice is to be explained with the help of Fig. 4.10.

Obviously, the exact imaging of the object plane is a precondition, i.e. the specimen has to be accurately focussed. The beam is "wobbled", i.e. it is periodically tilted forwards and backwards. Let us concentrate on a recognisable detail in the image of the specimen and we can see whether the image is exactly focussed or not. If not, we see a double image (Fig. 4.10a). It is important that we really concentrate on the recognisable detail and not on the edge of the beam. The double image has to be eliminated by adjustment of the objective focal length ("focus"). Now, we are looking at the edge of the beam. In the misaligned state it can be doubly seen (Fig. 4.10b). This is corrected by shifting of the pivot point along the optical axis (Fig. 4.10c—The setting control is often labelled as "pivot point X" and "pivot point Y", respectively.). There are two settings because of the two independent perpendicular directions x and y.

5. Alignment of the Rotation Centre The objective lens is the most important lens within the microscope. It has the strongest influence on the imaging quality. Hence, it is important that the ray bundle runs symmetrically to the optical axis of the objective. How do we control that?

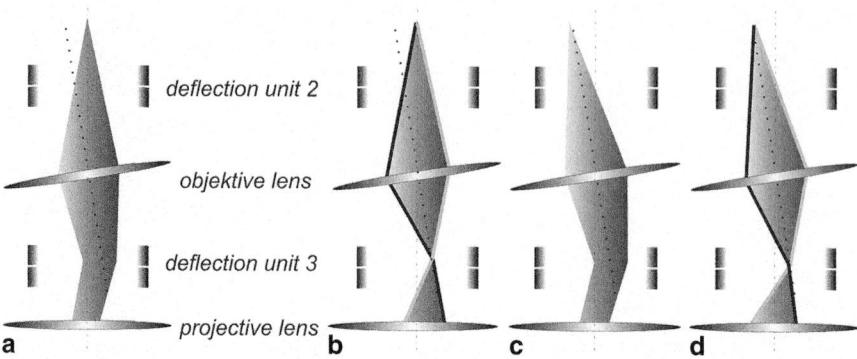

Fig. 4.11 Ray paths to illustrate the deflections of the ray bundle from the optical axis of the objective lens. **a** Misaligned state, focussed image. **b** Misaligned state after change of the objective focal length. **c** Aligned state, focussed image. **d** Aligned state after change of the objective focal length

Fig. 4.12 Images on the screen during wobbling the objective focal length to align the rotation centre. **a** Misaligned. **b** Aligned

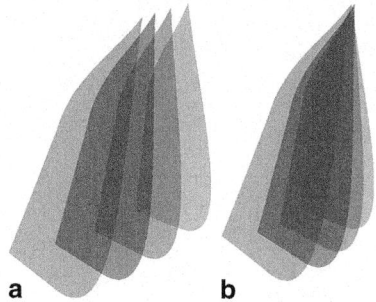

Let us think about the ray path also in this case (Fig. 4.11). The deviation between the symmetry axis of the ray bundle and the centre line (this should be the optical axis) of the objective lens is illustrated by the tilt of the objective lens. The image is focussed, the image shift on the screen is corrected by the deflection unit 3 (Fig. 4.11a). The change of the objective focal length leads to an unsharp image including a change of its size, for magnetic lenses it slightly rotates. At unchanged setting of the deflection unit 3 the centre of the image on the screen is shifted (Fig. 4.11b). The alignment is done by interaction of the deflection units 2 and 3: The beam is tilted to the centre line of the objective lens by deflection unit 2 and the resulting image shift on the screen is corrected by unit 3 (Fig. 4.11c). Contrary to the misaligned state the centre of the image stays at the same place of the screen, now (Fig. 4.11d).

Let us come to the practical procedure (Fig. 4.12): Firstly, we move the specimen in a position that we see a recognisable detail in the centre of the screen. A kind of tip as assumed in Fig. 4.12 is beneficial. After switching on the wobbler for the objective focal length ("rotation centre") we see in the misaligned state (Fig. 4.12a)

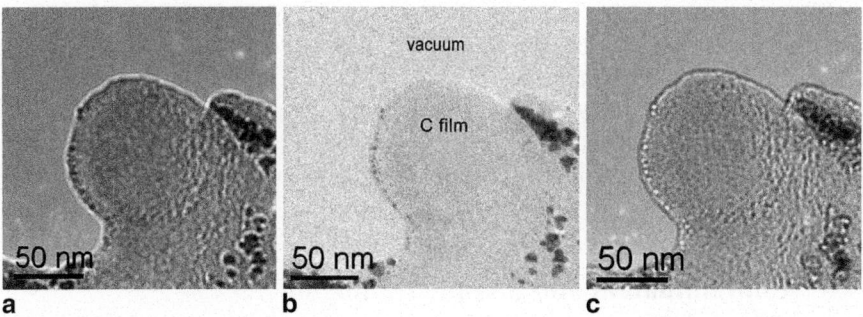

Fig. 4.13 TEM micrograph of the edge of a carbon film. **a** Underfocus. **b** Focus. **c** Overfocus

that the image of the tip periodically moves away from the centre of the screen. The magnification changes synchronously, the image "breathes". Additionally, the image slightly rotates. After successful alignment the image of the tip stays in the centre and rotates slightly around the tip (Fig. 4.12b). This is the reason to name the alignment step "rotation centre": With correct alignment the centre of the image rotation stays at the same place on the screen.

Alternatively to the objective lens current the acceleration voltage can be periodically changed ("voltage centre"). In this case the focal length of the objective lens is wobbled by the variation of the electron wavelength (details see [1]).

4.4 Focussing the Image—Sharpness and Contrast

In the previous paragraph we repeatedly wrote about the focussing of the image. But, when is the image sharp? The beginner often confuses sharpness with contrast. As a rule, an unsharp but high-contrast picture seems to be sharper than a sharp but low-contrast picture. There is a certain validity to this, since a sharp picture without any contrast does not make sense: We cannot see any details in this picture. On the other hand, we need a sharp image if we want to see tiny structures, if we want to reach the resolution limit on the sub-nanometre scale.

To explain this discrepancy between sharpness and contrast in transmission electron microscopic images let us look at Fig. 4.13. It shows an electron microscopic micrograph of the edge of a thin carbon film with some small gold particles. The focussed image 4.13b shows in comparison with the other images of Fig. 4.13 only low contrast. Vacuum and carbon film are similar grey. On the other hand, in the defocussed images a bright or dark fringe parallel to the carbon film edge is visible increasing the image contrast. This fringe is a consequence of the wave character of the electrons. In a plane beneath the carbon film the intensity is not sharply confined but has maxima and minima because of wave interference (Sect. 10.1). These maxima and minima can only be seen in a plane beneath or above the carbon film, i.e. in the image which is not exactly focussed on the carbon film plane. Hence, we see the contrast-boosting fringes only in the defocussed image. The best focus is reached in the low-contrast image.

We can also observe that the imaging e.g. of carbon structures in the electron microscope can be difficult because of their low contrast. In this case a compromise is often necessary: We slightly defocus the image just to see something.

4.5 Contamination and Sample Damaging

> ...The development of recent electron microscopes was mainly a struggle against the undesirable consequences of the same properties of electron beams which just enable the sub-light-optical-microscopic resolution. E.g. the short matter wave—the basis for the high resolution—is coupled with the high electron energy that is undesirable because of the stress it induces in a specimen....(translated of [2])

This quote by Ernst Ruska describes that among the desirable, contrast-inducing interactions between electrons and sample (this will be closer described in Chap. 6) undesirable results of these interactions have to be considered. We want to deal with two of these problems: contamination and sample damaging.

Contamination: The contamination, more detailed: the hydrocarbon contamination, is a process occurring in all instruments working with electron beams. Especially it disturbs in transmission electron microscopy. Although a "good" vacuum is given in the surrounding area of the specimen (pressure $< 10^{-3}$ Pa) there is a great number of gas molecules (particle density ca. 3×10^8 mm^{-3} at room temperature and 10^{-3} Pa). If only 0.01 % of them are hydrocarbon molecules there would be 30,000 per mm^3. The gas molecules touch the inner parts of the microscopic column as well as the sample surface and stay there for a short dwell time. The result is a gas cover on the sample surface with a contribution of hydrocarbons. There is a dynamic equilibrium between incoming molecules and molecules flying away. The details depend on pressure and temperature. At higher wall (and sample) temperature the dwell time of the molecules on the surface becomes shorter and the thickness of the gas cover decreases. At higher pressure in the chamber the flow of molecules onto the surface increases and the cover thickness grows.

At first, we imagine that the sample is illuminated over a large area. In transmission electron microscopy "large area" means e.g. an area of a few square micrometres. This is the precondition in Fig. 4.14. In this figure we see the dynamic equilibrium of incoming and outgoing hydrocarbon molecules on the sample. These molecules can also be hit by the beam electrons. In the case of two hydrocarbon molecules being close together on the sample surface the electron bombardment leads to a cross-linking of the molecules resulting in the formation of solid hydrocarbons, the contamination layer.

For an evaluation of the thickness growth by contamination we have to consider that the contamination layer grows on the top as well as on the bottom of the sample. The growth rate can reach up to some nanometres per second. To reduce the

4.5 Contamination and Sample Damaging

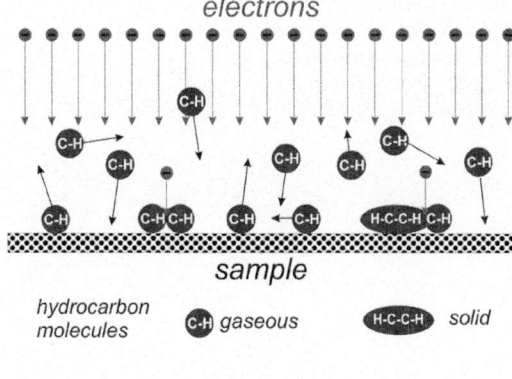

Fig. 4.14 Contamination at "large-area" illumination of the sample by electrons

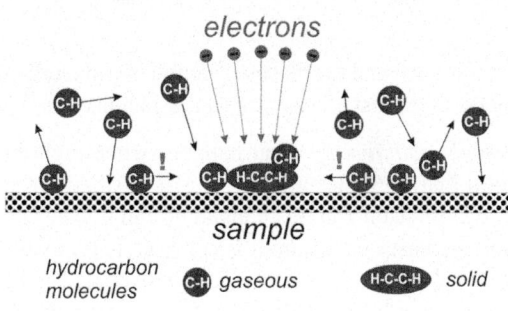

Fig. 4.15 Contamination at "probe-like" sample illumination by electrons. The molecules denoted by exclamation marks move over the surface to the place of contamination

contamination rate either the mean dwell time of the gaseous hydrocarbon molecules on the sample surface has to be reduced or/and the partial pressure of the hydrocarbons must be depleted. The dwell time depends on the surface temperature, heating of the sample would be one possibility. Unfortunately, the specimens are often sensitive to heat (e.g. in connection with embedding in epoxy). Additionally, especially at higher magnifications we need stable temperatures to avoid a thermal specimen drift. Hence, this possibility is rarely used, instead the partial pressure of the hydrocarbons is reduced by cooled copper plates surrounding the specimen [3]. Hydrocarbons and other gases (e.g. water vapour) are frozen at the plates, they work like a "poor" cryo pump. The cooling is done from outside the microscopic column by a Dewar vessel labelled "cold trap" in Fig. 4.1. Of course, the sample surface should be clean before inserting into the microscope, e.g. after cleaning by a plasmacleaner.

Another contamination mechanism exists at illumination of the sample by a small electron probe typical for the scanning electron microscope or for the scanning mode of the transmission electron microscope (see Chap. 8). In this case an additional possibility exists to deliver hydrocarbon molecules to the place of contamination: diffusion over the sample surface (Fig. 4.15).

The driving force of the surface diffusion is given by concentration differences between solid hydrocarbons at the probe position and mobile gaseous hydrocarbon molecules further away on the surface. This mechanism of deliver is very effective. The complete sample surface works as a kind of "antenna" to trap the molecules. Often the contamination rate at the probe position is much higher than for the large-area

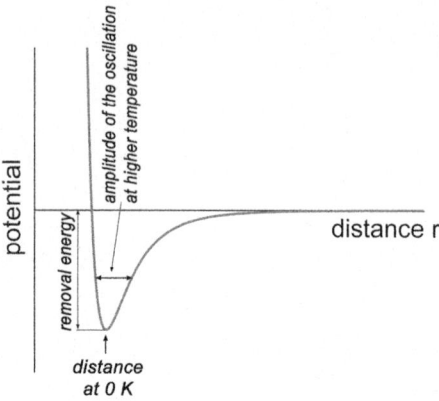

Fig. 4.16 Simplified potential model for explanation of the distances and the oscillation behaviour of the atoms in a solid

illumination and the plasmacleaning is eminently important before using the scanning mode. In almost all cases the specimen itself is the main source of contamination.

Sample damaging: What happens when high-energy electrons meet an atom of a solid that oscillates around its rest position? In principle, two effects are possible: The atom itself can be changed or/and it is pushed and the amplitude of the oscillation increases or the atom is completely removed from its rest position.

At electron energies used in transmission electron microscopes a change of an atom can only happen within its electron shell. We speak about two different kinds of electrons now: the electrons of the atomic shell and the electrons of the beam also named *primary electrons*. The shell electrons can be transferred from their original energy levels ("orbits") into higher ones or completely removed from the atomic shell. The transfer is called "ionisation" (cf. Chap. 9). The probability of such ionisation processes decreases with increasing energy of the primary electrons.

To understand the process of "pushing" an atom we have to think about the question: What is the reason for the special distance between the rest positions of the atoms in a solid? Simplifying we imagine that we had only two atoms which feel repulsive or attractive forces in dependence on the distance between both. In a mechanical model these forces can be created by a spring connecting the two atoms. Repulsive and attractive forces are equal in the state of equilibrium. For understanding of the oscillation it is useful to look at the potential affiliated to the forces. Figure 4.16 shows a simplified potential model. It is based on the overlay of repulsive and attractive forces dependent on the distance r.

At the temperature of 0 K the atoms do not significantly oscillate, their distance is determined by the position of the potential minimum. At higher temperatures our reference atom can move within a region of distances given by the width of the potential valley. An atom oscillates around its rest position: The higher the temperature is the broader the potential valley, i.e. the larger the amplitude of the oscillation. In this context it is noteworthy that the simple potential model gives an explanation of thermal expansion. This is caused by the asymmetric shape of the potential valley. The centre position of the range of motion shifts to larger distances at higher energies (temperatures, respectively).

Let us imagine that the energy is supplied not by heat but by primary electron impacts on atoms. Although the masses of atomic nuclei and electron are extremely different energy can be transferred from the electrons to the atoms leading to stronger oscillation. This effect is larger for light than for heavy atoms. Organic matter mainly consists of hydrogen, nitrogen, carbon, and oxygen, hence they are especially sensitive in this context.

If the impacts take place rapidly at a high number of atoms an increase of the temperature is measurable. It is difficult to evaluate the value of the temperature which can be reached. Beside the introduced energy and the sample mass, especially the heat dissipation plays an essential role and cannot be really determined (dependent on the sample geometry, heat contacts within the specimen and to the holder, heat radiation). At heat-sensitive samples the possible increase of the temperature has to be considered, perhaps the use of a cooling holder is helpful.

Finally, let us think about the case that the energy introduced by the electrons is larger than the removal energy drawn in Fig. 4.16. The atom reaches an energy level suitable for its "free motion". The sample melts if this effect pertains a high number of atoms. On the other hand, it can pertain only single atoms which are removed from their original position. There are such lattice vacancies within the real lattice whose number as a part of the thermodynamic equilibrium increases with increasing temperature. Additional vacancies can be obtained by primary electron impacts. Single atoms can be completely removed from the lattice. Often oxygen and halogens like fluorine or chlorine are involved.

It can be easily understood that such processes need a material-dependent minimum of introduced energy. On the other hand, the interaction needs a (short) time, i.e. at very high electron energies (and velocities, respectively) the probability of this kind of sample damage can be reduced.

Because of the mobility of the atoms within the crystal lattice electron beam induced defects can heal up. As consequence an accurate prognosis of possible damages is difficult.

Amorphous solids (e.g. bulk metallic glasses) are often thermodynamically metastable and a low energy input by the electron beam can lead to a limited crystallisation.

A review on radiation damage which can serve as starting point for further reading is published in [4].

References

1. Ishizuka, K., Shirota, K.: Lens-field center alignment for high resolution electron microscopy. Ultramicroscopy **65**, 71–79 (1996)
2. Ruska, E.: Das Entstehen des Elektronenmikroskops und der Elektronenmikroskopie. Nobel laureate lecture, 08. 12. 1986, Stockholm. Phys. Bl. **43**, 271 et seqq., http://ernst.ruska.de/daten_d/bibliothek/dokumente/999.nobelvortrag/vortrag.html. (1987)
3. Heide, H.G.: Die Objektverschmutzung im Elektronenmikroskop und das Problem der Strahlenschädigung durch Kohlenstoffabbau. Zeitschr. angew. Physik **15**, 116–128 (1963)
4. Egerton, R.F., Li, P., Malac, M.: Radiation damage in the TEM and SEM. Micron **35**, 399–409 (2004) and follow up publications

Chapter 5
Let us Switch to Electron Diffraction

Abstract One of the main advantages of the transmission electron microscope is the easy possibility to switch between the imaging of tiny structures in a thin specimen and the diffraction pattern of the same structures. Hence, we want to explain why different kinds of electron diffraction patterns are produced and which changes of the lenses are needed to see these patterns on the screen. Finally, we want to discuss which results can be obtained from the diffraction patterns regarding the materials science. For this purpose some basic knowledge about crystal structures is necessary and is explained in this chapter.

5.1 Why Diffraction Reflections?

Let us start with a short trip into the history of science. Until about 1910 two important questions in physics were not experimentally clarified:

1. What is the nature of the "X-rays" discovered by Wilhelm Conrad Röntgen[1] in 1895?
2. Do crystals exist with periodically arranged atoms?

On basis of Arnold Sommerfeld's[2] reflections Max von Laue[3] had the great idea to answer both questions by only one experiment: If the X-rays have wave character and the atoms are periodically arranged with distances not far from the order of the assumed X-ray wavelength after transmission through the crystal diffraction reflections have to be expected. Performing such an experiment with zinc blende W. Friedrich[4] und P. Knipping[5] observed single black spots on photo plate, i.e. they found diffraction reflections [1–3]. This experiment was repeated on a nickel crystal and with electrons by C. Davisson[6] und L.H. Germer[7] in 1927 [4].

---

[1] Wilhelm Conrad Röntgen, German physicist, 1845–1923, Nobel prize in physics in 1901 (first Nobel prize in physics).
[2] Arnold Sommerfeld, German physicist, 1868–1951.
[3] Max von Laue, German physicist, 1879–1960, Nobel prize in physics in 1914.
[4] Walter Friedrich, German biophysicist, 1883–1968.
[5] Paul Knipping, German physicist, 1883–1935.
[6] Clint Davisson, American physicist, 1881–1958.
[7] Lester Germer, American physicist, 1896–1971.

Fig. 5.1 Diffraction experiment at crystals. **a** Periodic arrangement of atoms in the crystal lattice. **b** Selected crystal planes ("lattice planes"). **c** Lattice planes act as semipermeable mirrors: The wave is partly reflected

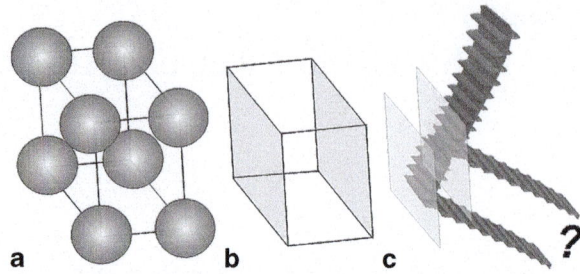

Fig. 5.2 Interference of electron waves after scattering at crystal planes. **a–c** Destructive interference (*cancellation*). **d–f** Constructive interference at changed wavelength (*intensification*). The pictures are snapshots of the propagating waves. At **b** and **e** a wave trough reaches the lattice plane on the *right side* and is reflected

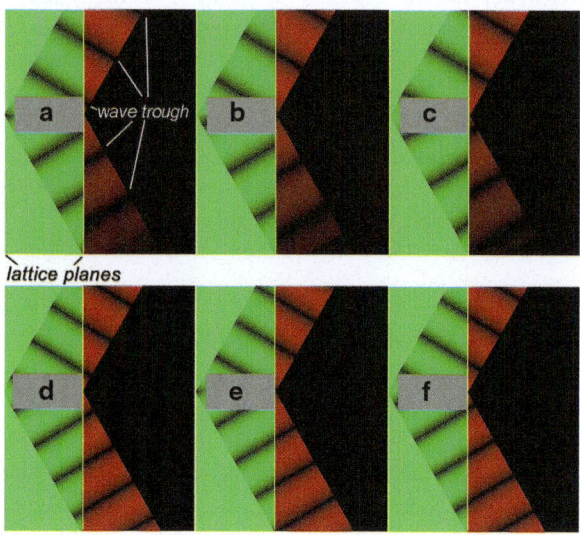

How did Laue get his idea? In Fig. 5.1a eight atoms are drawn periodically arranged in form of a lattice. The lattice is marked by lines between the atoms. We select two parallel planes (Fig. 5.1b) which are called atomic planes or *lattice planes*. We imagine that these planes are semipermeable and a wave (here: electron wave) incides at a flat angle. The wave is partly reflected at the front plane, partly at the following plane. This results in two waves propagating in the reflection direction. They interfere with each other (Fig. 5.1c).

We want to illustrate this interference with the help of Fig. 5.2. The pictures are snapshots of the waves partly reflected at the lattice planes with two different wavelengths. Parts b) and e) of Fig. 5.2 show the moment in which a wave trough reaches the lattice plane on the right side and gets partly reflected. The rest of the wave penetrates this plane and is reflected at the lattice plane on the left side. By doing so, it runs an additional distance, meets the firstly reflected part of the wave, and interferes (overlays) with it.

Commonly, both partial waves are phase-shifted against each other because of the additional distance which the second partial wave has to run ("path difference"),

5.1 Why Diffraction Reflections?

Fig. 5.3 Explanation of the calculation of the path difference during the diffraction of a wave at two lattice planes within a crystal

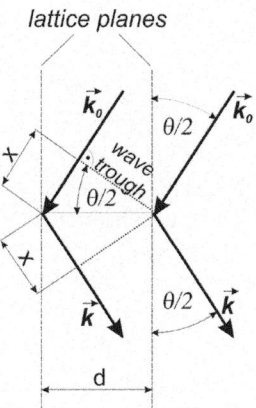

i.e. wave trough does not meet wave trough. The consequence is a reduction of the amplitude of the resulting wave. By change of the wavelength it is possible to reach that a wave trough meets a wave trough, i.e. an intensification of the resulting wave (constructive interference) occurs. Obviously, for maximal intensity the path difference has to be a whole-number multiple of the wavelength.

Let us think about the dependencies of the path difference between the two partial waves with the help of Fig. 5.3. We abstract: The waves are characterised by the vectors $\mathbf{k_0}$ and $\mathbf{k}$ (*wave number vectors*), which point in the propagation direction of the incident and the reflected wave, respectively. The distance between the lattice planes is labelled d, the angle between the wave number vector of the incident wave and the lattice plane is $\theta/2$. Following the law of reflection this angle is equal to the angle between the lattice plane and the wave number vector of the reflected wave.

As already shown in Fig. 5.2 the wave trough drawn in Fig. 5.3 is also directly reflected at the lattice plane on the right side. In the partial wave penetrating this lattice plane the wave trough must additionally run two times the distance x. Therefore the path difference is

$$\Delta s = 2 \cdot x \tag{5.1}$$

We already know that we get a constructive interference for path differences equal to a whole-number multiple of the wavelength λ. Using n as integer number the condition for a maximum of intensity is:

$$n \cdot \lambda = 2 \cdot x \tag{5.2}$$

Angles are equal if their arms are pairwise perpendicular to each other, i.e. the angle between the drawn wave trough and the line d is equal to $\theta/2$, too. The wave trough is perpendicular to the propagation direction. Herewith we read in Fig. 5.3:

$$x = d \cdot \sin \frac{\theta}{2} \tag{5.3}$$

and, respectively

$$n \cdot \lambda = 2 \cdot d \cdot \sin\frac{\theta}{2}. \tag{5.4}$$

This is the fundamental equation for diffraction and known as Bragg's[8] law (watch out: In X-ray diffraction the definition of the angle and therefore the equation are different). The model of the partly reflecting lattice planes was created by father and son Bragg [5, 6] in 1912. Bragg's law indicates the expected angles of the diffraction maxima ("diffraction reflections"). Normally, the wavelength is known and we can calculate the distances d between the lattice planes from the diffraction angles. The series of these distances is a "fingerprint" of the crystal.

Which order of these angles has to be expected in case of electron diffraction in the transmission electron microscope? We already learnt that the electron wavelength is only a few picometres for the transmission electron microscope; let us say typical 3 pm. Of course, the lattice distances depend on the material; we assume 0.3 nm for our evaluation. Using these values we get an angle of about 0.01 rad = 10 mrad ≈ 0.6° for the first diffraction maximum ($n = 1$). Contrary to X-ray diffraction it is possible to set

$$\sin\frac{\theta}{2} = \frac{\theta}{2} \tag{5.5}$$

at small angles and we get

$$n \cdot \lambda = d \cdot \theta \tag{5.6}$$

as the "base equation of electron diffraction".

5.2 Crystal Lattices and Lattice Planes

Let us turn the attention to the expected distances between the lattice planes within a crystal. The elements and compounds form different crystals, the specialists of materials science speak about different "phases". But be careful, please! We must not confuse these "materials phases" and the phase of a wave.

The differences between the crystals can be manifested in different shapes and different atomic distances. To classify crystals a unit cell is introduced, the complete crystal lattice is the result of the periodic arrangement of these unit cells. The shape and the size of the unit cell are described by three axes and three angles between these axes (Fig. 5.4).

---

[8] William Henry Bragg and William Lawrence Bragg: Australian/English physicists (father and son), 1862–1942 and 1890–1971, respectively, Nobel prize in physics in 1915.

5.2 Crystal Lattices and Lattice Planes

Fig. 5.4 Names of axes and angles in the unit cell

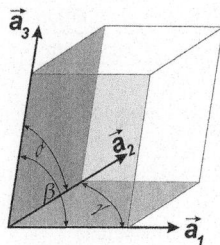

In this book the names of the three axes are a_1, a_2, and a_3, in the literature **a**, **b**, and **c** are also common. The angle α between a_2 und a_3 is opposite to the axis a_1 within the a_2-a_3 plane, the angle β is opposite to the axis a_2 within the a_1-a_3 plane, and the angle γ is opposite to the axis a_3 within the a_1-a_2 plane. Seven crystal systems result from symmetry regarding axes and angle relations (Table 5.1); the cubic system has got the highest symmetry, the triclinic system the lowest one.

Commonly, the symmetry properties of the crystal lattices are described by *space groups* (cf. Table 5.2).

Directions within the crystal are described by a linear combination of the three vectors a_1, a_2, a_3 and labelled by a triple of numbers in brackets:

[h k l] corresponds to the direction $h \cdot a_1 + k \cdot a_2 + l \cdot a_3$.

Therefore the crystal axes are labelled [100], [010], and [001].

Commonly, there are also atoms inside the unit cell and not only at the corners. To describe these positions we need either information about the packing of the unit cell (Table 5.3) or the positions of the atoms are calculated using the knowledge of the symmetry properties of the space groups and only some basic positions of the atoms. Determining the space group and the basic positions is the goal of crystal structure determination of an unknown phase.

In practice of electron diffraction the assignment of a measured diffraction pattern to a known phase is the most frequent purpose. Now, we want to explain with the help of a simple example how we can get the crystallographic data (distances of the unit cell and basic atomic positions) necessary for the calculation of diffraction patterns from the data usually given in the literature. We use sodium chloride (NaCl) as example. At first, we need a data basis or a reference with information about the phase NaCl. We can have a look into the "Pearson", a book (i.e. database) with summaries of the crystallographic data of known phases [8] or into the internet (e.g. "Inorganic Crystal Structures Database" [9]—licence is needed). For new, i.e. recently discovered phases, a look into the current scientific literature is often helpful.

Table 5.1 Definition of the seven crystal systems [7]

Nr.	Crystal system	Axis lengths	Axis angles
1	Cubic	$a_1 = a_2 = a_3$	$\alpha = \beta = \gamma = 90°$
2	Tetragonal	$a_1 = a_2 \neq a_3$	$\alpha = \beta = \gamma = 90°$
3	Orthorhombic	$a_1 \neq a_2 \neq a_3$	$\alpha = \beta = \gamma = 90°$
4	Hexagonal (also: trigonal with hexagonal axes)	$a_1 = a_2 \neq a_3$	$\alpha = \beta = 90°, \gamma = 120°$
5	Rhombohedral (trigonal with rhombohedral axes)	$a_1 = a_2 = a_3$	$\alpha = \beta = \gamma \neq 90°$
6	Monoclinic	$a_1 \neq a_2 \neq a_3$	$\alpha = \gamma = 90°, \beta \neq 90°$
7	Triclinic	$a_1 \neq a_2 \neq a_3$	$\alpha \neq \beta \neq \gamma$

Table 5.2 Correlation between space groups and crystal systems (the Pearson (Frederic Pearson Treadwell, American-Swiss chemist, 1857–1918)-Symbol is another name which immediately allows to read the crystal system—see also [8])

Crystal system	Space groups	PEARSON symbol
Cubic	No. 195–230 (quantity: 36)	cP, cI, cF
Tetragonal	No. 75–142 (quantity: 68)	tP, tI
Orthorhombic	No. 16–74 (quantity: 59)	oP, oI, oF, oC, oA, oB
Hexagonal	No. 168–194 (quantity: 27)	hP, hR
Rhombohedral (also: trigonal with hexagonal or rhombohedral axes)	No. 143–167 (quantity: 25)	hP, hR
Monoclinic	No. 3–15 (quantity: 13)	mP, mC, mA, mI
Triclinic	No. 1 and 2 (quantity: 2)	aP

For NaCl we find the following data:

Cell parameter: 5.64 Å; 5.64 Å; 5.64 Å; 90.0°; 90.0°; 90.0°			
Volume: 179.41 Å³			
Space group: Fm-3m (225); Pearson symbol: cF8			
Element	Wyckoff symbol	x	y	z
Na	4a	0.000	0.000	0.000
Cl	4b	0.500	0.500	0.500

Ralph Walter Graystone Wyckoff, American crystallographer, 1897–1994

Sometimes the units are omitted. In the first row we find the lengths of the axes **a₁, a₂, a₃** (also called lattice constants) and the angles α, β and γ between the axes. The volume of the unit cell resulting from the cell parameters is written in the second row. The third row includes the space group (F: face-centred) and the Pearson symbol which declares that NaCl has a cubic face-centred system with eight atoms within the unit cell.

How do we calculate the positions of these eight atoms? We use the data of the two last rows and the knowledge of the space group. The easiest way is to use a specialised computer program. But one can also use the "International Tables for Crystallography" [7]. In this book the calculation rules for the atomic positions are

5.2 Crystal Lattices and Lattice Planes

Table 5.3 Quantity and arrangement of the atoms within a unit cell for different kinds of packing

No.	Packing	Number of atoms	Atomic positions		Sketch of the unit cell
1	primitive (space groups P...)	1	0;0;0		
2	body-centred (space groups I...)	2	0;0;0 1/2; 1/2; 1/2		
3	face-centred (all faces) (space groups F...)	4	0;0;0 1/2; 0; 1/2	1/2; 1/2; 0 0; 1/2; 1/2	
4	base face-centred (footpoint C) (space groups C...)	2	0;0;0	1/2; 1/2; 0	
5	diamond lattice	8	0;0;0 1/2; 0; 1/2 1/4; 1/4; 1/4 3/4; 1/4; 3/4	1/2; 1/2; 0 0; 1/2; 1/2 1/4; 3/4; 1/4 1/4; 3/4; 3/4	
6	hexagonal close-packing of spheres (space groups R...)	2	2/3; 1/3; 1/4	1/3; 2/3; 3/4	

listed for the 230 space groups. Figure 5.5 shows one of the pages for the space group 225 as an example.

The line marked by A describes that the structure has a face-centred unit cell. The triples (0,0,0), (0, ½, ½), (½, 0, ½), and (½, ½, 0), respectively, must be added to each of the positions given in the next lines, i.e. we calculate four atomic positions from one value of the lines following A. The basic positions are given in the last two rows in the data set of NaCl. For sodium this is $x=0$, $y=0$, $z=0$. These values can be inserted into the calculation rules (1) to (48) written below the line A in Fig. 5.5. Doublings must be cancelled $(\bar{x} = -x)$. We see that all calculation rules lead to the

| CONTINUED | No. 225 | $Fm\bar{3}m$ |

Generators selected (1); $t(1,0,0)$; $t(0,1,0)$; $t(0,0,1)$; $t(0,\tfrac{1}{2},\tfrac{1}{2})$; $t(\tfrac{1}{2},0,\tfrac{1}{2})$; (2); (3); (5); (13); (25)

Positions

Multiplicity, Wyckoff letter, Site symmetry		Coordinates $(0,0,0)+\quad (0,\tfrac{1}{2},\tfrac{1}{2})+\quad (\tfrac{1}{2},0,\tfrac{1}{2})+\quad (\tfrac{1}{2},\tfrac{1}{2},0)+$ **A**	Reflection conditions
192	l 1	(1) x,y,z (2) $\bar{x},\bar{y},z$ (3) $\bar{x},y,\bar{z}$ (4) $x,\bar{y},\bar{z}$ (5) z,x,y (6) $z,\bar{x},\bar{y}$ (7) $\bar{z},\bar{x},y$ (8) $\bar{z},x,\bar{y}$ (9) y,z,x (10) $\bar{y},z,\bar{x}$ (11) $y,\bar{z},\bar{x}$ (12) $\bar{y},\bar{z},x$ (13) $y,x,\bar{z}$ (14) $\bar{y},\bar{x},\bar{z}$ (15) $y,\bar{x},z$ (16) $\bar{y},x,z$ (17) $x,z,\bar{y}$ (18) $\bar{x},z,y$ (19) $\bar{x},\bar{z},\bar{y}$ (20) $x,\bar{z},y$ (21) $z,y,\bar{x}$ (22) $z,\bar{y},x$ (23) $\bar{z},y,x$ (24) $\bar{z},\bar{y},\bar{x}$ (25) $\bar{x},\bar{y},\bar{z}$ (26) $x,y,\bar{z}$ (27) $x,\bar{y},z$ (28) $\bar{x},y,z$ (29) $\bar{z},\bar{x},\bar{y}$ (30) $\bar{z},x,y$ (31) $z,x,\bar{y}$ (32) $z,\bar{x},y$ (33) $\bar{y},\bar{z},\bar{x}$ (34) $y,\bar{z},x$ (35) $\bar{y},z,x$ (36) $y,z,\bar{x}$ (37) $\bar{y},\bar{x},z$ (38) y,x,z (39) $\bar{y},x,\bar{z}$ (40) $y,\bar{x},\bar{z}$ (41) $\bar{x},\bar{z},y$ (42) $x,\bar{z},\bar{y}$ (43) x,z,y (44) $\bar{x},z,\bar{y}$ (45) $\bar{z},\bar{y},x$ (46) $\bar{z},y,\bar{x}$ (47) $z,\bar{y},\bar{x}$ (48) z,y,x	$hkl: h+k, h+l, k+l = 2n$ $0kl: k,l = 2n$ $hhl: h+l = 2n$ $h00: h = 2n$ General:

Special: as above, plus

96	k ..m	$x,x,z\quad \bar{x},\bar{x},z\quad \bar{x},x,\bar{z}\quad x,\bar{x},\bar{z}\quad z,x,x\quad z,\bar{x},\bar{x}$ $\bar{z},\bar{x},x\quad \bar{z},x,\bar{x}\quad x,z,x\quad \bar{x},z,\bar{x}\quad x,\bar{z},\bar{x}\quad \bar{x},\bar{z},x$ $x,x,\bar{z}\quad \bar{x},\bar{x},\bar{z}\quad x,\bar{x},z\quad \bar{x},x,z\quad \bar{x},x,z\quad \bar{x},z,x$ $\bar{x},\bar{z},\bar{x}\quad x,\bar{z},x\quad z,x,\bar{x}\quad z,\bar{x},x\quad \bar{z},x,x\quad \bar{z},\bar{x},\bar{x}$	no extra conditions
96	j m..	$0,y,z\quad 0,\bar{y},z\quad 0,y,\bar{z}\quad 0,\bar{y},\bar{z}\quad z,0,y\quad z,0,\bar{y}$ $\bar{z},0,y\quad \bar{z},0,\bar{y}\quad y,z,0\quad \bar{y},z,0\quad y,\bar{z},0\quad \bar{y},\bar{z},0$ $y,0,\bar{z}\quad \bar{y},0,\bar{z}\quad y,0,z\quad \bar{y},0,z\quad 0,z,y\quad 0,z,\bar{y}$ $0,\bar{z},\bar{y}\quad 0,\bar{z},y\quad z,y,0\quad z,\bar{y},0\quad \bar{z},y,0\quad \bar{z},\bar{y},0$	no extra conditions
48	i $m.m2$	$\tfrac{1}{2},y,y\quad \tfrac{1}{2},\bar{y},y\quad \tfrac{1}{2},y,\bar{y}\quad \tfrac{1}{2},\bar{y},\bar{y}\quad y,\tfrac{1}{2},y\quad y,\tfrac{1}{2},\bar{y}$ $\bar{y},\tfrac{1}{2},y\quad \bar{y},\tfrac{1}{2},\bar{y}\quad y,y,\tfrac{1}{2}\quad \bar{y},y,\tfrac{1}{2}\quad y,\bar{y},\tfrac{1}{2}\quad \bar{y},\bar{y},\tfrac{1}{2}$	no extra conditions
48	h $m.m2$	$0,y,y\quad 0,\bar{y},y\quad 0,y,\bar{y}\quad 0,\bar{y},\bar{y}\quad y,0,y\quad y,0,\bar{y}$ $\bar{y},0,y\quad \bar{y},0,\bar{y}\quad y,y,0\quad \bar{y},y,0\quad y,\bar{y},0\quad \bar{y},\bar{y},0$	no extra conditions
48	g $2.mm$	$x,\tfrac{1}{4},\tfrac{1}{4}\quad \bar{x},\tfrac{1}{4},\tfrac{1}{4}\quad \tfrac{1}{4},x,\tfrac{1}{4}\quad \tfrac{1}{4},\bar{x},\tfrac{1}{4}\quad \tfrac{1}{4},\tfrac{1}{4},x\quad \tfrac{1}{4},\tfrac{1}{4},\bar{x}$ $\tfrac{1}{4},x,\tfrac{3}{4}\quad \tfrac{1}{4},\bar{x},\tfrac{3}{4}\quad x,\tfrac{1}{4},\tfrac{3}{4}\quad \bar{x},\tfrac{1}{4},\tfrac{3}{4}\quad \tfrac{3}{4},\tfrac{1}{4},x\quad \tfrac{3}{4},\tfrac{1}{4},\bar{x}$	$hkl: h = 2n$
32	f $.3m$	$x,x,x\quad \bar{x},\bar{x},x\quad \bar{x},x,\bar{x}\quad x,\bar{x},\bar{x}$ $x,x,\bar{x}\quad \bar{x},\bar{x},\bar{x}\quad x,\bar{x},x\quad \bar{x},x,x$	no extra conditions
24	e $4m.m$	$x,0,0\quad \bar{x},0,0\quad 0,x,0\quad 0,\bar{x},0\quad 0,0,x\quad 0,0,\bar{x}$	no extra conditions
24	d $m.mm$	$0,\tfrac{1}{4},\tfrac{1}{4}\quad 0,\tfrac{3}{4},\tfrac{1}{4}\quad \tfrac{1}{4},0,\tfrac{1}{4}\quad \tfrac{1}{4},0,\tfrac{3}{4}\quad \tfrac{1}{4},\tfrac{1}{4},0\quad \tfrac{3}{4},\tfrac{1}{4},0$	$hkl: h = 2n$
8	c $\bar{4}3m$	$\tfrac{1}{4},\tfrac{1}{4},\tfrac{1}{4}\quad \tfrac{3}{4},\tfrac{1}{4},\tfrac{1}{4}$	$hkl: h = 2n$
4	b $m\bar{3}m$	$\tfrac{1}{2},\tfrac{1}{2},\tfrac{1}{2}$ **B**	no extra conditions
4	a $m\bar{3}m$	$0,0,0$	no extra conditions

Symmetry of special projections

Along [001] $p4mm$
$a' = \tfrac{1}{2}a \quad b' = \tfrac{1}{2}b$
Origin at $0,0,z$

Along [111] $p6mm$
$a' = \tfrac{1}{6}(2a-b-c) \quad b' = \tfrac{1}{6}(-a+2b-c)$
Origin at x,x,x

Along [110] $c2mm$
$a' = \tfrac{1}{2}(-a+b) \quad b' = c$
Origin at $x,x,0$

679

Fig. 5.5 Page 679 of the "International Tables of Crystallography", Vol. A, for the space group 225 ([7]—explanations in the text)

5.2 Crystal Lattices and Lattice Planes

triple (0,0,0), i.e. they result in only one sodium atom. With consideration of the face centring the Na positions are:

$$(0,0,0)+(0,0,0)=(0,0,0),$$
$$(0,½,½)+(0,0,0)=(0,½,½),$$
$$(½,0,½)+(0,0,0)=(½,0,½), \text{ and}$$
$$(½,½,0)+(0,0,0)=(½,½,0).$$

It is possible to simplify the calculation by use of the Wyckoff symbols. For Na the Wyckoff position is *4a*. *4* is the number of the Na atomic positions and *a* characterises the symmetry. We find this symbol at the left edge of the lower line in box B in Fig. 5.5. In this line the position (0,0,0) is immediately readable. Analogously, the bottom line in box B shows $(½,½,½)$ for the chlorine atom and the four Cl positions are:

$$(0,0,0) + (½,½,½) = (½,½,½),$$
$$(0,½,½) + (½,½,½) = (½,0,0),$$
$$(½,0,½) + (½,½,½) = (0,½,0), \text{ and}$$
$$(½,½,0) + (½,½,½) = (0,0,½).$$

At a first view it is surprising that for the sum ½ + ½ = 0 is written. To understand that we have to consider what does e.g. the value x = ½ mean. This value is related to the axis length a_1, i.e. strictly spoken we have to write $x = ½ \cdot a_1$. Because of the translation property of the lattice the unit cell recurs and the atom at the position (1,0,0) is in the crystallographic sense equal to that at position (0,0,0). This is analogously valid for both other axes, i.e. the coordinates of the atomic positions are related to the axis lengths and are numbers between 0 and < 1.

The CIF format is a widespread standard for saving and exchanging of crystal structure data. These ASCII text files can be read by each text editor and are self-explained by the knowledge being taught in this paragraph.

For the description of the *lattice planes* we start with a freely chosen unit cell. A plane within it is described by a triple of integers (hkl) where 1/h, 1/k, and 1/l are the intersections of the plane (hkl) with the axes $\mathbf{a_1}$, $\mathbf{a_2}$, and $\mathbf{a_3}$ of the unit cell. If a plane is parallel to one of the axes the intersection of both is at infinity and the reciprocal value is equal to zero. For example, the (100) plane is the $\mathbf{a_2}$-$\mathbf{a_3}$ plane. Additional examples of lattice planes in the cubic, the tetragonal, and the hexagonal crystal system are shown in Fig. 5.6.

The indices h, k, l are also called Miller's[9] indices. Sometimes it is argued that the direction [hkl] is the normal of the plane (hkl), i.e. it is perpendicular to the plane characterised by equal indices. But be careful, please. Commonly, this is only valid in the cubic system.

---

[9] William Hallowes Miller, British crystallographer, 1801–1880.

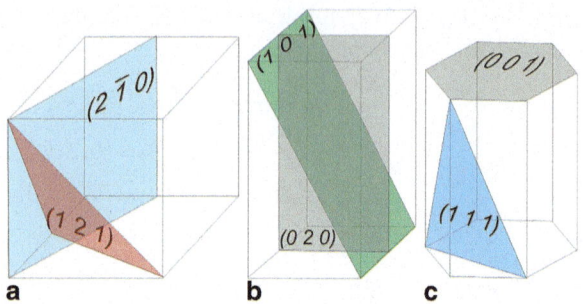

Fig 5.6 Examples for different lattice planes in the **a** cubic, **b** tetragonal, and **c** hexagonal crystal lattice

To calculate the diffraction angles the knowledge of the lattice plane distances is necessary. We consider the indexing of the lattice planes and specify the Eq. (5.6):

$$n \cdot \lambda = d_{hkl} \cdot \theta_{hkl}, \qquad (5.7)$$

i.e. each diffraction angle pertains to one lattice plane distance.

To calculate these distances d_{hkl} the shape of the unit cell has to be considered, i.e. different equations must be used in dependence on the crystal system. These equations are listed in Table 5.4; their basis is explained in Sect. 10.8.

Concerning the *hexagonal system* there is a speciality with respect to the indexing: Sometimes planes and directions are denoted by four indices. This is not really mandatory but this facilitates the recognition of equivalent crystallographic planes and directions in the hexagonal system. For understanding firstly we imagine a cubic system. The cube axes [100], [010], [001] are equivalent in the crystallographic sense. We can immediately see this: The indices include combinations of equal numbers. The three-fold indexing [uvw] in the hexagonal system shows another behaviour. The direction [110] is equivalent to the axes [100] and [010]. After introduction of a four-fold indexing [qrst] according to the rules

$$q = \frac{1}{3}(2 \cdot u - v), \quad r = \frac{1}{3}(2 \cdot v - u), \quad s = -\frac{1}{3}(u+v), \quad t = w \qquad (5.8)$$

we get $\left[2/3 \; -1/3 \; -1/3 \; 0 \right]$ and, after multiplication by 3, [2–1–1 0] instead of [100]. The direction [0 1 0] becomes [–1 2–1 0] and the direction [110] becomes [1 1–2 0]. We see there are (except for the sign) the same numbers. Further details of the indexing of hexagonal systems can be read e.g. in [10] and [11].

5.3 Selected Area and Convergent Beam Electron Diffraction

Let us come back to practical transmission electron microscopy. At first, we want to explain what has to be changed at the lenses to observe a diffraction pattern of a crystalline sample on the screen. Until now, we have seen an image of the

5.3 Selected Area and Convergent Beam Electron Diffraction

Table 5.4 Equations for the calculation of the lattice plane distances d_{hkl} in the seven crystal systems

Cubic: $a_1 = a_2 = a_3$

$$d_{hkl} = \frac{a_1}{\sqrt{h^2 + k^2 + l^2}}$$

Tetragonal: $a_1 = a_2$

$$d_{hkl} = \frac{1}{\sqrt{\frac{h^2 + k^2}{a_1^2} + \frac{l^2}{a_3^2}}}$$

Orthorhombic:

$$d_{hkl} = \frac{1}{\sqrt{\frac{h^2}{a_1^2} + \frac{k^2}{a_2^2} + \frac{l^2}{a_3^2}}}$$

Hexagonal: $a_1 = a_2$

$$d_{hkl} = \frac{1}{\sqrt{\frac{4}{3 \cdot a_1^2}(h^2 + h \cdot k + k^2) + \frac{l^2}{a_3^2}}}$$

Rhombohedral: $a_1 = a_2 = a_3$, $\alpha = \beta = \gamma$

$$d_{hkl} = \frac{1}{\sqrt{\frac{h^2}{a_1^2} + \left(-\frac{h \cdot \cot \alpha}{a_1} + \frac{k}{a_1 \cdot \sin \alpha}\right)^2 + \left(\frac{A}{a_1 \cdot \sin \varepsilon}\right)^2 + \frac{l^2}{a_1^2}}}$$

with

$$\cos \varepsilon = \cot \alpha \cdot \sqrt{2 \cdot (1 - \cos \alpha)}$$

and

$$A = h \cdot (\cot \alpha \cdot \sqrt{\cos^2 \varepsilon - \cos^2 \alpha} - \cos^2 \alpha) - k \cdot \frac{\sqrt{\cos^2 \varepsilon - \cos^2 \alpha}}{\sin \alpha} + 1$$

Monoclinic:

$$d_{hkl} = \frac{1}{\sqrt{\frac{h^2}{a_1^2} + \frac{k^2}{a_2^2} + \left(-\frac{h \cdot \cot \beta}{a_1} + \frac{1}{a_3 \cdot \sin \beta}\right)^2}}$$

Triclinic:

$$d_{hkl} = \frac{1}{\sqrt{\frac{h^2}{a_1^2} + C_{hk}^2 + C_{hkl}^2}}$$

with

$$C_{hk}^2 = -\frac{h \cdot \cot \gamma}{a_1} + \frac{k}{a_2 \cdot \sin \gamma}$$

and

$$C_{hkl}^2 = \frac{h \cdot \cot \gamma \cdot \sqrt{\cos^2 \varepsilon - \cos^2 \beta} - \cos \beta}{a_1 \cdot \sin \varepsilon} - \frac{k \cdot \sqrt{\cos^2 \varepsilon - \cos^2 \beta}}{a_2 \cdot \sin \gamma \cdot \sin \varepsilon} + \frac{1}{a_3 \cdot \sin \varepsilon}$$

aswellas

$$\cos \varepsilon = \frac{\sqrt{\cos^2 \alpha - 2 \cdot \cos \alpha \cdot \cos \beta \cdot \cos \gamma + \cos^2 \beta}}{\sin \gamma}$$

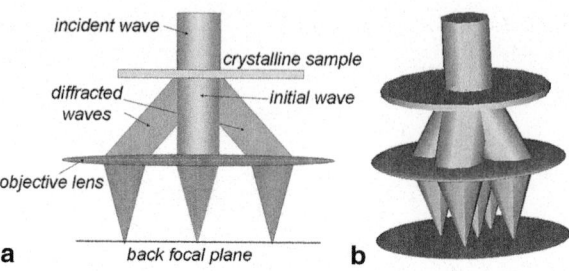

Fig. 5.7 Surrounding of the objective lens at selected area electron diffraction. **a** Section along the optical axis. **b** Perspective illustration

specimen created by multistage imaging. The focal lengths of the projective lenses are adjusted in a way to image the magnified real intermediate image on the screen.

We know from Sect. 5.1 that the diffraction maxima are observed under special angles. If we interpret the propagation direction of the electron wave as a "ray" in the sense of a geometric-optical treatment a diffraction maximum is represented by a bundle of rays which incide into the objective lens parallel to each other. Such parallel rays meet in the back-focal plane of the objective lens (cf. Sect. 2.7.2); they create a bright diffraction spot in this plane as illustrated in Fig. 5.7.

To observe the diffraction pattern the focal length of the projective lens system has to be set such that the back-focal plane of the objective is imaged on the observation screen. Often this setting is done by switching on an additional *diffraction lens*. The operator has only to press a button labelled "diffraction" on the panel.

Finally, we have to answer two questions: Which conditions must be fulfilled for "sharp" reflections and how can we determine the diffraction angle from the distance between reflection and the centre (i.e. the reflection of the initial wave) on the screen?

The condition for sharp reflections is already implied in Fig. 5.7: The crystalline sample is illuminated by a plane wave, i.e. with parallel illumination. This parallel illumination is set by a suitable excitation of the condenser system Then the diffraction pattern is focussed by the "diffraction focus" control. In practical work at a sufficiently well pre-aligned and stable microscope one has just to change the condenser setting ("intensity" or "brightness") until a sharp diffraction pattern is observed (cf. Sect. 2.7.1).

To determine the distances of the lattice planes by means of Bragg's law we need the diffraction angle θ. Assumed a parallel ray bundle incides against an angle θ to the optical axis into the objective lens with the focal length f_{Obj} then the rays meet in the back-focal plane in the distance $r_f = \theta \cdot f_{Obj}$ from the optical axis (cf. Fig. 2.18). The projective lens system additionally magnifies this distance by M_{Pro}:

$$r = \theta \cdot f_{Obj} \cdot M_{Pro}, \quad \text{respectively} \quad \theta = \frac{r}{f_{Obj} \cdot M_{Pro}}. \tag{5.9}$$

5.3 Selected Area and Convergent Beam Electron Diffraction

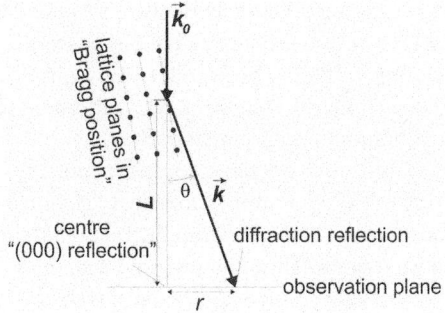

Fig. 5.8 Explanation of the term "camera length L". For electron diffraction in the transmission electron microscope $\theta \ll 1$ is valid

In practice, the product of the objective focal length and the magnification by the projective lenses is called *camera length L*:

$$\theta = \frac{r}{L}, \qquad (5.10)$$

and for all crystallographic reflection distances and diffraction angles, respectively:

$$\theta_{hkl} = \frac{r_{hkl}}{L}. \qquad (5.11)$$

Without any lenses between specimen and screen the camera length L would be the distance between those (Fig. 5.8).

The distance r from the central reflection (which gives the position of the initial, non-diffracted wave) is measured on the screen, on the photo plate or in the camera picture on the monitor. If we measure very precisely we will find that the values are systematically lower than the theoretically expected ones because of the spherical aberration of the objective lens. The deviation is the greater the larger the diffraction angle is.

In principle, the camera length can be calculated if the focal length of the objective lens and the magnification of the projective lens system are known. It is displayed on the monitor after switching to the diffraction mode. On the other hand, we see that the camera length depends on the focal length of the objective, i.e. under certain circumstances it can be variable. Therefore in practical work another route is usual: The camera length is measured using the diffraction diagram of a well-known crystalline substance. In this context it is useful to have a look back for the specified "base equation of electron diffraction" (5.7). We set $n=1$ (first diffraction maximum) und consider (5.11):

$$\lambda = d_{hkl} \cdot \frac{r_{hkl}}{L} \quad \text{and} \quad \lambda \cdot L = d_{hkl} \cdot r_{hkl}, \quad \text{respectively.} \qquad (5.12)$$

Not only the camera length but especially the product of camera length and wavelength is important. Obviously, we are able to calculate the wavelength from the accelerating voltage. However, if we calibrate at a standard anyway we can directly determine the product and avoid possible uncertainties of the precision of the accelerating voltage. The product $\lambda \cdot L$ is also called *apparatus constant* with the dimension Å mm = 0.1 nm mm. We will come back to the practical realisation of such a calibration measurement in Sect. 5.4.

Until now we have spoken about the "parallel illumination" of the specimen. But just the possibility to select "smallest areas" as sources of signal is one of the advantages of the electron diffraction in the transmission electron microscope. How does it work? Illuminating only a small area of the specimen would be the simplest way. Doing so we had to concentrate the ray bundle on the specimen by use of the condenser 2 lens with the result that the specimen is no longer parallel illuminated, the reflections become unsharp.

If we want to parallel illuminate only a small area of the sample we need either a third condenser lens or (and this is currently the more usual way) an aperture limiting the field of view. It seems to be obvious to position this aperture directly on the top of the specimen. In this case the aperture hole needs to be less than 1 μm to select such small areas. In practice it is difficult or impossible to realise both things. The specimen is moved along the optical axis to set the eucentric height, the aperture holder would have to be moved accordingly. Thinking about contamination it also is difficult to control holes with a diameter of less than 1 μm.

It is better to place this aperture in the first real intermediate plane. There the aperture limits the field of view, too. Geometrically, there is more space and the aperture hole is "backwards" demagnified onto the specimen by the objective lens. With an assumed magnification of 100 by the objective a hole diameter of 0.1 mm is small enough to select a specimen detail with 1 μm in size. The name of this method results out of this selection method of the investigated area: "*Selected Area Electron Diffraction*" (SAED). The method was proposed by H. Boersch already in 1936 [12].

Unfortunately, with decreasing diameter of the aperture the total intensity of the electrons in the observation plane decreases, too. It could become difficult to observe the diffraction pattern, and the specific setting of a desired orientation between crystal lattice and electron beam is nearly impossible. Additionally, the field of view can be displaced if the lenses are switched between imaging and diffraction mode because of the spherical aberration and the tolerances during the alignment of the electron-optical units. This displacement can reach up to more than 10 nm. If the selection area is less than about 50 nm the correlation between the diffractive area and the image of this area becomes unsure. Commonly, for these reasons the lower limit of the selected area is about 50 nm.

Selected area electron diffraction is possible on polycrystalline structures (*ring diagrams*) as well as on single crystals (*point diagrams*). To observe point diagrams the crystal must be larger than the area selected by the aperture and must extend across the complete specimen thickness, i.e. the source of the diffraction has to be only one crystal, in fact. There are also intermediate stages, whose appearance depend on the ratio of the size of the selected area and the size of the crystallites, i.e. on the number n of the crystallites contributing to the diffraction pattern. A large

5.3 Selected Area and Convergent Beam Electron Diffraction

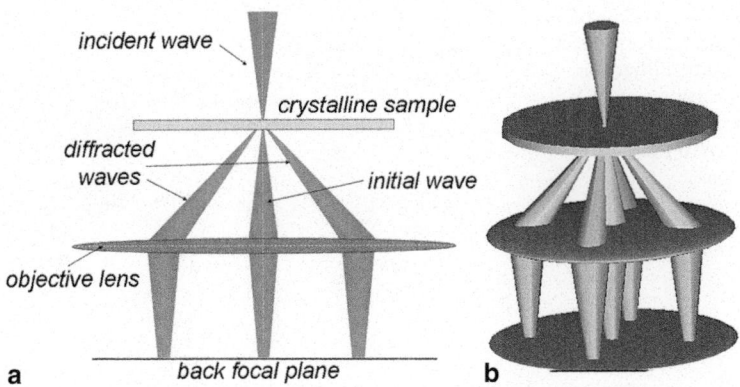

Fig. 5.9 Surrounding of the objective lens at convergent beam electron diffraction with low convergence angle. **a** Section along the optical axis. **b** Perspective illustration

number ($n \gg 10$) connected with statistic-uniformly distributed orientations of the crystallites results in ring diagrams with closed rings. Symmetrically distributed intensity fluctuations on single rings point to preferred crystallographic orientations (*textures*—cf. Sect. 6.3). With reduction of the number n the rings are no longer closed and they ultimately dissolve into individual reflections. For number n between 2 and 10 overlays of some point diagrams have to be expected making the analysis of the pattern more difficult.

What can we do to investigate "true nanostructures" by means of electron diffraction?

We relax the constraint of parallel illumination and focus the ray bundle on the specimen (*Convergent Beam Electron Diffraction, CBED*). The spot size can be influenced by the focal length of the condenser 1 lens. Hence this setting is called "spot size". Instead of the "sharp" diffraction reflections we observe "diffraction disks" whose diameter depends on the illumination aperture (angle of convergence). This is determined by the condenser 2 aperture (Fig. 2.16). At first, we imagine a very small convergence angle of only a few milliradians (Fig. 5.9).

In this case the deviations of the incident directions relative to the Bragg angle are smaller than the *ecitation error at electron diffraction* (we will come back to this topic later Sect. 10.9), i.e. at sufficiently small convergence angles the positions of the diffraction reflections are given by the centre of the diffraction disks. The pattern can be indexed like a "normal" point diagram obtained by selected area electron diffraction. This method is called "*Nano Beam Electron Diffraction*" (NBED). Ring diagrams do not make sense using this method because they result from many crystals anyway. There is a problem with the stacking of crystals if the crystals are smaller than the thickness of the specimen: In the projection plane we quasi hit only one crystal but in the three-dimensional reality more than one with different orientations and/or phases. We get a number of overlaid point diagrams and the indexing of the resulting diffraction pattern becomes very difficult or completely impossible.

What happens when we use larger convergence angles, i.e. when we use a larger condenser 2 aperture? Let us have a look at Fig. 5.10 to answer this question.

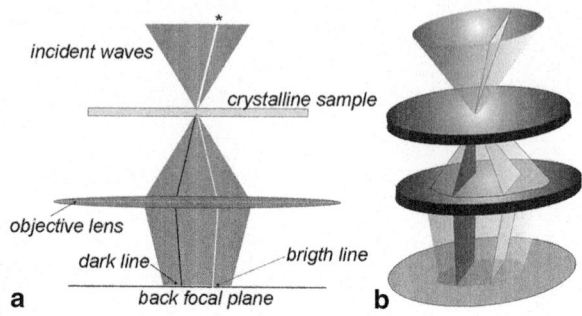

Fig. 5.10 Surrounding of the objective lens at convergent beam electron diffraction with large convergence angle. **a** Section along the optical axis. **b** Perspective illustration

Firstly, we observe a much larger diffraction disk in the centre. As speciality this disk includes an "inner structure" now. This can be explained by the following model: Because of the larger convergence angle we offer a cone of waves with very different incident angles to the crystal lattice. One of those directions (it is marked by * in Fig. 5.10a) fulfils Bragg's law with the direction of the selected lattice planes, i.e. an intensity rich diffraction maximum is generated. The line marked by * represents a plane that is perpendicular to the drawing plane in Fig. 5.10a. This line is the intersecting line of the drawing plane with the plane containing all the electron rays which fulfil Bragg's law (Fig. 5.10b). Therefore we observe a pair of lines: a dark line in the straight continuation of the "Bragg direction" and a bright line in the distance corresponding to the Bragg angle θ (cf. Sect. 5.4). These lines are named "Kikuchi[10] lines" or "Kikuchi bands" [13].

5.4 What Can We Learn from Selected Area Diffraction Patterns?

Let us consider the aim of electron diffraction for materials science. Which kind of results is useful? Answering this question we come back to the selected area electron diffraction and we discuss the ring diagrams which are produced when many small crystallites participate in the diffraction of polycrystalline structures and the crystal orientations are uniformly distributed.

5.4.1 Radii in Ring Diagrams

Each materials phase has a kind of "fingerprint" in the sequence of possible lattice spacings given by the crystal structure and which can be calculated by the equations in Table 5.4. Following Bragg's law (5.12) we are able to calculate the sequence

---

[10] Seishi Kikuchi, Japanese physicist, 1902–1974, discovered and explained the described bands in 1928.

5.4 What Can We Learn from Selected Area Diffraction Patterns?

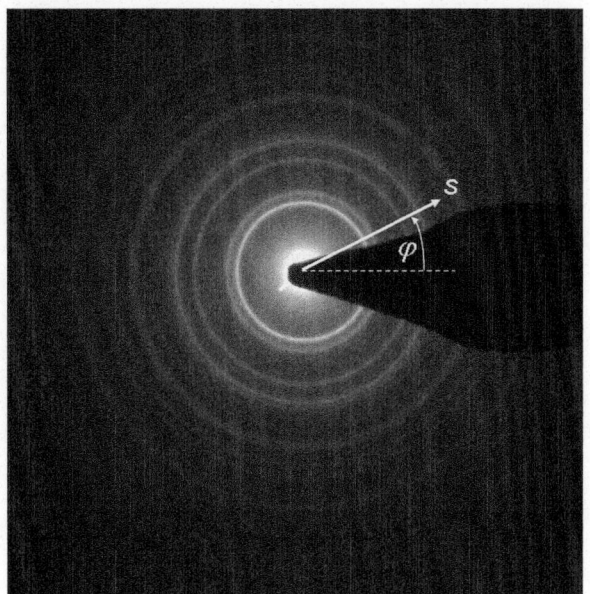

Fig. 5.11 Electron diffraction diagram of a fine-crystalline gold sample with overlaid polar coordinate system. Often the (000) reflection in the centre is so bright that it outshines the micrograph, even the CCD camera can be damaged! Hence, before recording the photo a small tip ("beam stopper") above the observation screen or the camera is inserted to cover this bright spot. In the figure this tip is inserted from the right side

of lattice spacings d_{hkl} from the measured ring radii r_{hkl} provided that the apparatus constant $\lambda \cdot L$ and the indexing are known.

Therefore we have to delve into the determination of the apparatus constant. For that purpose we need an electron-transparent specimen of a well-known substance with undeformed crystal lattice even in an extremely thin state. Often a sample containing tiny gold islands (10–20 nm in size) on a carbon support film is used. Such specimens can be bought as test specimens from suppliers for electron microscopic accessories. But be careful, please! Often the islands are not made of pure gold but are a gold-palladium alloy with slightly changed lattice constant.

Figure 5.11 shows an electron diffraction diagram of such a sample. On photographs the ring radii can be measured with a ruler or by a photometric method. At measurements on magnified pictures the magnification must be considered while the diffraction angles and the lattice spacings are calculated. The apparatus constant can be interpreted as the magnification of diffraction pictures, i.e. its value depends on the setting of the microscope and on a possible additional magnification of the observed picture.

If digital image processing is used, then it is advantageous to utilise a polar coordinate system s, φ in the image as shown in Fig. 5.11 and to sum the pixel brightness I_{Pix} in dependence on s for $\varphi = 0 \ldots 2\pi$:

$$I(s) = \int_0^{2\pi} I_{Pix}(s, \varphi) \, d\varphi. \tag{5.13}$$

We denote the radial coordinate with s to point out that s is the modulus of the scattering vector and we name the function $I(s)$ "radial brightness distribution". Following Eq. (5.13) we get Fig. 5.12 from Fig. 5.11.

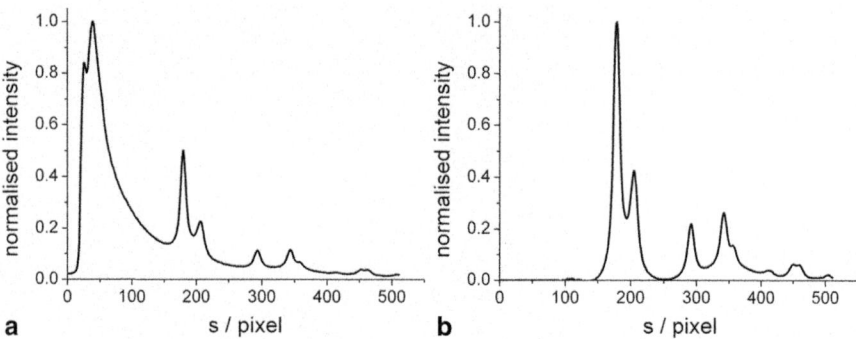

Fig. 5.12 Radial brightness distribution of the diffraction ring diagram of gold extracted from Fig. 5.11 (pixel size: 24 μm, camera length: 380 mm, accelerating voltage: 300 kV, i.e. wavelength $\lambda=1{,}97$ pm). **a** With background. **b** After background subtraction. Normalised intensity: maximal intensity is equal to 1

Table 5.5 Comparison between the measured sequence of the radii in Fig. 5.11 as "fingerprints" and the expectations for a cubic primitive unit cell

h	k	l	$h^2+k^2+l^2$	$179\cdot\sqrt{h^2+k^2+l^2}$	Measured s/Pixel
1	0	0	1	179	179
1	1	0	2	253	206
1	1	1	3	310	292
2	0	0	4	358	342
2	1	0	5	400	357
2	1	1	6	438	413
2	2	0	8	506	450
2	2	1	9	537	461

We can read the sequence of the measured ring radii (in pixels) in the background corrected Fig. 5.12b: 179, 206, 292, 342, 357, 413, 450, 461. Gold has a cubic face-centred (fcc) lattice (space group 225 with the basic positions 0,0,0). To determine the theoretical sequence of the ring radii firstly we consider the combination of lattice constant and apparatus constant as proportionality factor and set it to 1/(179 pixels), that is the reciprocal of the first measured radius. To get the theoretical radii we have to consider the (hkl) indices of the lattice planes. The needed equation follows from Table 5.4 for the cubic system and after variation of the indices h, k, and l we get the values listed in Table 5.5.

We see that the calculated and the measured sequences are completely different. Where is the mistake? We forgot the inner structure of the unit cell, the face-centring. Before we continue the determination of the apparatus constant we have to think about the influence of the inner packing of the unit cell on the diffraction pattern. As seen in Fig. 5.11 the intensities of the rings are very different. What is the reason of these differences? We are going to deal with this question in the following paragraphs, too.

Fig. 5.13 Diffraction at a cubic. **a** primitive. **b** face-centred unit cell

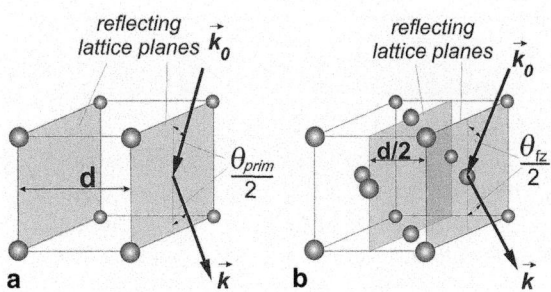

5.4.2 Rules for Forbidden Reflections

In Fig. 5.13 two cubic unit cells are illustrated. Within the primitive cell (part 5.13a) only the corners are occupied by atoms, i.e. there is only one atom in the unit cell. The remaining seven atoms drawn in Fig. 5.13a belong to other adjacent cells.

Following Bragg's law (5.6) the diffraction angle θ_{prim} of the 1st diffraction maximum ($n=1$, smallest diffraction angle) is given by

$$\theta_{prim} = \frac{\lambda}{d} \tag{5.14}$$

with the lattice spacing d of the primitive unit cell and the electron wavelength λ.

In the face-centred unit cell (Fig. 5.13b) additional atoms are inserted in the intersection points of the face diagonals. The distance of the reflecting lattice planes reduces to half of the distance in the primitive cell. The smallest diffraction angle is now given by

$$\theta_{fc} = \frac{\lambda}{d/2} = \frac{2 \cdot \lambda}{d}, \tag{5.15}$$

i.e. the first diffraction reflection appears, in comparison with the primitive cell, in double distance. The first reflection of the primitive cell is *forbidden*.

How can we conveniently calculate rules for such forbidden reflections? For this purpose we have to think about the fundamentals of the diffraction reflection intensities. Until now, we have calculated only the diffraction angles. We have to define the model of the reflecting lattice planes more precisely. In reality we have not only two but a variety of waves scattered at the lattice atoms and overlaid (Sect. 10.9).

The intensity of the diffraction reflections is determined by the scattering capability ("cross section for scattering") of the single atoms within the planes and the atomic occupation density of the planes responsible for a diffraction reflection. These influences are summarised in the *structure factor*. It depends on the special lattice plane and therefore, the structure factor has the index hkl indicating the lattice plane which is responsible for the reflection (hkl).

In Sect. 10.9 it is shown that the structure factor can be calculated by the equation

$$F_{hkl} = \sum_j f_j \cdot e^{-2\pi i (x_j \cdot h + y_j \cdot k + z_j \cdot l)} \qquad (5.16)$$

(f_j: atomic scattering factor of atom j, i: imaginary unit, x_j, y_j, z_j, coordinates of atom j within the unit cell). We come back to the problem of Sect. 5.4.1 and think about the structure factors of the cubic face-centred gold cell. There are four gold atoms with the atomic scattering factor f_{Au} on the places

j	x_j	y_j	z_j	j	x_j	y_j	z_j
1	0	0	0	3	0.5	0	0.5
2	0	0.5	0.5	4	0.5	0.5	0

Using these atomic positions in Eq. (5.16) we get:

$$F_{hkl} = f_{Au} \cdot \left(e^0 + e^{-2\pi i (\frac{k}{2} + \frac{l}{2})} + e^{-2\pi i (\frac{h}{2} + \frac{l}{2})} + e^{-2\pi i (\frac{h}{2} + \frac{k}{2})} \right)$$

$$F_{hkl} = f_{Au} \cdot \left(1 + e^{-\pi i (k+l)} + e^{-\pi i (h+l)} + e^{-\pi i (h+k)} \right). \qquad (5.17)$$

We consider Euler's[11] equation

$$e^{-i\varphi} = \cos\varphi - i \cdot \sin\varphi \qquad (5.18)$$

and

$$\cos(n \cdot 2\pi) = 1, \quad \sin(n \cdot 2\pi) = 0 \quad \text{and}$$

$$\cos\left(\left(n + \frac{1}{2}\right) \cdot 2\pi\right) = -1, \sin\left(\left(n + \frac{1}{2}\right) \cdot 2\pi\right) = 0 \qquad (5.19)$$

(n: integer number).

The sums (k+l), (h+l), and (h+k) must result in even integers to get a structure factor unequal to zero. This demand is fulfilled by the rule "either all h, k, l must be even or all h, k, l must be odd". This is the rule for the selection of reflections of the face-centred (on all sides) lattice including only one atomic type (Table. 5.6, 5.7).

Using this knowledge we are able to correct the Table 5.5:

We find that there is an excellent agreement between the measured and the calculated sequence. Using the known lattice constant of gold (d=0.4078 nm=4.078 Å) we can determine the apparatus constant $\lambda \cdot L$. A graphic solution is well suitable in this case. The pixel distances are plotted in dependence of the reciprocal lattice spacings (Fig. 5.14).

---

[11] Leonhard Euler, Swiss mathematician, 1707–1783.

5.4 What Can We Learn from Selected Area Diffraction Patterns?

Table 5.6 Structure factors of the face-centred unit cell (h, k, l=0…2)

h	k	l	$e^{-\pi\cdot i(k+l)}$	$e^{-\pi\cdot i(h+l)}$	$e^{-\pi\cdot i(h+k)}$	F_{hkl}/f_{Au}
1	0	0	1	−1	−1	1+1−1−1=0
1	1	0	−1	−1	1	1−1−1+1=0
1	1	1	1	1	1	1+1+1+1=4
2	0	0	1	1	1	1+1+1+1=4
2	1	0	−1	1	−1	1−1+1−1=0
2	1	1	1	−1	−1	1+1−1−1=0
2	2	0	1	1	1	1+1+1+1=4
2	2	1	−1	−1	1	1−1−1+1=0
2	2	2	1	1	1	1+1+1+1=4

Table 5.7 Comparison between the measured sequence of the radii in Fig. 5.11 "fingerprints" and the expectations for a cubic face-centred unit cell

h	k	l	$h^2+k^2+l^2$	$\dfrac{179}{\sqrt{3}}\cdot\sqrt{h^2+k^2+l^2}$	Measured/Pixel
1	1	1	3	179	179
2	0	0	4	207	206
2	2	0	8	292	292
3	1	1	11	343	342
2	2	2	12	358	357
4	0	0	16	413	413
3	3	1	19	451	450
4	2	0	20	462	461

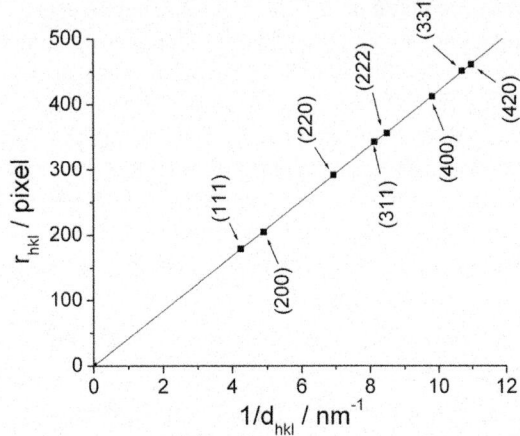

Fig. 5.14 Measured ring radii (*pixel*) against the reciprocal lattice spacings. The lattice planes, responsible for the measured points, are characterised by Miller's indices (hkl). The slope of the line of best fit is $m = 42.09$ pixel nm

Fig. 5.15 Electron diffraction diagram of a fine-crystalline gold film with inserted scale bar. The measured value is 5, the unit is 1/nm=nm⁻¹, i.e. the value is 5 nm⁻¹

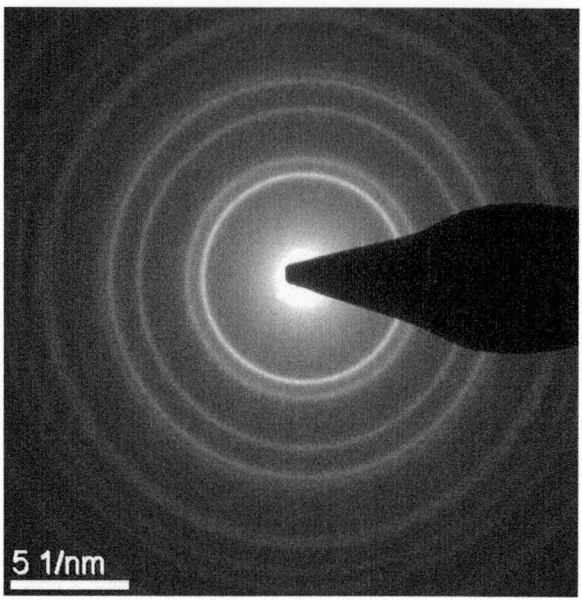

The slope of the graph is given by

$$m = \frac{r_{hkl}}{1/d_{hkl}} = d_{hkl} \cdot r_{hkl} = \lambda \cdot L \tag{5.20}$$

(cf. Eq. (5.12)) with the unusual dimensional unit pixel nm. To get a "true" unit of length the pixel size has to be considered.

In this case the pixel size is 24 µm, i.e. $\lambda \cdot L = 1010$ µm nm and, converted to a more common unit, $\lambda \cdot L = 10.10$ Å·mm, respectively. On the computer monitor the pixels are larger, e.g. 0.26 mm. Then, the apparatus constant would be 109.4 Å·mm. We see the dilemma: The apparatus constant depends on the pixel size. Similar to magnified pictures a scale bar on the diffraction pattern would be helpful. Its length also depends on the pixel size and is always correct. Because of the fact that the reflection distances are proportional to the reciprocal lattice spacings the unit of the scale bar is a reciprocal length, too. Of course, it is given by a real length in the picture (number of pixels). We preset a reciprocal distance $1/d$ and calculate the related length r of the scale in pixels using the knowledge of $\lambda \cdot L$:

$$r = \frac{1}{d} \cdot m. \tag{5.21}$$

In our case a scale bar representing a reciprocal distance of 5 nm⁻¹ has a length of 210.4 pixels. Now, the reciprocal distances of the reflections can be directly measured in the diffraction picture (Fig. 5.15).

The rules for forbidden reflections for the various types of packings listed in Table 5.3 are summarised in Table 5.8.

5.4 What Can We Learn from Selected Area Diffraction Patterns?

Table 5.8 Rules for forbidden reflections

No.	Packing	Number of atoms	Atomic positions	Rules for forbidden reflections
1	Primitive (space groups P...)	1	0;0;0	No forbidden reflections
2	Body-centred (space groups I...)	2	0;0;0 $1/2;1/2;1/2$	Sum $h+k+l$ odd
3	Face-centred (all faces space groups F...)	4	0;0;0 $1/2;1/2;0$ $1/2;0;1/2$ $0;1/2;1/2$	h, k, l is a mixture of even and odd numbers
4	Basis face-centred (basis face C = footpoint space groups C...)	2	0;0;0 $1/2;1/2;0$	Sum $h+k$ odd
5	Diamond lattice (space groups F...)	8	0;0;0 $1/2;1/2;0$ $1/2;0;1/2$ $0;1/2;1/2$ $1/4;1/4;1/4$ $3/4;3/4;1/4$ $1/4;3/4;3/4$ $3/4;1/4;3/4$	As No.3, additional cancellation if: $(h+k+l)/2, (3h+3k+l)/2,$ $(3h+k+3l)/2, (h+3k+3l)/2$ all odd
6	Hexagonal close-packing of spheres (space groups R...)	2	$2/3;1/3;1/4$ $1/3;2/3;3/4$	$h=k, l$ odd

5.4.3 Intensities of the Diffraction Reflections

In the kinematical model of electron diffraction (Sect. 10.9) the intensity of the diffraction reflections is influenced by two factors: the structure factor F_{hkl} known from the previous paragraph and the lattice factor G. The consequences of the atomic arrangement within the unit cell are summarised in the structure factor, the lattice factor includes the effects of the unit cell arrangement within the crystal. The intensity I of the electron waves is proportional to the product of the squares of the absolute values of structure and lattice factor (cf. Eqs. (10.167), (10.169), and (10.190) Sect. 10.9):

$$I(\mathbf{k}) \propto F_{hkl}^2 \cdot G^2 \quad \text{with}$$

$$F_{hkl}^2 = \left(\sum_j f_j \cdot e^{-2\pi \cdot i \cdot (h \cdot x_j + k \cdot y_j + l \cdot z_j)} \right)^2 \quad \text{and} \tag{5.22}$$

$$G^2 = \frac{1}{V_{EZ}^2} \cdot \frac{\sin^2(\pi \cdot u_1 \cdot M_1)}{(\pi \cdot M_1 \cdot u_1)^2} \cdot \frac{\sin^2(\pi \cdot u_2 \cdot M_2)}{(\pi \cdot M_2 \cdot u_2)^2} \cdot \frac{\sin^2(\pi \cdot u_3 \cdot M_3)}{(\pi \cdot M_3 \cdot u_3)^2}$$

with h, k, l as Miller's indices of the reflecting lattice planes, x_j, y_j, z_j as coordinates of the atom j within the unit cell and V_{UC} as volume of the unit cell. M_1, M_2, M_3 as number of the unit cells in $\mathbf{a}_1$-, $\mathbf{a}_2$-, and $\mathbf{a}_3$-direction together with V_{UC} characterise the crystal size. u_1, u_2, and u_3 are the components of the excitation error (i.e. deviation from the ideal diffraction condition)

$$\mathbf{u} = u_1 \cdot \mathbf{b}_1 + u_2 \cdot \mathbf{b}_2 + u_3 \cdot \mathbf{b}_3. \tag{5.23}$$

They are given by

$$u_1 = \frac{1}{M_1}, \quad u_2 = \frac{1}{M_2}, \quad u_3 = \frac{1}{M_3}. \tag{5.24}$$

The atomic scattering behaviour of atom j is considered in the atomic scattering factor f_j (Fig. 5.16).

We see that the atomic scattering factor depends not only on the element but also on the modulus of the scattering vector, i.e. the intensity decreases with increasing diffraction angle.

The lattice factor determines the shape of the diffraction reflections. If the crystal is very thin in one crystallographic direction (i.e. shaped plate-like) streak-like shaped reflections have to be expected. In ring diagrams the crystallite size determines the sharpness of the rings: the smaller the crystallites the broader the rings.

Finally, there are structures with many crystallites contributing to the diffraction pattern but their orientations are not uniformly distributed. They have a preferred orientation (*texture*). Dependent on the relation between preferred orientation and

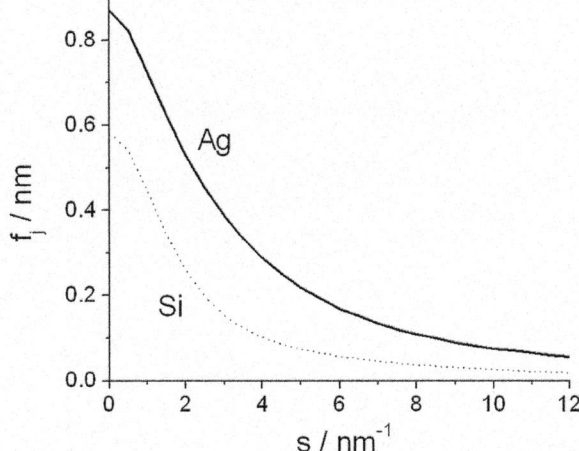

Fig. 5.16 Atomic scattering factor of silicon and silver in dependence on the modulus of the scattering vector $s = |\mathbf{k}-\mathbf{k}_0|$ following [14]

electron beam direction such a texture is evident as a periodic brightness fluctuation on single rings or permitted rings are missing in the diffraction pattern due to the absence of the according orientations (cf. Sect. 6.3).

5.4.4 Positions of Diffraction Reflections in Point Diagrams

After introducing the model "reciprocal lattice" the rule for the calculation of the diffraction reflection positions of a single crystal is given by the *Ewald construction* (Sect. 10.8). Let us try to explain these positions without this model.

The distance between reflection and centre can be calculated from the lattice spacings by Bragg's law (e.g. Eq. (5.12)). The respective equations are listed in Table 5.4. As resulting already from Fig. 5.1 the diffraction reflections are always perpendicular to the responsible lattice planes. Therefore, the angle ξ between two reflections is equal to that one between the related lattice planes (h_1, k_1, l_1) and (h_2, k_2, l_2).

The calculation is simple for the cubic system because the lattice plane normal [hkl] has the equal indexing as the lattice plane (hkl) itself. For the other crystal systems we use a transformation into an orthogonal system as already applied for finding the equations in Table 5.4 and as described in Sect. 10.8 (results see Table 5.9).

We need a reference reflection with the indices (h_1, k_1, l_1) and are able to calculate the angle ξ to each other reflection (h_2, k_2, l_2) with the knowledge of the unit cell (axis lengths and axes angles) and the equations of Table 5.9. Within the polar coordinate system the position of a diffraction reflection is completely determined by distance (radius) and angle.

Table 5.9 Equations for the calculation of the angle ξ between two point reflections (h_1,k_1,l_1) and (h_2,k_2,l_2) in a diffraction pattern

Cubic:	$a_1=a_2=a_3, \alpha=\beta=\gamma=90°$ $\varepsilon=90°$	$\cos\xi = \dfrac{h_1 \cdot h_2 + k_1 \cdot k_2 + l_1 \cdot l_2}{\sqrt{(h_1^2+k_1^2+l_1^2)\cdot(h_2^2+k_2^2+l_2^2)}}$
Tetragonal:	$a_1=a_2 \neq a_3$ $\alpha=\beta=\gamma=90°$ $\varepsilon=90°$	$\cos\xi = \dfrac{h_1 \cdot h_2 + k_1 \cdot k_2 + \left(\dfrac{a_1}{a_3}\right)^2 l_1 \cdot l_2}{\sqrt{\left(h_1^2+k_1^2+\left(\dfrac{a_1}{a_3}\right)^2 l_1^2\right)\cdot\left(h_2^2+k_2^2+\left(\dfrac{a_1}{a_3}\right)^2 l_2^2\right)}}$
Orthorhombic	$a_1 \neq a_2 \neq a_3$ $\alpha=\beta=\gamma=90°$ $\varepsilon=90°$	$\cos\xi = \dfrac{h_1 \cdot h_2 + \left(\dfrac{a_1}{a_2}\right)^2 \cdot k_1 \cdot k_2 + \left(\dfrac{a_1}{a_3}\right)^2 \cdot l_1 \cdot l_2}{\sqrt{\left(h_1^2+\left(\dfrac{a_1}{a_2}\right)^2 \cdot k_1^2+\left(\dfrac{a_1}{a_3}\right)^2 l_1^2\right)\cdot\left(h_2^2+\left(\dfrac{a_1}{a_2}\right)^2 \cdot k_2^2+\left(\dfrac{a_1}{a_3}\right)^2 l_2^2\right)}}$
Hexagonal:	$a_1=a_2 \neq a_3$ $\alpha=\beta=90°$ $\gamma=120°$ $\varepsilon=90°$	$\cos\xi = \dfrac{h_1 \cdot h_2 + \dfrac{1}{2}(h_1 \cdot k_2 + h_2 \cdot k_1) + k_1 \cdot k_2 + 3 \cdot \left(\dfrac{a_1}{a_3}\right)^2 \cdot l_1 \cdot l_2}{\sqrt{\left(h_1^2+h_1 \cdot k_1+k_1^2+3 \cdot \left(\dfrac{a_1}{a_3}\right)^2 l_1^2\right)\cdot\left(h_2^2+h_2 \cdot k_2+k_2^2+3\cdot\left(\dfrac{a_1}{a_3}\right)^2 l_2^2\right)}}$
Rhombo- hedral:	$a_1=a_2=a_3$ $\alpha=\beta=\gamma \neq 90°$	$\cos\xi = \dfrac{h_1 \cdot h_2 + F_1(h_1,h_2,k_1,k_2) + F_2(h_1,h_2,k_1,k_2,l_1,l_2)}{\sqrt{(h_1^2+F_3(h_1,k_1)+F_4(h_1,k_1,l_1))\cdot(h_2^2+F_3(h_2,k_2)+F_4(h_2,k_2,l_2))}}$

with

$F_1(h_1,h_2,k_1,k_2) = h_1 \cdot h_2 \cdot \cot^2\alpha - (h_1 \cdot k_2 + h_2 \cdot k_1)\dfrac{\cot\alpha}{\sin\alpha} + \dfrac{k_1 \cdot k_2}{\sin^2\alpha}$

Table 5.9 (continued)

$$F_2(h_1, h_2, k_1, k_2, l_1, l_2) =$$

$$= \frac{1}{\sin^2 \varepsilon} \left[h_1 \cdot h_2 \cdot \left(\frac{\cos \alpha}{2 \cdot \cos \frac{\alpha}{2}} \right)^2 + k_1 \cdot k_2 \cdot \left(\cot \alpha \cdot \tan \frac{\alpha}{2} \right)^2 + l_1 \cdot l_2 + \right.$$

$$+ (h_1 \cdot k_2 + h_2 \cdot k_1) \cdot \frac{\cos \alpha \cdot \cot \alpha \cdot \tan \frac{\alpha}{2}}{2 \cdot \cos \frac{\alpha}{2}} -$$

$$\left. - (h_1 \cdot l_2 + h_2 \cdot l_1) \cdot \frac{\cos \alpha}{2 \cdot \cos \frac{\alpha}{2}} - (k_1 \cdot l_2 + k_2 \cdot l_1) \cdot \cot \alpha \cdot \tan \frac{\alpha}{2} \right]$$

$$F_3(h, k) = h^2 \cdot \cot^2 \alpha - 2 \cdot h \cdot k \cdot \frac{\cot \alpha}{\sin \alpha} + \frac{k^2}{\sin^2 \alpha}$$

$$F_4(h, k, l) = \frac{1}{\sin^2 \varepsilon} \left[h^2 \cdot \left(\frac{\cos \alpha}{2 \cdot \cos \frac{\alpha}{2}} \right)^2 + k^2 \cdot \left(\cot \alpha \cdot \tan \frac{\alpha}{2} \right)^2 + l^2 + \right.$$

$$+ h \cdot k \cdot \frac{\cos \alpha \cdot \cot \alpha \cdot \tan \frac{\alpha}{2}}{\cos \frac{\alpha}{2}} -$$

$$\left. - h \cdot l \cdot \frac{\cos \alpha}{\cos \frac{\alpha}{2}} - 2 \cdot k \cdot l \cdot \cot \alpha \cdot \tan \frac{\alpha}{2} \right]$$

as well as

$$\cos \varepsilon = \cot \alpha \cdot \sqrt{2 \cdot (1 - \cos \alpha)}$$

Table 5.9 (continued)

Monoclinic:	$a_1 \neq a_2 \neq a_3$ $\alpha = \gamma = 9°$ $\beta \neq 90°$ $\varepsilon = \beta$	$\cos \xi = \dfrac{\dfrac{h_1 \cdot h_2}{a_1^2 \cdot \sin^2 \beta} + \dfrac{k_1 \cdot k_2}{a_2^2} - \dfrac{(h_1 \cdot l_2 + h_2 \cdot l_1) \cos \beta}{a_1 \cdot a_3 \cdot \sin^2 \beta} + \dfrac{l_1 \cdot l_2}{a_3^2 \cdot \sin^2 \beta}}{\sqrt{F_5(h_1, k_1, l_1)} \cdot F_5(h_2, k_2, l_2)}$ with $F_5(h, k, l) = \dfrac{h^2}{a_1^2 \cdot \sin^2 \beta} + \dfrac{k^2}{a_2^2} - \dfrac{2 h \cdot l \cdot \cos \beta}{a_1 \cdot a_3 \cdot \sin^2 \beta} + \dfrac{l^2}{a_3^2 \cdot \sin^2 \beta}$
Triclinic:	$a_1 \neq a_2 \neq a_3$ $\alpha \neq \beta \neq \gamma$	$\cos \sphericalangle (\mathbf{B_1}, \mathbf{B_2}) = \dfrac{\mathbf{B_1} \cdot \mathbf{B_2}}{|\mathbf{B_1}| \cdot |\mathbf{B_2}|}$ $\mathbf{B_1} \cdot \mathbf{B_2} = h_1 h_2 b_{1x}^2 + (h_1 b_{1y} + k_1 b_{2y}) \cdot (h_2 b_{1y} + k_2 b_{2y}) + (h_1 b_{1z} + k_1 b_{2z} + l_1 b_{3z})(h_2 b_{1z} + k_2 b_{2z} + l_2 b_{3z})$ $|\mathbf{B_1}| \cdot |\mathbf{B_2}| = \sqrt{(h_1 b_{1x})^2 + (h_1 b_{1y} + k_1 b_{2y})^2 + (h_1 b_{1z} + k_1 b_{2z} + l_1 b_{3z})^2}$ $\cdot \sqrt{(h_2 b_{1x})^2 + (h_2 b_{1y} + k_2 b_{2y})^2 + (h_2 b_{1z} + k_2 b_{2z} + l_2 b_{3z})^2}$ $b_{1x} = \dfrac{1}{a_1}, b_{1y} = \dfrac{-\cot \gamma}{a_1}, b_{1z} = \dfrac{\cot \gamma \sqrt{\cos^2 \varepsilon - \cos^2 \beta} - \cos \beta}{a_1 \cdot \sin \varepsilon}$ $b_{2y} = \dfrac{1}{a_2 \cdot \sin \gamma}, b_{2z} = \dfrac{-\sqrt{\cos^2 \varepsilon - \cos^2 \beta}}{a_2 \cdot \sin \gamma \cdot \sin \varepsilon}$ $b_{3z} = \dfrac{1}{a_3 \cdot \sin \varepsilon}$ with $\cos \varepsilon = \dfrac{\sqrt{\cos^2 \alpha - 2 \cos \alpha \cos \beta \cos \gamma + \cos^2 \beta}}{\sin \gamma}$

5.4 What Can We Learn from Selected Area Diffraction Patterns?

Fig. 5.17 Indexing of a dot-shaped diffraction diagram: The indices of diffraction reflections are these ones of the reflection lattice planes. The intersection line of all reflecting lattice planes is named "zone axis"

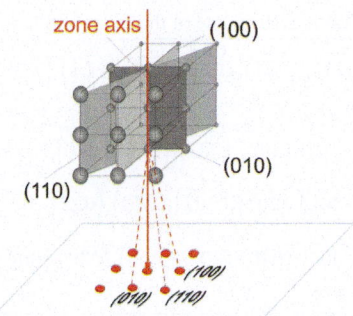

Fig. 5.18 Point diagram of silicon (measured diffraction pattern). Distances between the reflections: $s_1 = s_4 = 5.13$ nm^{-1}, $s_2 = s_5 = 5.11$ nm^{-1}, $s_3 = s_6 = 7{,}21$ nm^{-1}, angle between s_1 and s_2: 89.9°, angle between s_2 and s_3: 45.0°

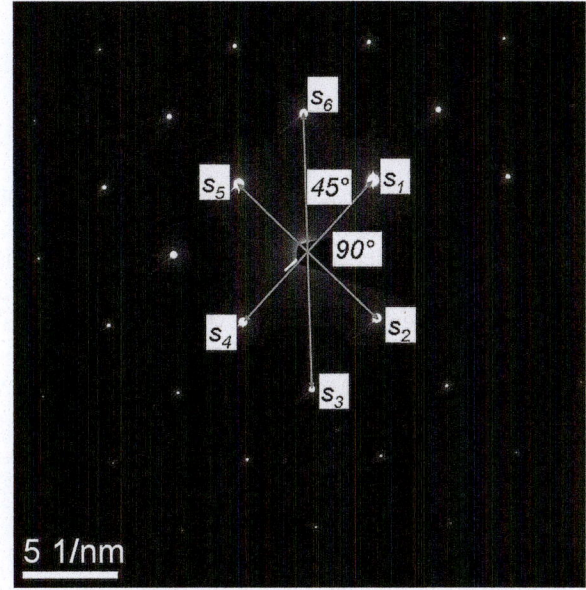

5.4.5 Indexing of Diffraction Reflections

With the knowledge presented in the previous paragraphs of Chap. 5 it seems to be simple to index the diffraction reflections: The diffraction reflections have the same indices as the responsible lattice planes (Fig. 5.17).

We want to demonstrate at an example how the indexing can be practically done. We use a diffraction pattern of silicon for this purpose (Fig. 5.18).

At first, it has to be considered which diffraction reflections of silicon agree with the distances s_1, s_2, and s_3. Silicon has a cubic crystal system with diamond structure, i.e. a reflection appears only if all indices h, k, l are either even or odd. From this quantity of indices the following are additionally excluded: (h+k+l)/2, (3h+3k+l)/2, (3h+k+3 l)/2, and (h+3k+3 l)/2 are odd integers (cf. Table 5.8).

104   5 Let us Switch to Electron Diffraction

Table 5.10 Some of the reciprocal lattice spacings of silicon

$(111),(11\bar{1}),(1\bar{1}1),(\bar{1}11),(1\bar{1}\bar{1}),(\bar{1}1\bar{1}),(\bar{1}\bar{1}1),(\bar{1}\bar{1}\bar{1})$	$\dfrac{1}{d_{hkl}} = 3.19\,\text{nm}^{-1}$
$(220),(202),(022),(2\bar{2}0),(\bar{2}20),(\bar{2}\bar{2}0),$ $(20\bar{2}),(\bar{2}02),(\bar{2}0\bar{2}),(0\bar{2}2),(02\bar{2}),(0\bar{2}\bar{2})$	$\dfrac{1}{d_{hkl}} = 5.21\,\text{nm}^{-1}$
$(400),(040),(004),(\bar{4}00),(0\bar{4}0),(00\bar{4})$	$\dfrac{1}{d_{hkl}} = 7.37\,\text{nm}^{-1}$

The axis length of the unit cell (lattice constant) of silicon is $a_1 = 0.543$ nm. From Table 5.4 we take the equation

$$\frac{1}{d_{hkl}} = \frac{\sqrt{h^2 + k^2 + l^2}}{a_1} \tag{5.25}$$

for the cubic crystal system. Considering this, the reciprocal lattice spacings listed in Table 5.10 are possible among others.

The deviations between the measured reciprocal distances and the calculated ones for (200) as well (400) are less than 2%. This is an acceptable tolerance for selected area electron diffraction. The next step is to calculate the angles between different reflections and to select such pairs of indices resulting in 90° and 45°, respectively. We take the needed equation from Table 5.9:

$$\cos\xi = \frac{h_1 \cdot h_2 + k_1 \cdot k_2 + l_1 \cdot l_2}{\sqrt{(h_1^2 + k_1^2 + l_1^2)\cdot(h_2^2 + k_2^2 + l_2^2)}}. \tag{5.26}$$

For an angle $\xi = 90°$ is $\cos\angle(\mathbf{s}_1,\mathbf{s}_2)=0$, i.e. $h_1\cdot h_2 + k_1\cdot k_2 + l_1\cdot l_2 = 0$. Since s_1 and s_2 both amount to about 5.21 nm^{-1} we use the indices from the second row of Table 5.10. One of the indices is zero for all the listed reflections. To fulfil the demand mentioned above

$$h_1\cdot h_2 = -k_1\cdot k_2 \quad \text{or} \quad h_1\cdot h_2 = -l_1\cdot l_2 \quad \text{or} \quad k_1\cdot k_2 = -l_1\cdot l_2$$

has to pertain. That is fulfilled for the reflection pairs

$$(220) \text{ or } (\bar{2}\bar{2}0) \text{ and } (2\bar{2}0) \text{ or } (\bar{2}20),$$
$$(202) \text{ or } (\bar{2}0\bar{2}) \text{ and } (20\bar{2}) \text{ or } (\bar{2}02),$$
$$(022) \text{ or } (0\bar{2}\bar{2}) \text{ and } (02\bar{2}) \text{ or } (0\bar{2}2).$$

5.4 What Can We Learn from Selected Area Diffraction Patterns?

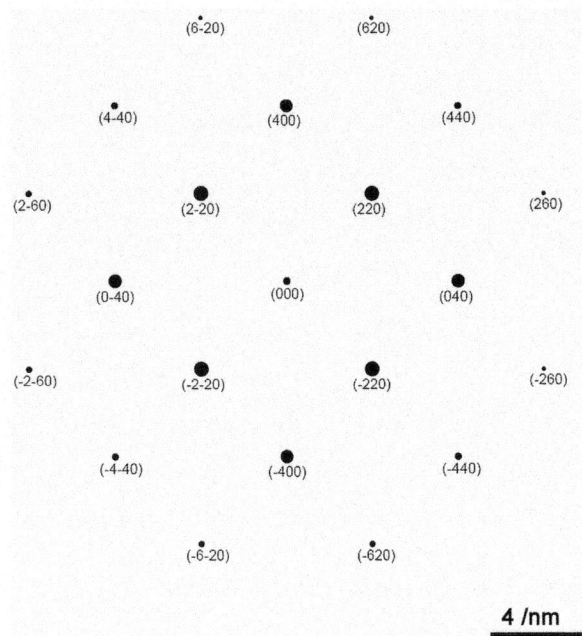

Fig. 5.19 Calculated and completely indexed diffraction pattern of silicon, zone axis [001]

We use (randomly) the first pair and index s_1 with (220) and s_2 with ($\bar{2}$20). The angle to s_3 should be 45°, i.e.

$$\cos \sphericalangle (s_2, s_3) = \frac{2 \cdot (-h_2 + k_2)}{\sqrt{(4+4)(h_2^2 + k_2^2)}} = \frac{-h_2 + k_2}{\sqrt{2} \cdot \sqrt{(h_2^2 + k_2^2)}} = \frac{1}{\sqrt{2}} \quad (5.27)$$

Obviously, this requirement is satisfied for $h_2 = -4$ und $k_2 = 0$. Therewith the three selected reflections are indexed, the ones on the opposite side are mirror reflections and just have opposite signs. Performing all the calculations and seeking for possible indices is a typical job for today's computers. There are some specialised programs available for this task (e.g. Java-EMS [15] or ELDISCA [16]). Commonly, the result is a completely indexed diffraction diagram (Fig. 5.19).

The appearance of a point pattern depends on the crystal orientation (zone axis). Finally, we determine the zone axis of the pattern shown in the Figs. 5.18 and 5.19. We know the indices of at least two reflections: (220) and ($\bar{2}$20). The components of the zone axis in the crystal system are (Sect. 10.8):

$$k_1 \cdot l_2 - k_2 \cdot l_1, h_2 \cdot l_1 - h_1 \cdot l_2 \text{ and } h_1 \cdot k_2 - h_2 \cdot k_1, \quad (5.28)$$

i.e. in our case [008] and, after cancelling of the common multiples of all single indices, [001].

For practitioners useful overviews about indexed diffraction patterns with often used zone axes of the closest-packed crystal systems are represented e.g. in [16].

Fig. 5.20 Sketch for explanation of the Kikuchi lines

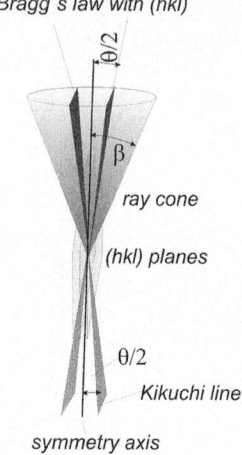

5.5 Kikuchi- and HOLZ-lines

Let us remember: To get a sharp diffraction pattern in selected area electron diffraction a parallel illumination is necessary, i.e. only one electron incident direction is offered to the crystal lattice. We now come back to the convergent beam electron diffraction with "large" angle of convergence, i.e. we offer the lattice a variety of incident directions. We assume that the symmetry axis of the ray cone is positioned parallel to the lattice plane (hkl) within the crystal (Fig. 5.20).

The lattice planes (hkl) "search" within the manifold directions for such ones which satisfy Bragg's law. These planes are positioned parallel to the (hkl) lattice planes in the direction of the normal of the drawing plane in Fig. 5.20. They are drawn in black in this figure. After reflection at the lattice plane (hkl) these planes are inclines at an angle $\theta/2$ (θ: Bragg angle) to the symmetry axis of the ray cone. The reciprocal lattice vectors are perpendicular to the lattice planes having equal indices. The length of these vectors characterises the Bragg angle, i.e. the half length the half Bragg angle. This is a simple possibility to construct a Kikuchi pattern in context with a point diagram. As an example we use the Si point diagram known from Fig. 5.19 with zone axis [001] and with limit to the inner reflections with high intensity.

In Fig. 5.21a the method is demonstrated with the (040) reflection: The (040) Kikuchi line is the perpendicular bisector of the connecting line between the (000) and the (040) reflection. Accordingly, other Kikuchi lines are added in Fig. 5.21b. The intensities of the Kikuchi lines correspond to these of the point reflections used for their construction.

Kikuchi lines in this form can only arise if the incident directions demanded by the Bragg angle are present within the ray cone, i.e. the half Bragg angle must be less than the opening angle of the ray cone (illumination aperture or angle of con-

5.5 Kikuchi- and HOLZ-lines

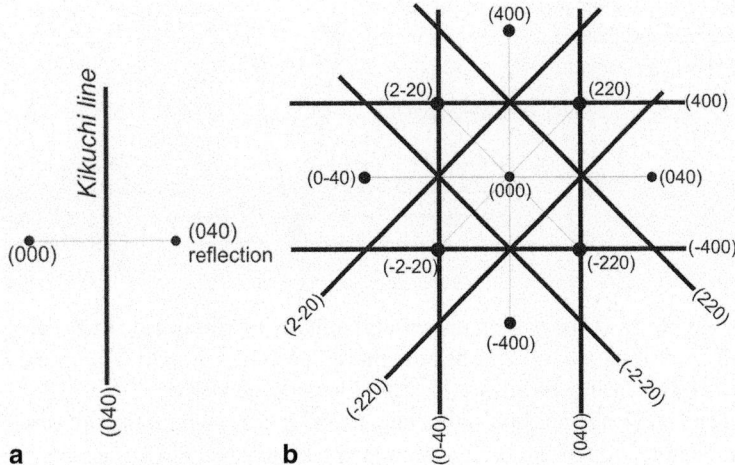

Fig. 5.21 Construction of Kikuchi line pattern for silicon, zone axis [001]. **a** Construction of the (040)-line from the (040) reflection. **b** Complete Kikuchi pattern of the inner reflections

Fig. 5.22 Change of the Kikuchi line pattern after sligthly tilting the crystal

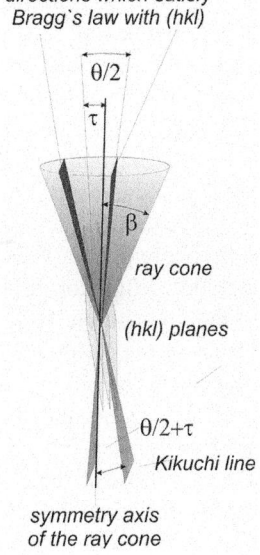

vergence β): $\theta/2 < \beta$ must be valid. One can never get Kikuchi patterns using a very small condenser 2 aperture!

What happens if the lattice planes (hkl) are inclined against the symmetry axis of the ray cone (Fig. 5.22)? In practice this means that the crystal is slightly tilted by an angle τ of a few mrad against the predefined zone axis (e.g. [001]).

We realise that the symmetry point of the Kikuchi pattern (pole) shifts away from the centre. The direction of this movement depends on the tilt direction. At our

Fig. 5.23 Convergent illumination and reciprocal lattice (*section view*)

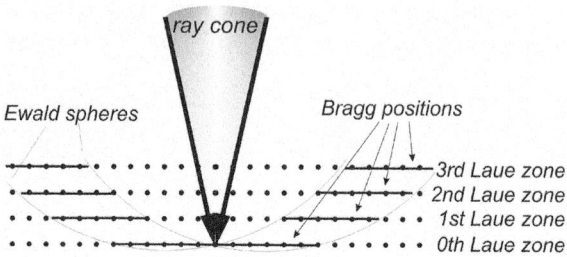

example in Fig. 5.21 the pattern moves horizontally by tilting around the axis $\mathbf{a}_1$ and vertically by tilting around $\mathbf{a}_2$. The amount of the shift is proportional to the angle τ (Fig. 5.22). Increasing τ leads to two possible consequences:

1st The half Bragg angle is only "one-sided" located within the ray cone. In the straight continuation of the Bragg-orientated lattice plane the diffracted intensity is missing and a dark line arises parallel to the bright line in the angle distance θ. This circumstance was described in Sect. 5.3 as a general case of the convergent beam electron diffraction with large angle of convergence.

2nd Another ("low-indexed") set of lattice planes reaches the vicinity of the Bragg position. A new pole arises that is not located in the centre of the screen. It is possible to shift this pole into the centre by tilting the crystal around the "right" crystal axis. On the other hand, it means that it is easily possible to orientate a crystal into the direction of a low-indexed zone axis very accurately by doing this (Fig. 7.6b).

Let us illustrate this by means of the model of the reciprocal lattice (Sect. 10.8 and Fig. 5.23, respectively).

In the section the marginal incident directions of the ray cone with their Ewald spheres are drawn. We realise that between the two spheres a region exists which includes diffraction reflections ("Bragg positions" of lattice planes) contributing to the line pattern. Especially, we also realise that this region extends to multiple planes of the reciprocal lattice. These planes are called "Laue zones". The widely used abbreviations are "ZOLZ" for the zero order Laue zone, "FOLZ" for the first order Laue zone, and in general, "HOLZ" for high order Laue zone. In Fig. 5.23 we see also that the HOLZ lines appear at comparatively large diffraction angles which can be outside of the observation screen. But we know that the diffraction pattern can be shifted on the screen by a slight tilt of the sample. In practice, the sample is tilted by a few milliradians against the low-indexed zone axis, i.e. a higher indexed axis is selected. As a rule, we cannot observe pairs of lines as in the Kikuchi pattern but only single dark lines (Fig. 5.24).

Because of the comparably large diffraction angles of the HOLZ lines minute changes of the lattice constants result in a measurable influence on the line patterns. In principle it is possible to determine lattice constants with an accuracy of better than 0.01% in areas of some nanometres by use of this method. However, monocrystalline areas with low mechanical strain gradients are needed and the quantitative interpretation of the measurements is not straightforward.

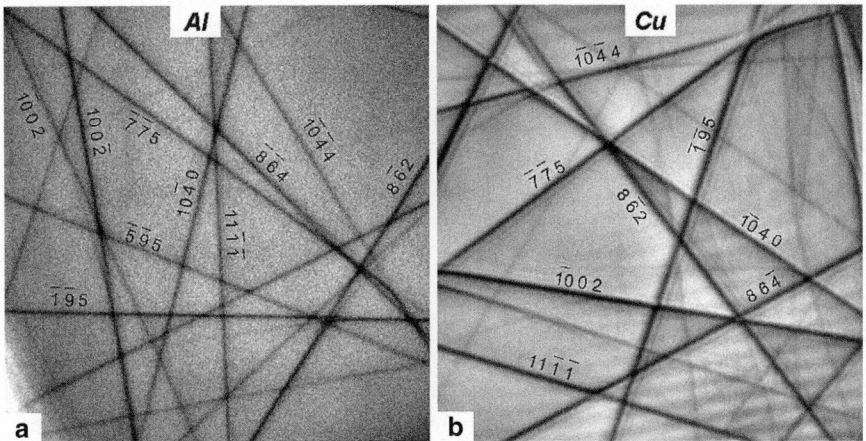

Fig. 5.24 HOLZ line pattern of **a** Aluminium, **b** Copper. Zone axis close to [136]. Especially at copper line bendings are visible (*top right*) which cannot be explained using the kinematical model. (Micrographs recorded by M. Hofmann)

In Fig. 5.24b we see line bendings at intersections of several lines coming from the interaction between the electron waves within the crystal. Such interactions between the waves but also the interactions of the diffracted waves with the crystal lattice are not considered in the kinematical model. Consequently, especially at lattices with heavy elements the kinematical model is not precise enough to reach the accuracy mentioned above. We need a model that considers the interactions between the diffracted waves and the crystal lattice. It is given by the dynamical model that is harder to understand and the calculation is more time consuming. It requires the solution of Schrödinger's[12] equation (this is the base equation of matter waves Sect. 10.2) within the crystal potential [17]. Beside the already mentioned line bending this has additional consequences:

The electron wavelength depends on the crystal potential felt by the wave and, therefore, on the propagation direction within the crystal, too. A positive crystal potential shortens the wavelength. Ewald's construction (Sect. 10.8) is no longer a sphere with constant radius but its radius depends on the crystallographic direction. The exact positions of the diffraction reflections are slightly shifted against the result of the kinematical approximation. Generally, it does not play any role for the interpretation of selected area electron diffraction diagrams, however it does so for the accurate measurement of lattice constants by means of HOLZ line patterns.

The dynamical model includes the multiple scattering of electron waves within the crystal. The intensity oscillates between diffracted and initial waves while the wave is penetrating the crystal. Considering this, the intensity of the diffraction reflections depends on the sample thickness, too.

It is possible to reduce the influence of the dynamical effects and that of the excitation error on the diffraction pattern. The Ewald sphere is not retained but it

---

[12] Erwin Schrödinger, Austrian physicist, 1887–1961, Nobel prize in physics in 1933.

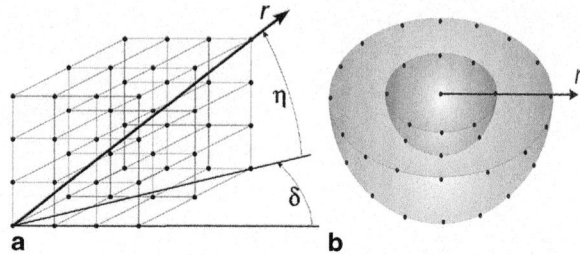

Fig. 5.25 Atomic arrangement in a **a** crystalline and **b** amorphous solid

"nutates". In this way a variety of reciprocal lattice points is hit in the same kind within the temporal mean equivalent to identical excitation conditions for all diffraction reflections. In practice this can be reached by a precession movement of the electron beam in front of the specimen. To get a "stable" diffraction pattern in the observation plane this movement has to be compensated behind the specimen ("Precessing Electron Diffraction"—PED [18]).

The (more accurate) dynamical model enables additional conclusions: The intensity distribution within the diffraction disks (not closer explained here) in Convergent Beam Electron Diffraction (CBED) is strongly influenced by the scattering properties of the crystal lattice. In principle one has the chance to measure the electronic structure of the solids, e.g. the electron density [19].

5.6 Amorphous Samples

We come back to selected area electron diffraction, i.e. to parallel illumination. The observed diffraction diagram depends strongly on the size of the crystallites. For crystallites which completely cover the size of the selected area we observe point diagrams. By continuously decreasing the sizes of the crystallites diagrams arise whose sharp maxima finally switch over to diffuse rings.

We want to examine the extreme case to demonstrate the possibilities of the characterisation of *amorphous substances* in the transmission electron microscope. We have to consider that from the viewing point of characterisation the transition from the fine-crystalline to the amorphous state is smooth and the measured result depends on the method. The term "X-ray amorphous" for materials with crystallite sizes of less than about 5 nm emphasises that.

The result of the measurement, the dependence of the intensity on the scattering angle is called *scattering curve* (regarding the radial brightness distribution of ring diagrams—cf. Sect. 5.4.1). The kind of the atoms and their arrangement within the solid are essential for the appearance of the scattering curve. In our model we want to restrict ourselves to the influence of the atomic arrangement with only one sort of atoms.

Single crystals are characterised by a periodic arrangement of atoms. It follows that only discrete atomic distances exist (we ignore the thermal motion) and these atomic distances depend on the direction (Fig. 5.25a).

5.6 Amorphous Samples

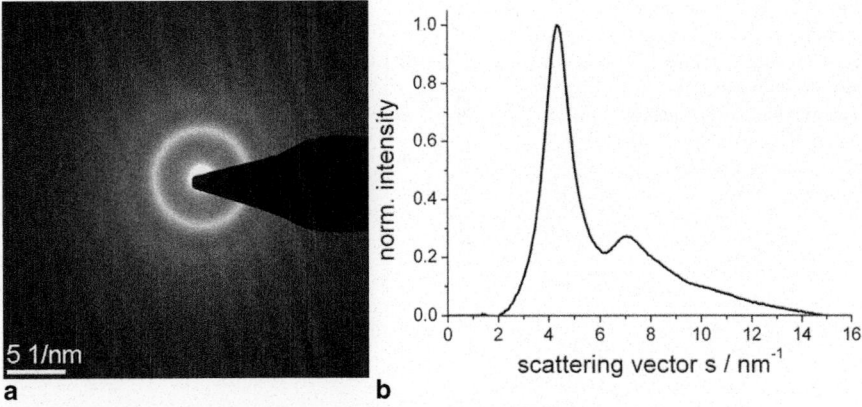

Fig. 5.26 Diffraction diagram of an amorphous film (*thinned bulk metallic glass*) and radial brigthness distribution extracted from it after background subtraction (*scattering curve*)

The distribution of the atoms ρ depends on the distance r and its direction characterised by the angles δ and η:

$$\rho = \rho(r, \delta, \eta), \tag{5.29}$$

i.e. it is anisotropic.

If the periodic arrangement of atoms is eliminated (that can be reached by an increase of the temperature above the melting point, by different force fields in the surrounding of the atoms or by special conditions during rapid solidification of the melt) the discrete distances and the anisotropy vanish in average (Fig. 5.25b). Then the atomic distribution can be written as:

$$\rho = \rho(r) \tag{5.30}$$

Nevertheless, there are still interatomic forces with the result that preferred atomic distances and therewith modulations of the atomic density distribution have to be expected even after the elimination of the crystal lattice. The *long-range order* as typical for a crystal is lost and substituted by a *short-range order*.

Figure 5.26 shows a diffraction pattern of a thin amorphous sample.

For interpreting this measurement we have to understand how its fundamentals. The scattering curve $I(s)$ is the result of the overlay of all electron waves scattered at the atomic arrangement. The intensity is given by

$$I(s) \sim |\psi(s)|^2 \text{ with } \psi(s) = \sum_{m=1}^{N} \sum_{n=1}^{N} f_m \cdot f_n \cdot \frac{\sin(2 \cdot \pi \cdot s \cdot r_{mn})}{2 \cdot \pi \cdot s \cdot r_{mn}} \tag{5.31}$$

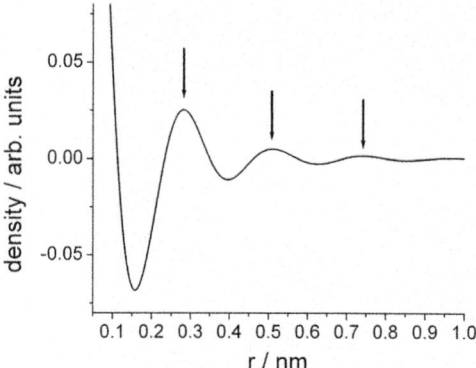

Fig. 5.27 Radial density function of the scattering curve of an amorphous sample shown in Fig. 5.26. The maxima of the density function (marked by *arrows*) are at 0.28 nm, 0.51 nm, and 0.74 nm

f_m, f_n cross sections (atomic scattering factors) of the scattering centres m and n,

$s = |\mathbf{k} - \mathbf{k}_0|$ modulus of the scattering vector, this is the difference of the wave number vectors of the diffracted and the initial wave,

r_{mn} modulus of the difference vector between the centres of scattering m and n

(Sect. 10.10).

Assuming just one kind of atoms the distribution function $\rho(r)$ can be calculated from the background corrected scattering curve $I(s)$:

$$\rho(r) = \frac{2}{r} \cdot \int_0^\infty I(s) \cdot s \cdot \sin(2 \cdot \pi \cdot s \cdot r) \cdot ds. \tag{5.32}$$

$\rho(r)$ is also termed *radial density function*.

Regarding the numerical solution of the integral the upper bound of integration depends on the maximal scattering angle acquired during the measurement. Because of its finiteness at small distances r oscillations of $\rho(r)$ can exist as a consequence of the calculation procedure being not typical for the sample. To avoid this the function $I(s)$ is often multiplied by a damping factor before starting the integration.

The result of such a calculation for the example in Fig. 5.26 is plotted in Fig. 5.27. It shows preferred atomic distances in amorphous matter by the maxima of the radial density function.

References

1. Ewald, P.P.: Die Entdeckung der Röntgeninterferenzen vor zwanzig Jahren und zu Sir William Braggs siebzigstem Geburtstag. Naturwissenschaften **29**, 527–530 (1932)
2. Friedrich, W.: Erinnerungen an die Entdeckung der Interferenzerscheinungen bei Röntgenstrahlen. Naturwissenschaften **36**, 354–356 (1949)

3. Schulze, G.E.R.: M. v. Laue und die Geschichte der Röntgenfeinstrukturuntersuchung. Krist. Techn. **8**, 527–543 (1973)
4. Davisson, C., Germer, L.H.: The scattering of electrons by a single crystal of nickel. Nature **119**, 558–560 (1927)
5. Bragg, W.L.: The specular reflection of X-rays. Nature **90**, 410 (1912)
6. Bragg, W.H., Bragg, W.L.: The reflections of X-rays by crystals. Proc. Royal Soc. London **88**, 428–438 (1913)
7. Hahn, T. (ed.): International tables for crystallography, vol. A, Space-Group Symmetry. Kluwer Academic Publishers, Dordrecht (1996)
8. Villars, P., Calvert, L.D.: Pearson's handbook of crystallographic data for intermetallic phases. American Society of Metals (ASM Intern.), Ohio (1991)
9. http://icsd.fiz-karlsruhe.de/icsd/. Accessed 25 Oct 2013
10. Otte, H.M., Crocker, A.G.: Crystallographic formulae for hexagonal lattices. Phys. Stat. Sol. **9**, 441–450 (1965)
11. Nicholas, J.: The simplicity of Miller-Bravais indexing. Acta. Crystallogr. **21**, 880–881 (1966)
12. Boersch, H.: Über das primäre und sekundäre Bild im Elektronenmikroskop. Ann. Phys. **26**, 631–644 (1936) and **27**, 75–80 (1936)
13. Kikuchi, S.: Japan. J. Phys. 5 (1928), 83 und Phys. Z. 31 (1930), 777, cited in: Möllenstedt, G.: My early work on convergent-beam electron diffraction. Phys. Stat. Sol. (a) **116**, 13–22 (1989)
14. Doyle, P.A., Turner, P.S.: Relativistic Hartree-Fock X-ray and electron scattering factors. Acta. Crystallogr. **A24**, 390–397 (1968)
15. Stadelmann, P.: EMS—a software package for electron diffraction analysis and HREM image simulation in materials science. Ultramicroscopy **21**, 131–146 (1987), used Java-EMS-Version: 6.6201U2011
16. Thomas, J., Gemming, T.: ELDISCA C#—a new version of the program for identifying electron diffraction patterns. In: Luysberg, M., Tillmann, K., Weirich, T. (eds.) EMC 2008, Aachen, vol. 1, pp. 231–232. Springer-Verlag, Berlin (2008)
17. von Laue, M.: Materiewellen und ihre Interferenzen. Akad. Verl.-Ges. Geest & Portig, Leipzig (1948)
18. Vincent, R., Midgley, P.A.: Double conical beam-rocking system for measurement of integrated electron diffraction intensities. Ultramicroscopy **53**, 271–282 (1994)
19. Deininger, C., Mayer, J., Rühle, M.: Determination of the charge-density distribution of crystals by energy-filtered CBED. Optik **99**, 135–140 (1995)

Chapter 6
Why Do We See Any Contrast in the Images?

Abstract If we draw two white points on a white paper we cannot see these points. The white pencil does not create any contrast on the white paper. In transmission electron microscopy we need not only a good resolution limit but also an image contrast. This contrast is the result of the interaction between the beam electrons (primary electrons) und the transmitted specimen. Different to the light optical microscopy where the attenuation of the amplitude of the light wave is mainly responsible for the contrast, the absorption of electrons within the sample (i.e. the attenuation of the amplitude of the electron wave) plays only a secondary role in transmission electron microscopy. Here, the different scattering (deflection) of the electrons acts as the dominant reason for contrast. Later, we will see that, especially at very high magnifications, phase shifts of the electron waves can be responsible for the contrast, too.

6.1 Elastic Scattering of Electrons Within the Sample

We assume that the beam electrons interact only with the atomic nuclei while they are transmitting the sample (Coulomb interaction). Because of the extremely different masses of nucleus and electron (the proton mass is 1,836 times larger than the electron mass) the position of the nucleus does not change by interaction with one electron. This is comparable with the collision of a billiard ball with the edge of the billiard table. Therefore we speak about elastic interaction, the magnitude of the velocity and therewith the energy of the electron do not change. The probability of the deflection (deflection angle θ) of an electron after Coulomb interaction with an atomic nucleus and subsequent scattering into the solid angle segment $d\Omega$ characterised by θ (cf. Fig. 6.1—Rutherford's[1] differential cross section $d\sigma/d\Omega$ of scattering $\rightarrow$ Sect. 10.11) is given by

$$\frac{d\sigma}{d\Omega} = \left(\frac{1}{4\pi \cdot \varepsilon_0} \cdot \frac{Z \cdot e^2}{4 \cdot E_0}\right)^2 \cdot \frac{1}{\sin^4 \frac{\theta}{2}} \tag{6.1}$$

---

[1] Ernest Rutherford, New Zealand/British physicist, 1871–1937, Nobel prize in chemistry in 1908

J. Thomas, T. Gemming, *Analytical Transmission Electron Microscopy*,
DOI 10.1007/978-94-017-8601-0_6, © Springer Science+Business Media Dordrecht 2014

Fig. 6.1 For explanation of the differential cross section

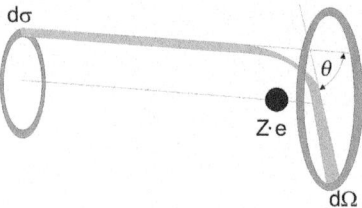

(ε_0: permittivity of vacuum, Z: atomic number, e: elementary electric charge, E_0: energy of the primary electrons). We neglect the partial shielding of the electric potential of the atomic nucleus by the electron shell. The total scattering of the electrons is also influenced by the atomic density within the sample. The number N of atoms within the volume V is given by

$$N = N_A \cdot \rho \cdot V \tag{6.2}$$

(N_A: Avogadro's constant, ρ: density).

The scattering ability of a single atom is described by its cross section σ for scattering. We want to find out all electrons whose scattering angle is larger than an assumed acceptance angle α. From Eq. (6.1) it follows after integration (Sect. 10.12):

$$\sigma(\alpha) = \left(\frac{Z \cdot e^2}{8 \cdot \varepsilon_0 \cdot E_0}\right)^2 \cdot \frac{1}{\pi \cdot \tan^2 \frac{\alpha}{2}}. \tag{6.3}$$

The magnitude of the scattering also depends on the sample thickness, of course. As shown in Sect. 10.12 the number N_E of electrons deflected by less than the angle α after travelling the path s within the sample can be written as

$$N_E = N_{E,0} \cdot e^{-s/\Lambda_{el}}. \tag{6.4}$$

$N_{E,0}$ is the number of incident electrons, Λ_{el} is a statistical value known as *mean free path of elastic scattering*. Using our simple model it is

$$\Lambda_{el} = \left(\frac{8 \cdot \varepsilon_0 \cdot E_0}{Z \cdot e^2}\right)^2 \cdot \frac{\pi \cdot \tan^2 \frac{\alpha}{2}}{N_A \cdot \rho}. \tag{6.5}$$

6.2 Mass Thickness and Diffraction Contrast

Transferring the different scattering ability into an image contrast the more strongly scattered electrons have to be removed from the ray path. This happens by an aperture whose radius defines the acceptance angle. Hence, this aperture has to be

6.2 Mass Thickness and Diffraction Contrast

Fig. 6.2 Arrangement and operation principle of the contrast aperture

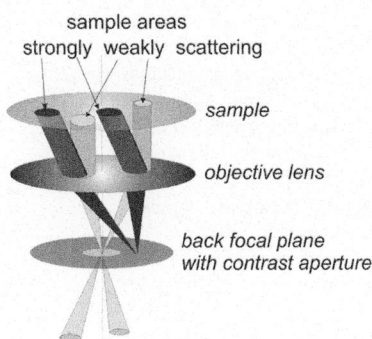

positioned in an angle-selective plane, i.e. in the back-focal plane of the objective lens (objective aperture—cf. Sect. 2.7.2 and Fig. 6.2). Because of its contrast-enhancing effect it is named *contrast aperture*, too.

Its effect can be seen in Fig. 6.2: Weakly scattered electrons move almost parallel to the optical axis. After the lens they meet close to the focal point within the back-focal plane and penetrate the contrast aperture. Strongly scattered electrons hit the aperture plate, are removed from the ray path and do not contribute to the image brightness. The more strongly scattering sample areas appear dark in the image.

We want to evaluate by means of the Eqs. (6.4) and (6.5) which contrast can be expected for small gold islands, only some 10 nm in size, on a thin carbon film. We define the image contrast K_{1-2} between two image areas 1 and 2 by the brigthnesses H_1 and H_2 in these areas:

$$K_{1-2} = 2 \cdot \left| \frac{H_1 - H_2}{H_1 + H_2} \right| \tag{6.6}$$

The quantity of incident electrons per time and area is responsible for the image brightness. Using Eq. (6.4) we write:

$$H_1 = H_{Au} = N_{E,0} \cdot e^{-s_{Au}/\Lambda_{el,Au}}$$
$$H_2 = H_C = N_{E,0} \cdot e^{-s_C/\Lambda_{el,C}} \tag{6.7}$$

respectively

$$K_{Au-C} = 2 \cdot \left| \frac{e^{-s_{Au}/\Lambda_{el,Au}} - e^{-s_C/\Lambda_{el,C}}}{e^{-s_{Au}/\Lambda_{el,Au}} + e^{-s_C/\Lambda_{el,C}}} \right| \tag{6.8}$$

Simplifying the problem we assume uniform sample thicknesses $s_{Au} = s_C = s$ and calculate the contrast K_{Au-C} in dependence on the sample thickness s for two different diameters d of the contrast aperture (Fig. 6.3). For the acceptance angle

Fig. 6.3 Expected contrast within the electron microscopic image of gold islands on carbon film at a primary electron energy of 200 keV and two different diameters d of the contrast aperture in dependence on the sample thickness s

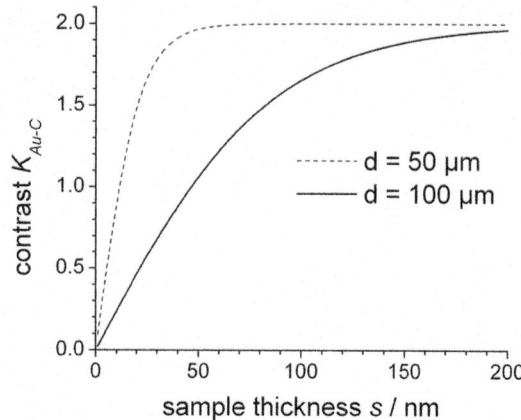

$$\alpha = \frac{d}{2 \cdot f} \approx \frac{d}{2 \cdot 1\,\text{mm}} \qquad (6.9)$$

is valid (Fig. 2.18).

Comparing the calculation shown in Fig. 6.3 with experimental data we use the electron microscopic image in Fig. 6.4. It shows just such gold islands on a carbon film. The contrast between carbon and gold of 0.72 detected in this image is reached at a sample thickness of 30 nm as read from Fig. 6.3. This is a plausible thickness.

This good agreement shows that our simple model is well suitable to explain this kind of contrast. From Eq. (6.5) and Fig. 6.3 we can see which actions are helpful to enhance the contrast without any changes of the specimen: decrease of the accelerating voltage (i.e. the primary electron energy) and decrease of the diameter of the contrast aperture (i.e. acceptance angle α).

The German name of this contrast mechanism is "Streuabsorptionskontrast" (scattering absorption contrast). It summarises the two reasons: scattering of electrons within the specimen and absorption of scattered electrons at the contrast aperture. In English the term *mass thickness contrast* is common practice.

For crystalline samples the various deflections of the electrons can be also reached by diffraction (cf. Chap. 5). We conceive a polycrystalline sample with differently orientated grains. The crystal lattices of the grains include different angles with the incident electrons. If the angle between lattice planes and electron incident direction fulfils Bragg's law (5.6) diffraction maxima with high intensity are produced. The electrons scattered into these diffraction maxima are removed from the ray path by the contrast aperture. The grains being in "Bragg position" appear dark in the image (Fig. 6.5). We speak about *diffraction contrast*. It allows the distinction of crystallites in electron microscopic images.

Because of the excitation error permissible at electron diffraction strict compliance of Bragg's condition is not necessary to get diffraction reflections

6.2 Mass Thickness and Diffraction Contrast

Fig. 6.4 Transmission electron microscopic image of a carbon film with gold islands ($E_0 = 200$ keV, diameter of the contrast aperture: 100 μm). The mean value of the image brightness was measured within the two white-shaped rectangles: It amounts to 82 in the upper rectangle (*gold*) and 175 in the lower one (*carbon*). Using Eq. (6.6) it follows a contrast of 0.72

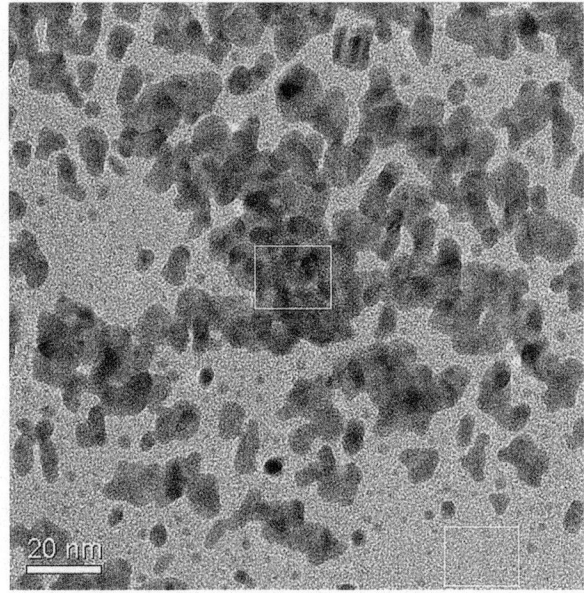

Fig. 6.5 Transmission electron microscopic micrograph of a thin polycrystalline Ni/NiO-layer. The grains (crystallites) can be clearly distinguished by diffraction contrast

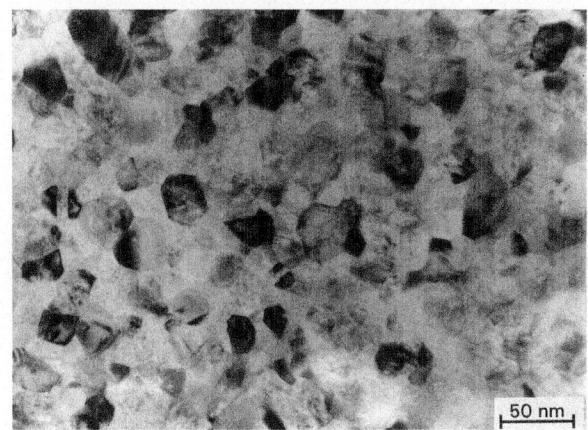

(cf. Sect. 5.4.3 as well as Sect. 10.9). Hence, a tolerance of the Bragg position also exists depending on the size of the crystallites.

Contrary to the mass thickness contrast the diffraction contrast depends on the orientation, i.e. it changes while the specimen is being tilted. Thereby the operator can distinguish between both kinds of contrast: He has to tilt the sample stage (it is possible on all transmission electron microscopes with side-entry holders at least around one axis). If the contrast changes it is diffraction contrast and therefore the sample must be crystalline. If we compare both contrasts we observe that the diffraction contrast is much stronger (in fact it is the strongest contrast mechanism in transmission electron microscopy).

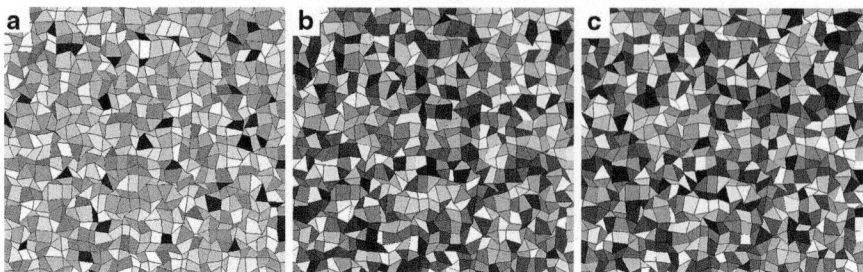

Fig. 6.6 Calculated brightfield and darkfield images of a polycrystalline layer without any preferential orientation of the grains. **a** Brightfield image, mean brightness: 165. **b** Darkfield image 1, mean brightness: 126. **c** Darkfield image 2, mean brightness: 124

6.3 Brightfield and Darkfield Imaging

So far, we have assumed that the contrast aperture is adjusted concentrically to the optical axis, i.e. no or only slightly scattered electrons pass the aperture. This case is named *brightfield imaging*. The Figs. 6.4, 6.5 show such brightfield images.

What happens when we shift the contrast aperture sidewards and in this way the slightly scattered electrons remove from the ray path? Instead of them at least a part of the strongly scattered electrons pass the contrast aperture (*darkfield imaging*). For a non-crystalline sample we get a simple inverse contrast and, hence no additional information about the sample. This circumstance changes for crystalline samples. Let us firstly assume that all grain orientations within a thin layer have equal statistical probability. In other words, there is no preferred orientation. Using a computer model we calculated the expected brightfield image shown in Fig. 6.6a. After shifting of the contrast aperture to two different positions outside of the optical axis we get the darkfield images 1 and 2 shown in Figs. 6.6b, c.

As expected the mean brightness within the darkfield images is significantly lower than that within the brightfield image. It does not substantially vary in between both darkfield images. This changes when the grains are preferentially orientated (texture).

Under this condition the images of Fig. 6.7 were calculated. The setting of the contrast aperture is the same as in Fig. 6.6. Contrary to Fig. 6.6 the mean brigthnesses of both darkfield images in Fig. 6.7 are significantly different.

By means of a small contrast aperture single reflections can also be selected within the diffraction pattern. In the corresponding darkfield image these grains arise bright which are responsible for the selected reflection. A uniform distribution of the orientations leads to similar mean brigthnesses in the darkfield images independent on the selected reflection (cf. Figs. 6.6b, c). That is different for textured layers. When a reflection is selected that stems from the preferentially orientated grains a larger quantity of grains looks bright in the darkfield images than in the case of the selection of a reflection stemming from grains with non-preferential

6.3 Brightfield and Darkfield Imaging

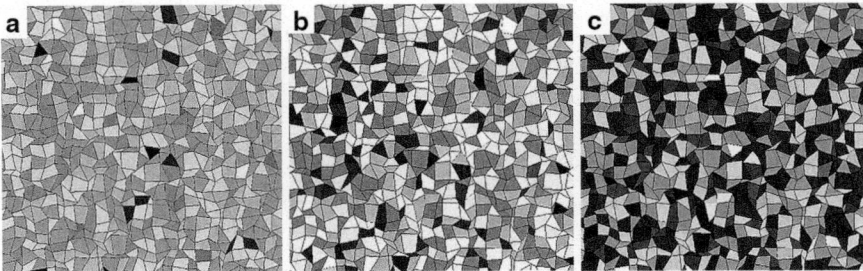

Fig. 6.7 Calculated brightfield and darkfield images of a polycrystalline layer with preferential orientation of the grains. **a** Brightfield image, mean brightness: 160. **b** Darkfield image 1, mean brightness: 149. **c** Darkfield image 2, mean brightness: 99

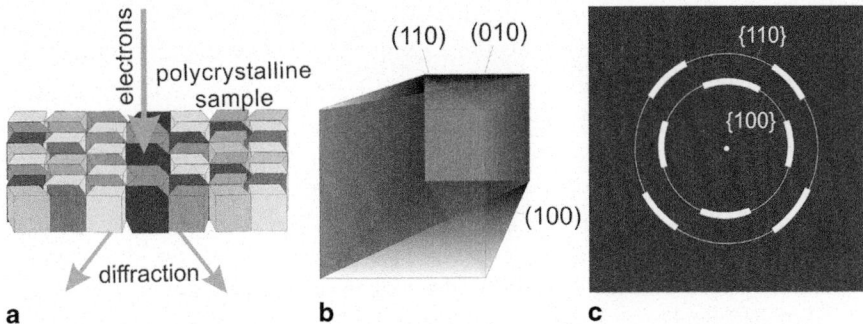

Fig. 6.8 Thin layer with <001>—preferential orientation. **a** Sketch of a textured layer. **b** Position of the diffracting lattice planes. **c** Diffraction pattern of this layer (schematically)

orientations (cf. Figs 6.7b and 6.7c). Also, the textures influence directly the diffraction patterns as shown for a simple example in Fig. 6.8c.

In the example shown in Fig. 6.8 we assume cubic crystallites whose [001] axes point upwards and which are randomly rotated around this axis by only a few degrees (Fig. 6.8a). Therefore, the intensities of the diffraction rings are not uniformly distributed but concentrated on symmetrically arranged azimuthal areas (Fig. 6.8c). When the contrast aperture is adjusted on a position with high intensity in the diffraction pattern a large number of grains appears bright in the darkfield image. The systematic change of the setting of the contrast aperture allows to identify textures in polycrystalline structures by darkfield imaging. Unfortunately, it is difficult to adjust the contrast aperture mechanically with the necessary accuracy and reproducibility.

In fact, the same effect as by adjusting the contrast aperture can be obtained by tilting of the electron beam (Fig. 6.9). By doing so, it is important that the point at which the beam is tilted around lies exactly within the sample plane. We remember: The adjustment of this point ("pivot point") was a part of the alignment control of

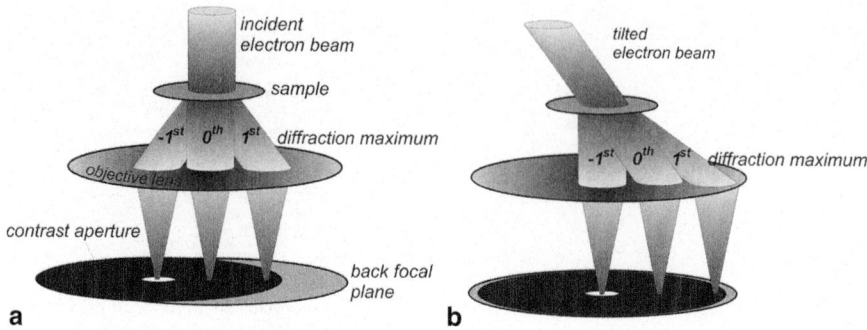

Fig. 6.9 Two possibilities to use the -1st diffraction reflection for darkfield imaging. **a** Shift of the contrast aperture. **b** Tilt of the electron beam

the microscope as described in Sect. 4.3. In practice, the beam tilting happens by means of electromagnetic deflection units.

Beside the higher precision during the setting the beam-tilt method has two additional advantages:

1. As evident from Fig. 6.9b at the beam-tilt method the contrast aperture remains concentrically adjusted to the optical axis. Rays close to the optical axis are used for the darkfield imaging, too. That leads to a minimisation of the imaging aberrations. This is also the reason why the method of shifting the aperture is called "dirty darkfield".
2. By a suitable activation of the beam deflection units it is possible to guide the electron beam on a conical envelope where its tip is located in the sample plane (*conical darkfield*). This has the same effect like moving the contrast aperture on a ring (that seems to be mechanically impossible). The radius of the ring is determined by the cone angle of the beam guidance.

Therewith an additional method for phase analysis is introduced. We assume that two different materials phases have crystallographic differences so that at least one ring of the diffraction diagram is accounted to only one of the phases and it is possible to select just this one by the contrast aperture. On the one hand this is a question of the differences in the crystallographic structures on the other hand it is a question of the contrast aperture size.

A mixture of nickel and nickel oxide is an example which is suitable to separate both phases by conical darkfield imaging. Both of them generate cubic phases with lattice constants of 0.35 nm and 0.42 nm, respectively.

Therewith the ring with the smallest diameter in the diffraction diagram can be matched to the nickel oxide having the larger lattice constant (Fig. 6.10). By setting a cone angle of 8 mrad it is possible to select the innermost ring with the contrast aperture and to record a conical darkfield micrograph (Fig. 6.10b). In this darkfield image the NiO crystallites arise bright.

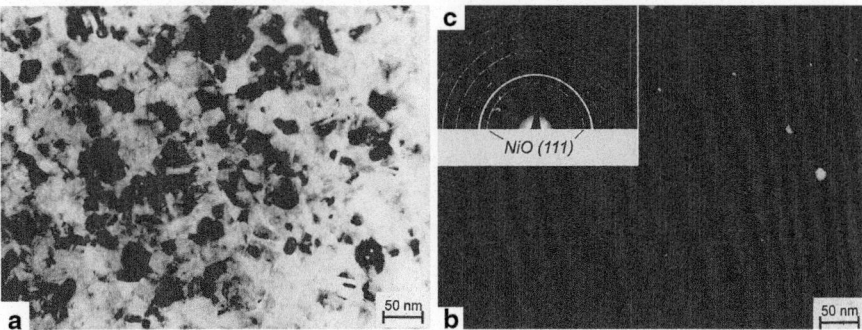

Fig. 6.10 Phase analysis of a Ni/ NiO mixture by conical darkfield imaging; diameter of the contrast aperture: 50 μm. **a** Brightfield image. **b** Conical darkfield image with inserted diffraction diagram. **c** The (111) ring of NiO is labelled

6.4 Bending Contours, Dislocations, and Semicoherent Particles

We have seen that the relation between the orientations of lattice planes and the direction of the incident electron beam is very important for the creation of diffraction contrast. That kind of contrast is the base for many electron microscopic investigations of the lattice structure within the samples. In other words, all deviations of the lattice structure leading to localised changes of the lattice plane orientations induce diffraction contrasts. In this paragraph we want to discuss some of such deviations from the ideal lattice structures together with their influences on the contrast of transmission electron microscopic images.

At first, we imagine a single crystal as specimen prepared with a thickness of only about 100 nm in reality. It is not surprising when this veritably ultra-thin crystal is bent (Fig. 6.11).

The lattice planes, in Fig. 6.11 only these orientated close to the Bragg position are drawn, follow the bend of the crystal, i.e. their orientation fluctuates a bit. Hence, there are small sample areas with lattice planes whose angle to the incident electron beam is equal to the Bragg angle. The part of the electron beam which hits this area is scattered into the direction, given by the Bragg angle, with higher intensity than for the other sample areas; its intensity into the straightforward direction is reduced.

The scattered intensity is removed from the ray path by the contrast aperture and the areas being in Bragg position appear dark in the image. In Fig. 6.11 this is illustrated by the dark line in the image plane. As already mentioned in Sect. 6.2 the Bragg condition must be only approximately fulfilled because of the excitation error of the electron diffraction. Therefore we do not observe sharp but "smeared" bending contours. Additionally, we cannot expect that the mechanical strains, intrinsic or generated by the electron microscopic preparation, lead to a bend of the

Fig. 6.11 Origin of the bending contours: Bent sample with different angles between lattice planes and incident electron beam. Result: dark line within the image plane

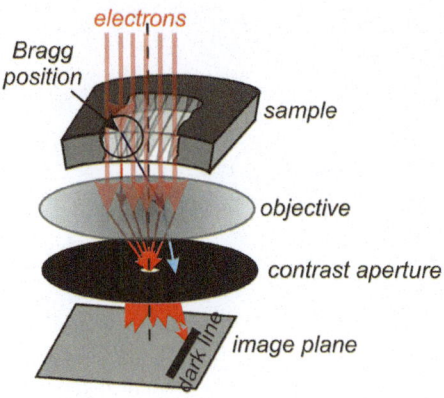

crystal only in one plane. Instead of such a simple bend mainly torsions of the crystal are the origin of curved *bending contours*.

Other sample areas reach the Bragg position when the sample or the electron beam is tilted. In the image the bending contours "run" over the sample (Fig. 6.12). The silicon crystal shown in the brightfield mode in Fig. 6.12 is dimpled and thinned by ion milling with argon ions to the generation of a hole (cf. Sect. 3.3). At the lower left in the pictures of Fig. 6.12 one can see a part of the hole edge (mass thickness contrast). With the help of this shape it can be seen that it is really the same sample area, only the bending contours have been shifted as consequence of sample tilting.

Dislocations are another possibility to generate localised lattice deformations. An edge dislocation is defined by an additional half-plane that is inserted into the lattice (Fig. 6.13). In the neighbourhood of the dislocation core (this is the starting point of the interjectional lattice plane) the atoms relax. A bending of the lattice planes in this area follows. In principle, it is the same behaviour as at the bending contours: For a suitable direction of the incident electron beam a part of the bent lattice planes is in a Bragg position, the electrons incidenting this position are more strongly scattered, and in the picture a dark line arises with the same direction as that of the interjectional lattice plane.

Contrary to the bending contours the lattice deformation is confined to the area around the dislocation core. The dark line does not move while the sample is being tilted. The line just disappears when the tilt angle is large enough. Using this behaviour the operator is able to distinguish between dislocation lines and bending contours. This emphasises the importance of the sample-tilt possibility during the work at the transmission electron microscope.

At an edge dislocation the lattice deformation arises only along a line and is confined of only one lattice plane direction. On the one hand this leads to the dependence of its visibility on the observation direction; on the other hand this allows to determine the position and the direction of the dislocation. Let us illustrate this with the help of Fig. 6.14.

We realise that the dislocation line should be observable in the direction $\mathbf{k}_{0,3}$ of the incident electron beam as a dark line with high contrast in the brightfield image.

6.4 Bending Contours, Dislocations, and Semicoherent Particles 125

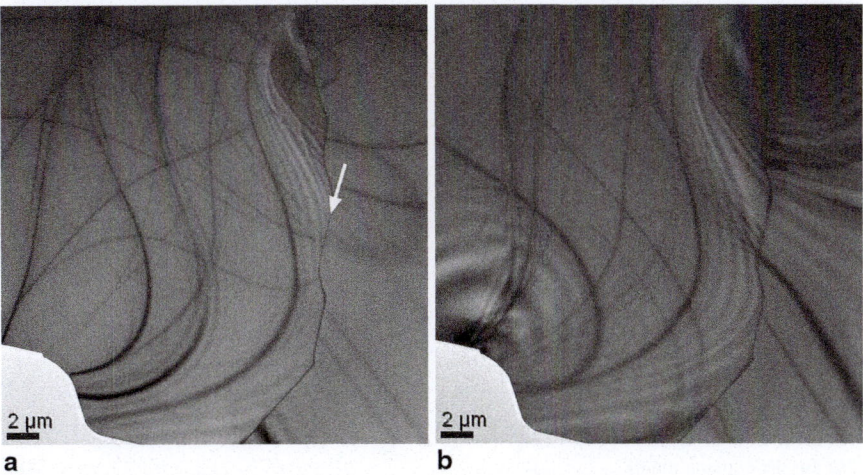

Fig. 6.12 Bending contours in a thinned silicon crystal. **a** Brightfield image. **b** Image of the same area after tilting the crystal by 2°. The crystal is wrenched, the crack line (white arrow) does not shift while the crystal is being tilted

Fig. 6.13 Generation of diffraction contrast at an edge dislocation. In this sketch only the sample area is drawn, arrangement of the objective lens and contrast aperture as displayed in Fig. 6.11

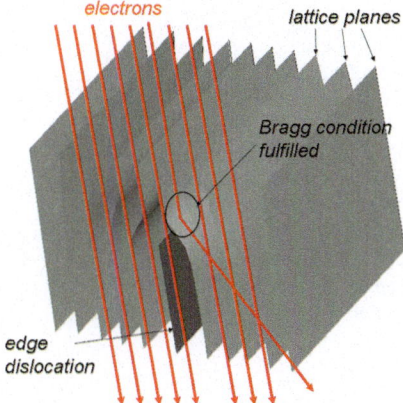

To discuss this circumstance furthermore we assume that we have a cubic crystal. In this case the three incident directions agree (we neglect the sign) with the crystal directions [100], [010], and [001]. Then, the [100] planes are curved. The *Burgers*[2] *vector* serves as the crystallographic description of the dislocation direction.

Its direction and its magnitude are determined by a "Burgers circulation" as shown in Fig. 6.15 relevant to our example. We look at the lattice projection in $k_{0,2}$ direction and choose the area around the dislocation core. We find that the Burgers vector is perpendicular to the dislocation line (here: crystal direction [100]).

---

[2] Johannes Martinus Burgers, Dutch physicist, 1895–1981

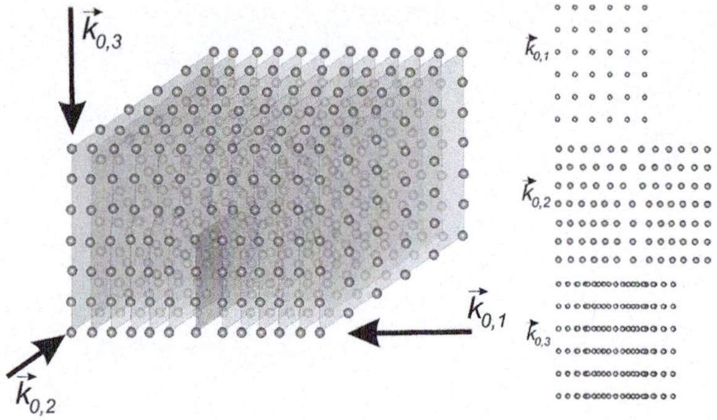

Fig. 6.14 Atomic arrangement of an edge dislocation in different projections: directions $\mathbf{k}_{0,1}$ and $\mathbf{k}_{0,2}$: no deformation, i.e. no different lattice plane orientations observable in these directions. Direction $\mathbf{k}_{0,3}$: shift of the atoms visible in this projection, i.e. variation of the lattice plane orientation observable

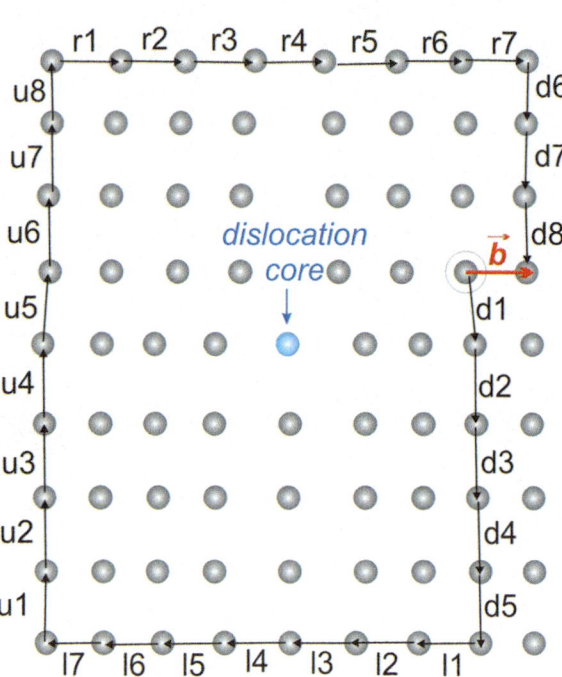

Fig. 6.15 Burgers circulation around the dislocation core for the explanation of the Burgers vector. The black framed atom is the starting point. From this atom we go five atoms downwards (arrows labelled by d1—d5), then seven atoms to the left (l1—l7), eight atoms upwards (u1—u8), and seven atoms to the right (r1—r7). We need the same number of steps downwards and upwards. Therefore, three steps downwards are missing. We add the steps d6—d8 for correction and reach the neighbour of the starting atom. The gap is closed by the Burgers vector **b**

This indicates an edge dislocation. Furthermore, we find that the scalar products of the three incident directions are

$$\mathbf{k}_{0,1} \cdot \mathbf{b} \neq 0, \ \mathbf{k}_{0,2} \cdot \mathbf{b} = 0, \ \mathbf{k}_{0,3} \cdot \mathbf{b} = 0, \text{ since } \mathbf{k}_{0,i} \cdot \mathbf{b} = |\mathbf{k}_{0,i}| \cdot |\mathbf{b}| \cdot \cos \sphericalangle (\mathbf{k}_{0,i}, \mathbf{b}) \quad (6.10)$$

The dislocation line disappears in darkfield images acquired "in the light" of the (010) and (001) reflection. The cross product of these two reflection directions delivers the direction of the Burgers vector: [100]. This is the basis of the determination of Burgers vectors by means of transmission electron microscopic images. In practice three important things have to be considered:

6.4 Bending Contours, Dislocations, and Semicoherent Particles

1. It must be possible to orientate the specimen into suitable crystallographic directions. Therefore, the use of a double tilt holder is absolutely necessary in such investigations.
2. Often it is required to draw the Burgers vector in an electron microscopic micrograph of the crystal. In this case the possible image rotation between imaging and diffraction mode has to be controlled and considered. For controlling suitable specimens are necessary, e.g. orthorhombic α-MoO_3 needles whose axes have the crystallographic [001] direction. Such specimens can be bought from companies selling electron microscopic accessories. We remember: The electrons move on spiral paths through the magnetic lenses. Switching between imaging and diffraction needs a change of the focal length of the projective lens system (imaging of the intermediate image plane or of the back-focal plane of the objective lens). This lens change leads to the image rotation mentioned above. In modern electron microscopes this image rotation is often corrected by a suitable interaction of the projective lenses, but this should be controlled for all magnifications and camera lengths.
3. We explained in Sect. 6.3 that darkfield images should be generated by tilting of the electron beam. This must be reconsidered for the investigation of dislocations because of the slightly varied incident direction as a consequence of the beam tilt. As seen before, the incident direction plays a crucial role in this kind of observations. Considering these circumstances the contrast aperture must be shifted to get the darkfield images, unless the beam tilt is wanted (*weak beam technique*).

The edge dislocation with a dislocation line almost parallel to the sample surface served us for the explanation of the contrast generation in principle. The Burgers vector is perpendicular to the dislocation line in this case. There are additional configurations of dislocations: the screw dislocation with a Burgers vector parallel to the dislocation line, and dislocations with a dislocation line which is inclined against the surface and ends within the sample.

The last case leads to bends of several orientated lattice planes with the result that these ends of the dislocation lines are observable at different incident directions of the electrons. Often they dominate in micrographs recorded without the selection of a special electron beam direction (Fig. 6.16).

Finally, let us discuss a third possibility of the generation of localised lattice deformations: semicoherent precipitates. Following our understanding these are particles which are embedded within a matrix and whose crystallographic structure and lattice constants are very similar. Figure 6.17 shows a sketch with the assumption that the lattice constant of the particle is less than that of the surrounding matrix.

The coffee-bean-like contrast in the transmission electron microscopic image which can be observed in Fig. 6.17 is typical for semicoherent segregations. Of course, the precondition for such observations is a suitable incident direction of the electron beam as this is true for all electron microscopic investigations using the diffraction contrast.

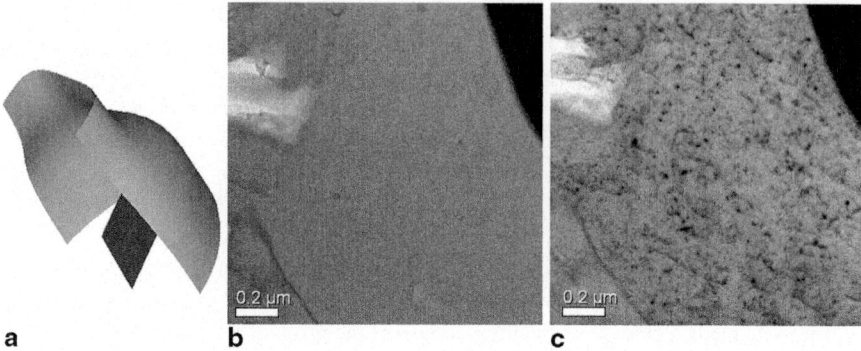

Fig. 6.16 Imaging of dislocations in the TEM. **a** Bend of lattice planes at the end of an edge dislocation. **b** TEM brightfield image of Al: dislocations are not visible. **c** The same sample area but sample tilted by 8°: dislocation lines and ends are visible

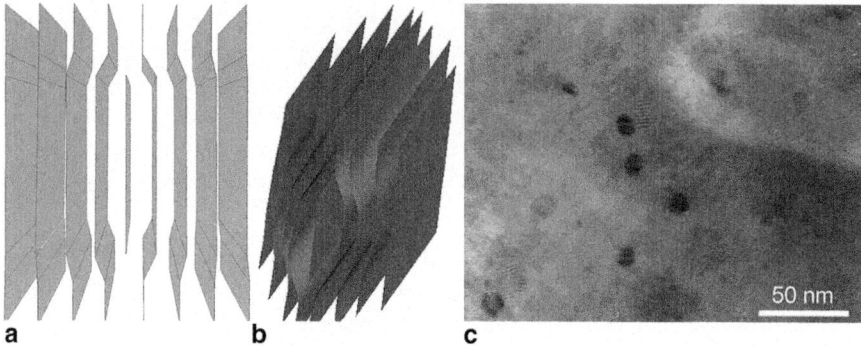

Fig. 6.17 Imaging of semicoherent precipitates in a TEM. **a** Adaption of the lattice planes by bends and insertion (edge dislocation). **b** Perspective illustration of the bends. **c** TEM micrograph of copper particles in a lead matrix (particles about 20 nm in size)

6.5 Thickness Contours, Stacking Faults, and Twins

We now address contrast phenomena in transmission electron microscopic images which cannot be explained by bent lattice planes connected with localised Bragg positions. For this purpose we have a look at Fig. 6.18. It shows the grain structure within a thinned copper layer. Beside the diverse contrast by localised bends of the lattice planes we observe dark stripes at the edge of a grain. It is the goal of the following discussion to understand the origin of these stripes.

We recall the diffraction of the electron wave at the crystal lattice. Thinking about Bragg's law we assumed a partial reflection of the electron wave at the lattice planes with subsequent interference of the partial waves (cf. Sect. 5.1). In that discussion we did not consider what can additionally happen with the diffracted wave within the crystal.

6.5 Thickness Contours, Stacking Faults, and Twins

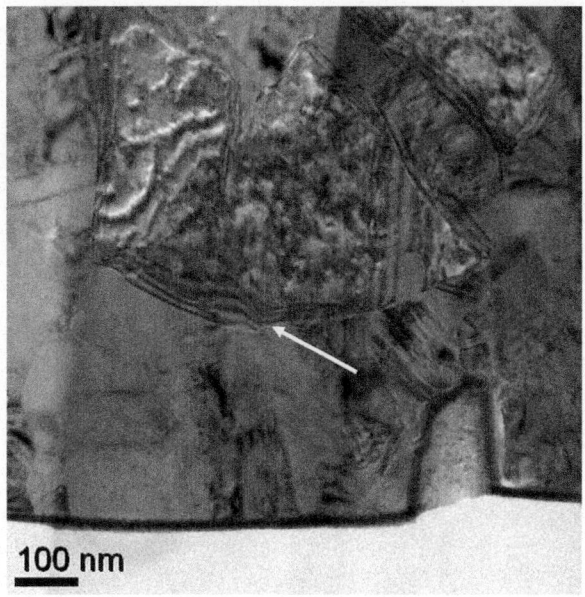

Fig. 6.18 Thickness contours (*arrow*) at the edge of a grain within a thin copper layer

To explain the stripes in Fig. 6.18 we must enhance our model and have a look at Fig. 6.19. We assume a crystal whose thickness increases wedge-shaped like at its edge (Fig. 6.19a). The lattice planes sketched in the opened section form the Bragg angle θ together with the incident electron wave, i.e. they are exactly in Bragg position. In our simple model the scattering capability should be equal for all lattice planes and the complete intensity of the electron wave should be reflected at one lattice plane. It reaches the neighboured plane and is reflected into the initial direction again (Fig. 6.19b). Because of the multiple reflection the intensity oscillates between diffracted and initial wave.

The length of one period is termed *extinction distance* L_{ext}. We realise that the sample thickness determines which of both waves has the higher intensity: the initial or the diffracted one. Obviously, within the thickness interval of the sample (i =0, 1, 2,..)

$$i \cdot L_{ext} \leq t < (i + \frac{1}{2}) \cdot L_{ext} \tag{6.11}$$

the diffracted wave has got more intensity and within the thickness interval

$$(i + \frac{1}{2}) \cdot L_{ext} \leq t < (i + 1) \cdot L_{ext} \tag{6.12}$$

the initial wave is the one with the higher intensity. If the diffracted intensity is removed from the ray path by the contrast aperture we get in the image bright and dark stripes called *thickness contours*.

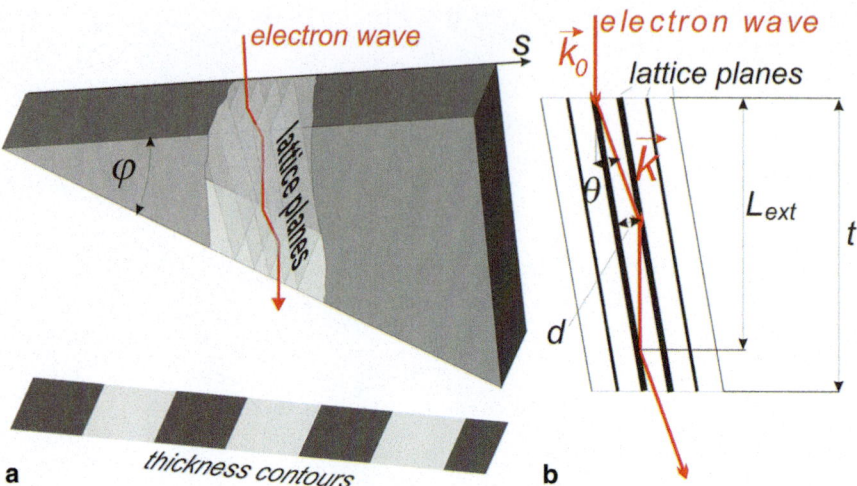

Fig. 6.19 For explanation of the thickness contours. **a** Crystalline sample with wedge-shaped thickness exactly orientated in Bragg position: Intensity of the electron wave oscillates between reflecting lattice planes. **b** Bragg angle θ, lattice distance d, extinction distance L_{ext}, and sample thickness t

According to Fig. 6.19b one obtains for small Bragg angles θ

$$\frac{L_{ext}}{2} = \frac{2 \cdot d}{\theta} \tag{6.13}$$

and following Bragg's law (5.6) for n=1 (1st diffraction order) with the electron wavelength λ

$$L_{ext} = \frac{4 \cdot d^2}{\lambda} \tag{6.14}$$

The lattice distances depend on the orientation. Hence, it is better to write d_{hkl} instead of d. Therewith the extinction distance depends on the orientation, too:

$$L_{ext,hkl} = \frac{4 \cdot d_{hkl}^2}{\lambda} \tag{6.15}$$

Following this simple geometric model the extinction distance, e.g. of the (111) lattice planes of copper (lattice constant: 0.362 nm) and 300 keV electrons (λ=0.00197 nm) amounts to 91 nm. In our model we neglect the scattering capability of the lattice planes (structure factor and volume of the unit cell) which influences the extinction distance, too. Hence, Eq. (6.15) gives only a rough estimate of the extinction distance.

By counting the dark stripes it is possible to evaluate the thickness of a crystal: We multiply the number of the stripes by the extinction distance. In reality the

6.5 Thickness Contours, Stacking Faults, and Twins

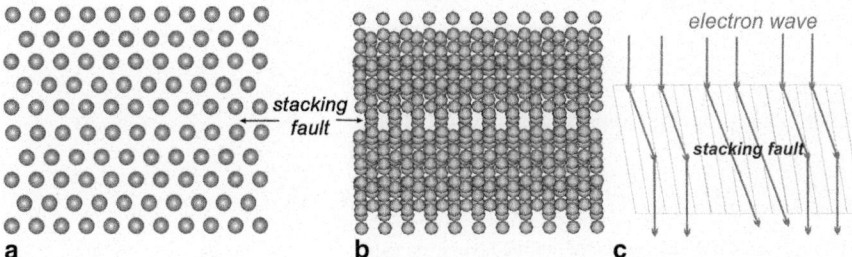

Fig. 6.20 Generation of contrast by a stacking fault. **a** Creation of a stacking fault. **b** Perspective illustration of a stacking fault. **c** Effect on the electron wave

electron wave is only partially reflected at the lattice planes. Therefore the thickness contours are more and more smeared at increasing crystal thickness.

Until now, in this paragraph we have assumed an undisturbed lattice. But, what happens if irregularities in the lattice planes are present with the consequence that the scattering capability is changed at localised positions? *Stacking faults* are typical examples for such a behaviour. For explanation of a stacking fault we assume that we have only one sort of atoms and these atoms are stacked "on gap" (Fig. 6.20a). The atomic arrangement is disturbed at the position marked by "stacking fault", the atoms are not stacked on gap. In the three-dimensional case the density of the atomic occupation of the lattice planes is changed (Fig. 6.20b). These deviations from the uniform atomic arrangement lead to additional phase shifts while the wave is propagating within the crystal lattice. We illustrate this by small gaps in the lattice planes with the consequence that their reflection behaviour is changed.

In our model one of the lattice planes is in Bragg position. Similar to the situation for the thickness contours the intensity oscillates between initial and scattered wave. In the undisturbed crystal area the intensity of the initial wave dominates, i.e. the crystal appears bright in the electron microscopic brightfield image. The gaps in the lattice planes lead to a redistribution of the intensity to the diffracted wave in the region of the stacking fault, i.e. this area appears dark in the image.

So far, we assume a stacking fault orientated almost perpendicular to the incident direction of the electron wave. Inclinations against this direction can lead to the imaging as thin dark stripe or additional thickness contours can arise at the edge of the stacking fault in the electron microscopic image (bright and dark stripes).

The shift of atoms connected with the stacking fault causes the alteration of the total energy of the crystal system. The difference to the origin state without stacking fault is named stacking fault energy. Because of the given interatomic forces only special configurations of stacking faults are preferred.

Let us discuss two of the different possibilities to stack atomic layers now (Fig. 6.21). The two possibilities differ by a half atomic distance to the side and by $\sqrt{3}/6$ atomic distances in the height. If it is switched from stacking variant (1) to variant (2) within the crystal mirrored lattice planes are generated (Fig. 6.22).

In Fig. 6.22a an example is emphasised by a black line. Such mirrored lattice planes are typical for *crystal twins*, the mirror planes are named *twinning planes*.

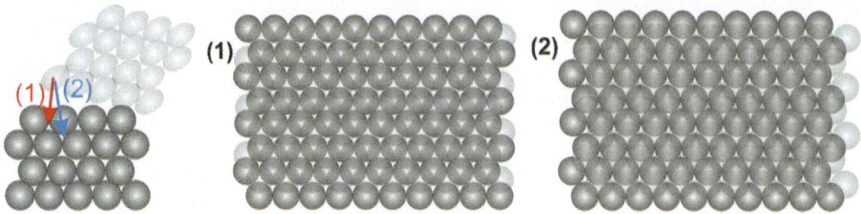

Fig. 6.21 Two possible stacks of atomic layers "on gap"

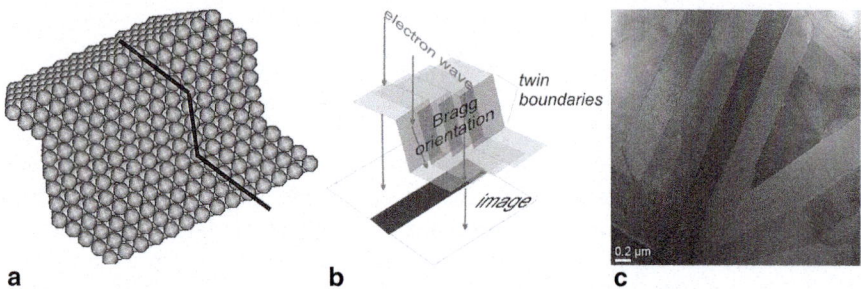

Fig. 6.22 Twins generated by stacking faults. **a** Perspective sketch. **b** Explanation of the resulting diffraction contrast. **c** Electron microscopic micrograph of twins in a Bi-Te alloy.

We discussed in the paragraph before how the crystal orientation influences the contrast in an electron microscopic image. We assume that the middle area of the lattice planes in Fig. 6.22b reaches the Bragg orientation by a suitable tilting of the specimen. After removing of the strongly scattered electrons by the contrast aperture this part of the twins appears dark in the image (Figs. 6.22b, c). Also in this case the contrast alternates when the specimen is tilted.

6.6 Moiré Patterns

Let us treat the case of two crystals on top of each other in the direction of the incident electrons (this direction is also called projection direction). The lattice planes represent periodic structures, their overlay can produce additional stripes, so-called "*Moiré patterns*"[3]. Three possible variants are sketched in Fig. 6.23.

The distances h between the Moiré stripes can be calculated by the equation

$$h = \frac{d}{2 \cdot \sin\left(\frac{\delta}{2}\right)} \tag{6.16}$$

---

[3] from the French "moirer": to marble

6.7 Magnetic Domains: Lorentz Microscopy

Fig. 6.23 Moiré patterns by overlay of two crystals. **a** Torsion with equal lattice plane distances. **b** Without torsion but with different lattice plane distances. **c** With torsion and different lattice plane distances

from the lattice spacing d and the torsion angle δ in the case of Fig. 6.23a, by

$$h = \frac{d_1 \cdot d_2}{|d_1 - d_2|} \qquad (6.17)$$

from both lattice spacings d_1 und d_2 in the case of Fig. 6.23b, and

$$h = \frac{d_1 \cdot d_2}{\sqrt{d_1^2 + d_2^2 - 2 \cdot d_1 \cdot d_2 \cdot \cos \delta}} \qquad (6.18)$$

from lattice spacings and torsion angle in the case of Fig. 6.23c (Sect. 10.13).

The electron microscopic image of Fig. 6.24 shows an example of Moiré fringes in practice. They must not be confused with possible larger lattice spacings.

Concerning the diffraction the overlaid crystals mean that strong diffraction reflections of the first crystal are able to generate additional reflections within the second crystal, i.e. the strong reflections of the first crystal can be initial waves for the second one. In principle that can also happen within a single crystal with larger thickness. This phenomenon is called "*double diffraction*". The resulting diffraction pattern can be understood as a linear combination of the diffraction patterns of the single diffraction events. Please note that double diffraction is also a possibility to excite reflections which are otherwise forbidden due to the kinematical extinction.

6.7 Magnetic Domains: Lorentz Microscopy

After discussion of the contrast generation by the orientation dependence of the electron wave interferences we come back to the particle model of the electrons. Let us think about the deflection of the electrons in a ferromagnetic sample and the possibility to use it for contrast generation in a transmission electron microscopic image.

Fig. 6.24 Moiré fringes (white mark) by overlay of two TiN crystals in projection direction (transmission electron microscopic image)

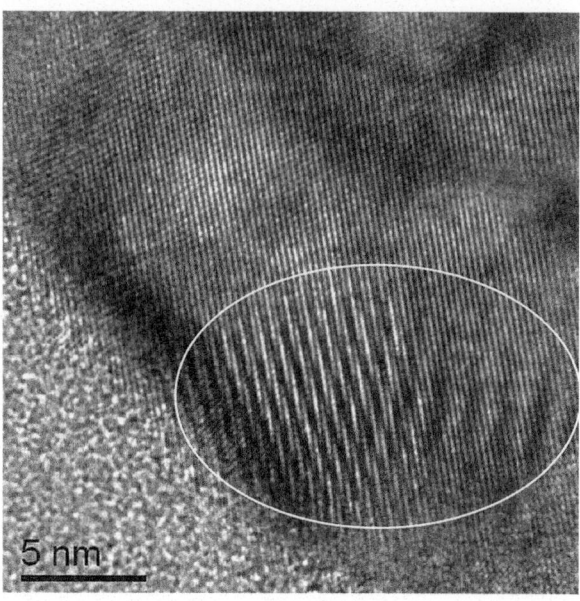

A ferromagnetic sample consists of domains ("*Weiss*[4] *domains*") where the atomic magnetic moments (elementary magnets) have equal direction. The domains are separated by *Bloch*[5] *walls*. The direction of the magnetisation rotates within a Bloch wall. The magnetisations of neighboured domains are often antiparallel.

A detail of such a ferromagnetic sample with antiparallel aligned domains is drawn in Fig. 6.25. The electrons follow the Lorentz force while they are passing the specimen:

$$\mathbf{F} = -e \cdot \mathbf{v} \times \mathbf{B} \qquad (6.19)$$

with the elementary electric charge e, the electron velocity **v**, and the magnetic induction **B**. This equation is already known from Sect. 2.2 where the function of rotational-symmetric magnetic fields as electron lenses was explained. From there the direction of the Lorentz force is also known. The direction of the magnetic induction is equal to that one of the magnetisation. This is considered in Fig. 6.25.

The deflection of the electrons within the sample generates a modulation of the electron density in a plane 1 below the specimen. In the area of Bloch walls separating domains with the magnetisation backwards on the left side and forwards on the right side the electron density is reduced. This is marked by black stripes in plane 1. Bloch walls between domains with forwards directed magnetisation on the left and backwards directed magnetisation on the right create a higher electron density

---

[4] Pierre-Ernest Weiss, French physicist, 1865–1940
[5] Felix Bloch, Swiss physicist, 1905–1983, Nobel prize in physics in 1952

6.7 Magnetic Domains: Lorentz Microscopy

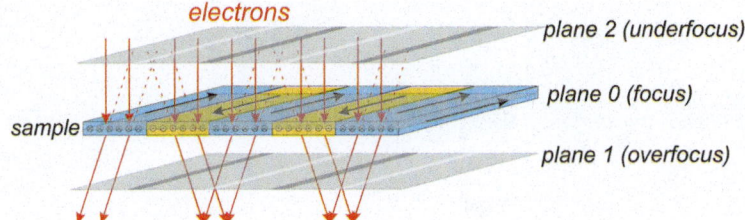

Fig. 6.25 Electron density in different planes after deflection on antiparallel directed magnetic domains within a ferromagnetic sample. The black arrows mark the direction of the magnetisation. Additional explanations in the text

in plane 1 symbolised by bright stripes. The electron density does not modulate in plane 0 (object plane).

As we already know, in the electron microscope different planes can be imaged on the screen by changing the lens focal lengths. In this way it is possible to image the plane 0 (corresponding to the exact focus) as well as the plane 1 (corresponding to the overfocus) on the observation screen. In the first case the domain boundaries are invisible, in the second one the different electron densities generate a contrast, the domain boundaries are brought out by dark and white lines.

There is an additional consequence of the variation of the lens focal length: It is also possible to image the plane 2 above the sample plane 0. The back projection of the directions of the electron deflection results in a contrast inversion in this plane. While increasing the lens current from underfocus to overfocus firstly we see the domain walls as bright and dark lines. They are invisible in the sharp image and arise again with inverse contrast as dark and bright lines.

There is a serious problem when this *Lorentz microscopy* is to be performed in practice. Normally, the sample is located within the high magnetic field of the objective lens. The adjustments of the magnetisation in the sample follow the magnetic field of the lens and Bloch walls disappear. To avoid this a special objective lens (*Lorentz lens*) is necessary with a significantly lower magnetic field surrounding the specimen. Unfortunately, in doing so the spherical aberration of a rotational-symmetric lens increases and the resolution limit degrades.

Another possibility is to turn off the conventional objective lens or at least to operate with a drastically reduced objective coil current. Doing so is simply possible by working in the low magnification range. In this case the (low) magnification is mainly reached by the projective lenses leading to dramatic consequences for the resolution limit, too. The advantage compared to a Lorentz lens is to save the costs of a Lorentz lens and the changes of the microscope setup.

The electron microscopic pictures of magnetic domains within nanoscaled cobalt bars (Fig. 6.26) were acquired in the "low-mag" mode of the objective lens [1]. The geometry of the specimen is a consequence of the cross-section preparation of a Co/Cu/Co layer stack on silicon substrate.

The upper Co layer (Co2) is partly oxidised and non-ferromagnetic. The Co layer directly on the substrate (Co1) shows the expected behaviour: In the focussed

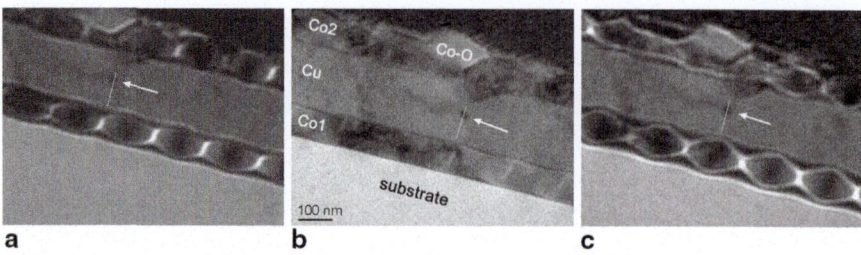

Fig. 6.26 Lorentz microscopy on a cross-section of a Co/Cu/Co layer stack. **a** Underfocus. **b** Focus. **c** Overfocus. The arrow marks the identical sample position

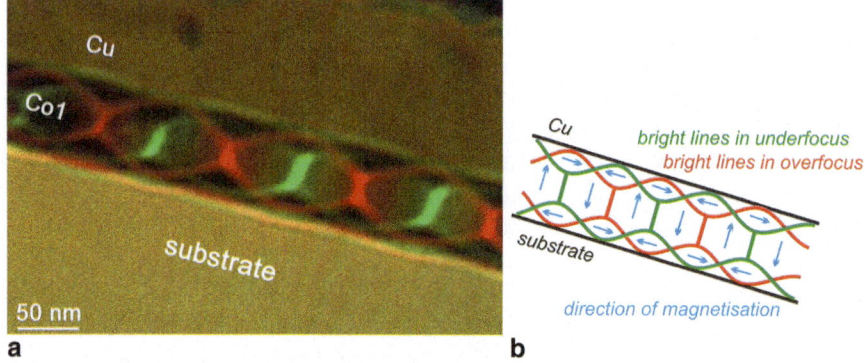

Fig. 6.27 Explanation of the image contrast in Fig. 6.26. **a** Overlay of the green (*underfocus*) and red (*overfocus*) coloured pictures of Figs. 6.26a and 6.26c. **b** Sketch with determination of the magnetisation directions within the domains

micrograph (Fig. 6.26b) the domain boundaries are invisible. In the underfocussed image (Fig. 6.26a) the domain wall at the arrow position arises bright and in the overfocussed image (Fig. 6.26c) it becomes dark. The contrast inverts during the transition from underfocus to overfocus.

Following the basics as sketched in Fig. 6.25 and the observed image contrast (Fig. 6.26) the magnetisation directions in the domains can be determined. The result is shown in Fig. 6.27.

References

1. Brückner, W., Thomas, J., Hertel, R., Schäfer, R., Schneider, C.M.: Magnetic domains in a textured Co nanowire. Journ. Magn. Magn. Mat. **283**, 82–88 (2004)

Chapter 7
We Increase the Magnification

Abstract Let us now focus on the imaging of structures with sizes close to the resolution limit of a transmission electron microscope. The atomic distances within the crystal lattices are in this order. The investigation of grain boundaries and other interfaces on the atomic scale are topics in this context, too. Additionally, we discuss the effects arising at the high-magnified and high-resolved imaging of amorphous materials. In order to do this we have to consider an additional contrast mechanism: the phase contrast.

7.1 Imaging of Atomic Columns in Crystals: Phase Contrast

Let us start with an example. We have a look at the unit cell of a silicon crystal from different directions. Silicon forms a diamond lattice described by the space group 227 ($Fd\bar{m}3$) with the atomic basic position 0.25, 0.25, 0.25. Illustrating this we imagine two cubic face-centred lattices of which the second one is shifted by a quarter of the space diagonal of the first one (Fig. 7.1a). The lattice constant of silicon amounts to 0.543 nm.

At an arbitrary viewing direction on the crystal lattice as shown in Fig. 7.1b we cannot see any systematic arrangement of the atoms. The atoms are overlapped in the projection direction, single atomic columns cannot be observed. This changes if we tilt the crystal and look into the direction of a cube edge on the crystal as shown in Fig. 7.2a, for instance.

The electron microscopic micrograph (HRTEM image, this is the abbreviation of High Resolution Transmission Electron Microscopic image) in Fig. 7.2b exactly corresponds to the lattice projection along the cube edge of the Si unit cell (Fig. 7.2c).

From Chap. 5 we know that the cube edges are crystallographically labelled [100], [010] or [001]. Negative signs of the number 1 are also possible. In a cubic crystal all these axes are crystallographically equal. If just one of these directions is meant then one writes <100>. We see that the indices within the crystallographic description of this direction are low. Such *low-indexed directions* are advantageous for imaging since the projected atomic distances are comparably large in this case.

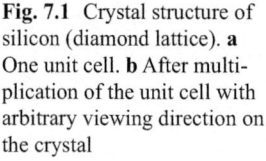

Fig. 7.1 Crystal structure of silicon (diamond lattice). **a** One unit cell. **b** After multiplication of the unit cell with arbitrary viewing direction on the crystal

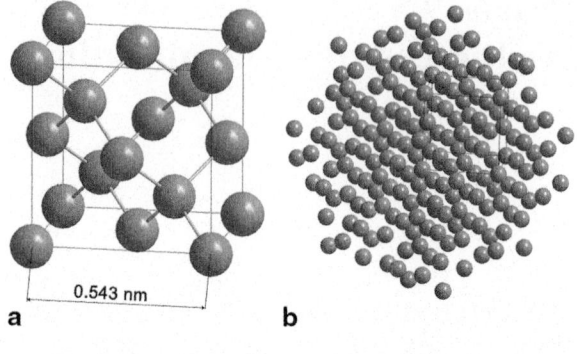

Fig. 7.2 Interpretation of a high resolution electron microscopic micrograph of silicon. **a** Silicon lattice in [100] direction. **b** HRTEM image of Si. **c** HRTEM image with overlaid unit cell of Si in [100] projection

Additionally, it must be noted that the image shown in Fig. 7.2b can only be obtained with especially thin high quality samples.

Now, let us think about the reason of the image shown in Fig. 7.2b: Why can we see the atomic columns in the transmissions electron microscopic image, what contrast mechanism is responsible?

Two kinds of contrast are known from Chap. 6: mass thickness and diffraction contrast. Mass thickness contrast needs differences of the density and/or the specimen thickness including the contrast enhancement by the contrast aperture; diffraction contrast needs different orientations of the crystal lattice towards the direction of the incident electron beam or localised differences in scattering abilities of lattice planes. Neither of them is fulfilled during imaging of the silicon crystal. There must exist an additional contrast mechanism, important especially for the high resolution imaging in transmission electron microscopy.

To understand this contrast mechanism we remember the wave character of the electrons and consider the interaction between the electron wave and the periodically arranged atoms within the crystal lattice. From Schrödinger's equation (Sect. 10.2) it follows that the wavelength λ_{Cr} of the electron wave within a potential Φ can be calculated by

7.1 Imaging of Atomic Columns in Crystals: Phase Contrast

Fig. 7.3 Illustration of the lattice potential. **a** Definition of the coordinate system. **b** Sketch of the crystal potential within the x-y plane

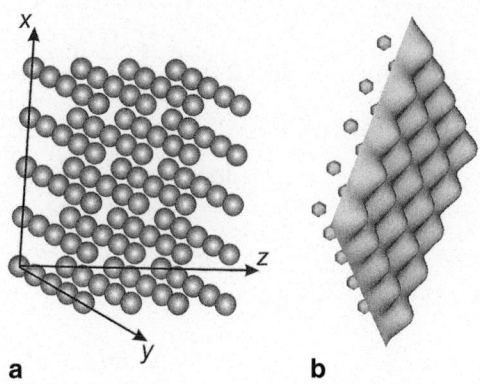

$$\lambda_{Cr} = \frac{h}{\sqrt{p^2 + 2 \cdot m \cdot e \cdot \Phi}} \quad (7.1)$$

(h: Planck's constant, p: momentum of the electron, m: electron mass, e: elementary electric charge). In non-relativistic approximation the correlation between momentum p, energy E, and acceleration voltage U_0 is given by

$$p^2 = 2 \cdot m \cdot E = 2 \cdot m \cdot e \cdot U_0, \quad (7.2)$$

i.e. with consideration of the position-dependence of the potential $\Phi(x, y, z)$ within the periodic crystal lattice (Fig. 7.3):

$$\lambda_{Cr} = \frac{h}{\sqrt{2 \cdot m \cdot e \cdot (U_0 + \Phi(x, y, z))}}. \quad (7.3)$$

Outside of the crystal ($\Phi=0$) the wavelength amounts to

$$\lambda = \frac{h}{\sqrt{2 \cdot m \cdot e \cdot U_0}}. \quad (7.4)$$

Following equation (7.3) the wavelength decreases within the crystal (positive potential) resulting in a phase shift $d\phi$ in a thin crystal layer of thickness dz:

$$d\phi = 2 \cdot \pi \cdot \left(\frac{dz}{\lambda_{Cr}} - \frac{dz}{\lambda} \right) = 2 \cdot \pi \cdot \frac{dz}{\lambda} \cdot \left(\frac{\lambda}{\lambda_{Cr}} - 1 \right). \quad (7.5)$$

With Eqs. (7.3) and (7.4) it follows

$$d\phi = 2\cdot\pi\cdot\frac{dz}{\lambda}\cdot\left(\sqrt{\frac{2\cdot m\cdot e\cdot(U_0+\Phi(x,y,z))}{2\cdot m\cdot e\cdot U_0}}-1\right) \quad (7.6)$$

$$d\phi = 2\cdot\pi\cdot\frac{dz}{\lambda}\cdot\left(\sqrt{1+\frac{\Phi(x,y,z)}{U_0}}-1\right).$$

For the usual accelerating voltages in transmission electron microscopy $U_0 \gg \Phi(x, y, z)$ is valid and, with it,

$$\frac{\Phi(x,y,z)}{U_0} \ll 1. \quad (7.7)$$

We expand the radical term in Eq. (7.6) as a Taylor[1] series at

$$\frac{\Phi(x,y,z)}{U_0} = 0 \quad (7.8)$$

up to the second term and get

$$d\phi = 2\pi\cdot\frac{dz}{\lambda}\cdot\left(1+\frac{1}{2}\frac{\Phi(x,y,z)}{U_0}-1\right) = \frac{\pi}{\lambda\cdot U_0}\cdot\Phi(x,y,z)\cdot dz. \quad (7.9)$$

The phase shift ϕ in a specimen with thickness t can be obtained by integration:

$$\phi = \frac{\pi}{\lambda\cdot U_0}\cdot\int_0^t \Phi(x,y,z)\cdot dz. \quad (7.10)$$

Assuming the potential depends only on x and y and is constant along the coordinate z it follows

$$\phi(x,y) = \frac{\pi\cdot t}{\lambda\cdot U_0}\cdot\Phi(x,y). \quad (7.11)$$

This means the phase shift reflects the periodicity of the crystal. The plane wave incident into the crystal becomes phase-modulated and this phase modulation reflects the periodicity of the crystal lattice in the exit wave (Fig. 7.4).

How can we visualise this phase modulation? The recording medium, screen, photo plate or CCD camera, displays only changes of the wave amplitude, not of its phase. From Chap. 1 we know that the image can be understood as an interference

---

[1] Brook Taylor, British mathematician, 1685–1731.

7.1 Imaging of Atomic Columns in Crystals: Phase Contrast 141

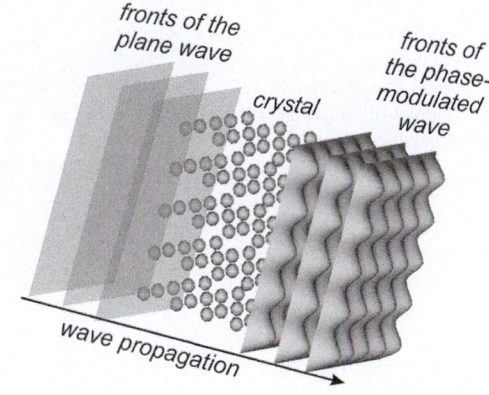

Fig. 7.4 Phase modulation by the crystal potential while the incident electron plane wave is penetrating a crystal lattice

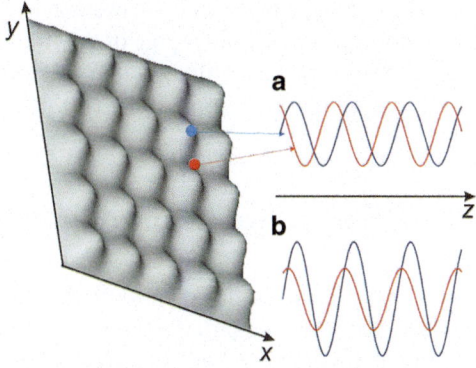

Fig. 7.5 Transformation of the phase modulation into an amplitude modulation by interference with a diffracted wave. **a** Examples for an area with strong (*blue*) and weak (*red*) phase shift of the phase-modulated exit wave. **b** Changed amplitudes (i.e. intensities) after interference with a wave in phase with the *blue* area: The amplitude of the *blue* wave is amplified

figure of the waves incident into the lens. For the creation of an image we need at least a second (diffracted) wave. The diffracted wave also experiences a phase modulation but a modified one because of the other direction in relation to the crystal. Simplifying, we assume in our model that the diffracted wave is a plane wave. Dependent on the phase of this second wave we get a constructive interference with the phase-modulated exit wave at special positions of x, y. Assuming the phase difference is equal to a whole-number multiple of the wavelength just for the modulation at the atomic column positions the wave part at these positions experiences a constructive interference with amplified amplitude and, therefore, these positions arise bright in the image contrary to the interstices (Fig. 7.5). Another phase of the second wave would create another contrast.

Since the direct interaction between electron wave and crystal creates a (initially invisible) phase modulation this kind of contrast mechanism is termed *phase contrast*. In dependence on the additional phase shift of the diffracted wave the positions of the atomic columns or even the interstices between them can show a higher image brightness. According to Eq. (7.11) the phase shift within the object exit wave also depends on the specimen thickness t, i.e. the specimen thickness influences the phase contrast, too.

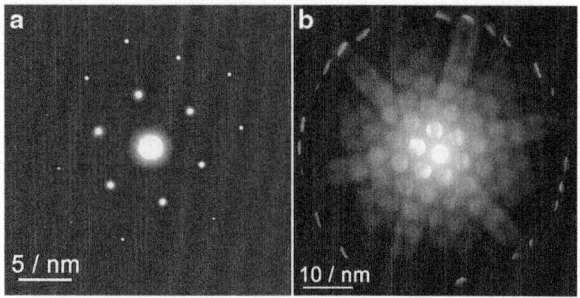

Fig. 7.6 Diffraction pattern of Si at low-indexed incident direction (zone axis [111]). **a** Parallel illumination. **b** Convergent illumination with Kikuchi pattern

In the electron microscopic practice the imaging of crystal lattices needs a very thin (<50 nm) specimen not covered by amorphous layers and free from lattice defects eventually created by the electron microscopic specimen preparation. The low-indexed incident direction is preferably adjusted in the diffraction mode using a convergent illumination. A pole of the Kikuchi pattern is shifted into the centre of the screen by systematic tilting of the specimen (Fig. 7.6).

Of course, such high resolution investigations need a double-tilt specimen holder.

7.2 Contrast Transfer by the Objective Lens

How can we reach a suitable phase shift of the diffracted wave to transfer the phase modulation into an amplitude modulation? We remember: An ideal optical imaging by a lens is achieved when all the diffracted waves overlay in the image space without any additional phase shift by the lens. However, we want to get a special phase shift of the diffracted wave, i.e. we have to use the lens in another way than necessary for ideal imaging.

In Fig. 7.7a the ray path close to the objective lens is drawn for the ideal imaging. In this case the optical paths are equal for both waves: the wave 1 running through the centre of the lens and the diffracted wave 2. There is no phase shift between them in the Gaussian image plane.

The deviation from the ideal ray path leads to different optical paths of both waves originating a phase difference in their cross over (Fig. 7.7b). This deviation can be generated by the spherical aberration of the lens but also by defocussing, i.e. by a change of the focal length. The latter one can be done by the operator and therefore it is possible to create the special phase shift between the waves needed for the transfer of the phase modulation into an amplitude modulation and, in so far, for the visualisation of the crystal lattice.

Let us demonstrate this possibility with the help of an image simulation using the software JEMS by P. Stadelmann [1] often used for such purposes (Fig. 7.8). The sample is a copper crystal accurately observed from the top (crystallographic [001] direction).

7.2 Contrast Transfer by the Objective Lens 143

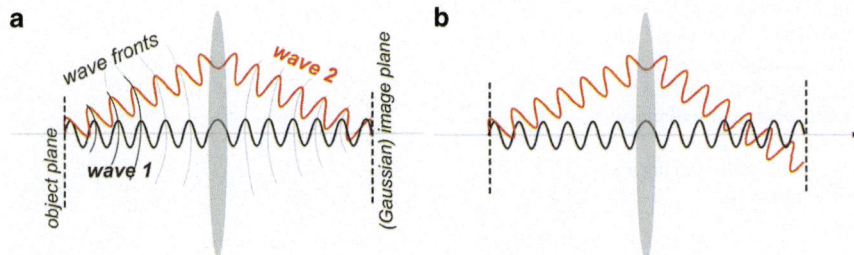

Fig. 7.7 Wave interferences after transition through a lens. **a** For ideal imaging (no phase shift in the Gaussian image plane). **b** Waves do not meet in the Gaussian image plane (phase shift caused by the lens)

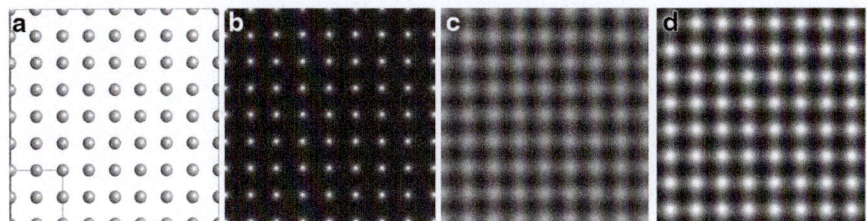

Fig. 7.8 Influence of the focus on the phase contrast. **a** Cu lattice in [001] projection (4 × 4 unit cells). **b** Projected potential. **c** Calculated HRTEM image (U_0 = 300 kV, C_S = 1.2 mm, C_C = 2 mm, ΔE = 1 eV, t = 11 nm, defocus Δf = 79 nm). **d** As (c) but Δf = 99 nm

The projected potential (Fig. 7.8b) reflects exactly the positions of the copper atomic columns. At a defocus of 79 nm just these positions arise bright in the simulated high resolution (HRTEM) image (Fig. 7.8c), contrary, at a defocus of 99 nm they appear dark (Fig. 7.8d).

Besides the dependencies of the phase shift ϕ on the adjustable lens focal length (defocus Δf, i.e. difference of actual focal length and that needed for focussing within the Gaussian image plane) and on the spherical aberration (characterised by its aberration coefficient C_S) an influence of the angle θ between diffracted wave and optical axis has to be considered:

$$\phi = \phi(\theta, C_S, \Delta f). \tag{7.12}$$

This is mathematically treated in Sect. 10.14. At this point we want to plausibly explain this fact.

If the phase shift ϕ is set for a given diffraction angle θ in a way that the path difference between initial and diffracted wave is equal to a whole-number multiple of the electron wavelength λ in the image, amplifying, i.e. high image brightness, arises. At a path difference equal to an odd-numbered multiple of $\lambda/2$ destructive interference happens, i.e. we get low image brightness.

Fig. 7.9 Contrast transfer in dependence on the detail distance: **a** At phase shift leading to maximal amplification by constructive interference. **b** Intermediate phase. **c** At phase shift leading to destructive interference

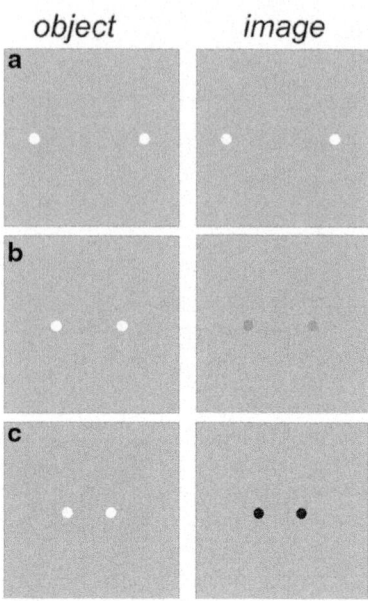

From Bragg's law (5.6) it follows that the angle θ is inversely proportional to the lattice spacing distance d. We would like to generalise this message: The angle θ is inversely proportional to the distance d between two object details which are to be imaged:

$$\theta \sim \frac{1}{d}. \tag{7.13}$$

The distance d can be interpreted as structure size. The term $1/d$ (reciprocal length) is named space frequency q (sometimes $q = 2\pi/d$ is used). The phase shift ϕ depends on the space frequency, too:

$$\phi = \phi(q, C_S, \Delta f). \tag{7.14}$$

What are the consequences for the imaging? We assume that for a chosen distance d_1 between two details and a given spherical aberration a focus is selected which leads to a phase shift ϕ originating a constructive interference of initial and diffracted wave. As consequence these details are displayed with excellent brightness and contrast. At another detail distance d_2 the diffraction angle θ is changed and therefore also the phase shift ϕ. The condition for the maximal amplification during the interference is no longer fulfilled; the contrast of these details is lower. The other extreme is a distance-dependent phase shift leading to a destructive interference. In other words: The image contrast depends on the structure sizes (i.e. distances between two neighbouring details) which are to be imaged. This dramatic consequence is illustrated in Fig. 7.9.

7.3 Wave-optical Interpretation of the Resolution Limit

Varying distances between the object details lead to completely different contrast relations. Without any proper understanding of these relationships the image interpretation becomes ambiguous:

▶ "Believe only what you see after you have understood why you see it!"

7.3 Wave-optical Interpretation of the Resolution Limit

The dependence of the image contrast on the structure size (i.e. on the space frequency) mentioned in the previous Sect. 7.2 is illustratively presented in a sufficient approximation by the linear phase contrast transfer function *CTF*. Using the knowledge about this function (Sect. 10.14)

$$CTF(q) = \sin\left(\frac{\pi}{2}\left(C_S \cdot \lambda^3 \cdot q^4 - 2 \cdot \Delta f \cdot \lambda \cdot q^2\right)\right) \cdot e^{-\pi^2 \cdot C_C^2 \cdot \left(\frac{\Delta E}{E_0}\right)^2 \lambda^2 \cdot q^4}. \quad (7.15)$$

(q: space frequency, C_S: coefficient of spherical aberration, λ: electron wavelength, Δf: defocus, i.e. difference of the actual focal length and the one needed for focussing in the Gaussian image plane, C_C: coefficient of chromatic aberration, ΔE: energy width of the electrons, E_0: primary electron energy) we would like to deal again with the definition of the resolution limit. The argument of the sine function shows that the influence of the spherical aberration can be reduced by a suitable defocus. The "best" property of the contrast transfer, i.e. the highest space frequency with simple interpretation of the image contrast can be obtained by a slightly defocussed objective lens. This reasoning defines the *Scherzer focus* (Sect. 10.15)

$$\Delta f_{Sch} = 1.2 \cdot \sqrt{C_S \cdot \lambda}. \quad (7.16)$$

An example of the contrast transfer function is shown in Fig. 7.10.

Space frequencies up to q_δ are transferred without any contrast inversion. There are no problems with image interpretation up to this point since the contrast transfer takes place without oscillations. Higher space frequencies are also transferred but with varying contrast depending on the space frequency, the contrast transfer function oscillates. Therefore, two different space frequencies can be defined as limitations:

The reciprocal value of q_δ is known as point resolution:

$$\delta_P = \frac{1}{q_\delta}. \quad (7.17)$$

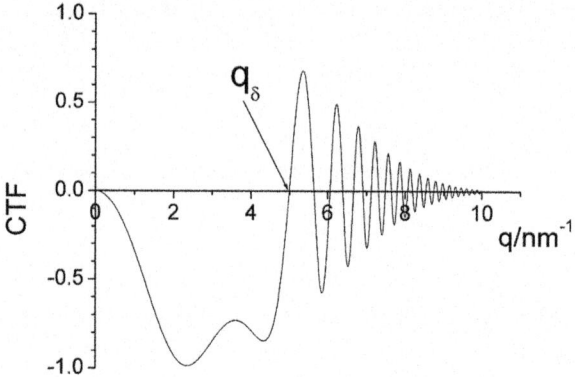

Fig. 7.10 Contrast transfer function for $E_0 = 300$ keV, $\Delta E = 0.7$ eV (Schottky field emission gun), $C_S = 1.2$ mm, $C_C = 1.5$ mm, $\Delta f = 58$ nm (Scherzer focus)

Following equation (7.15) the first zero is reached at:

$$C_S \cdot \lambda^3 \cdot q_\delta^4 = 2 \cdot \Delta f \cdot \lambda \cdot q_\delta^2$$
$$q_\delta = \sqrt{\frac{2 \cdot \Delta f}{C_S \cdot \lambda^2}} \tag{7.18}$$

and after use of the Scherzer focus (7.16):

$$q_\delta = \sqrt{\frac{2 \cdot 1.2 \cdot \sqrt{C_S \cdot \lambda}}{C_S \cdot \lambda^2}} = \sqrt{2.4 \cdot \sqrt{\frac{C_S \cdot \lambda}{C_S^2 \cdot \lambda^4}}} = \frac{\sqrt{2.4}}{\sqrt[4]{C_S \cdot \lambda^3}}, \tag{7.19}$$

i.e. the point resolution is given by

$$\delta_P = 0.645 \cdot \sqrt[4]{C_S \cdot \lambda^3}. \tag{7.20}$$

The radicand agrees with that of equation (2.18), the coefficient in front of the square root is smaller using the wave-optical approach (0.645 instead of 0.9). With the parameters mentioned in Fig. 7.10 we get a point resolution of 0.20 nm.

The second limitation is given by the space frequency which can be transferred independent on the oscillations. In our model the damping by the chromatic aberration is responsible for this limitation. There is a certain arbitrariness during the determination of this value. According to the usual practice we use the space frequency at which the amplitude of the contrast transfer function is reduced to $1/e^2 = 0.135$. The reciprocal value of this space frequency is termed *information limit* and can be calculated by

$$\delta_{lim} = 1.49 \cdot \sqrt{C_C \cdot \left(\frac{\Delta E}{E_0}\right) \cdot \lambda} \tag{7.21}$$

(Sect. 10.14). Following this equation the information limit for the parameter set of Fig. 7.10 amounts to 0.12 nm.

For a more accurate description of the contrast transfer our approximation explained here has to be completed by the interactive overlay of all the electron waves with different wave vectors, commonly considered by a transmission cross coefficient. This can be calculated numerically in software programs for HRTEM image simulations, a simple illustrative explanation of the details of this quite complex iteration seems to be impossible.

7.4 Periodic Distribution of Brightness in Pictures: Fourier Analysis

Obviously, for imaging of very small structures with sizes close to the resolution limit of the transmission electron microscope the space frequency plays an important role for the interpretation of image contrast. The name "frequency" is evocative of the harmonic analysis of oscillations, i.e. the notation of arbitrary periodic time functions with an oscillation period T by a sum of sine and/or cosine functions with different coefficients A_k and B_k as well as multiples of a basic frequency $\omega = 2\pi/T$ in the argument of the trigonometric functions:

$$f(t) = A_0 + \sum_{k=1}^{n} A_k \cdot \cos(k \cdot \omega \cdot t) + \sum_{k=1}^{n} B_k \cdot \sin(k \cdot \omega \cdot t). \quad (7.22)$$

The coefficients A_k and B_k are called *Fourier*[2] *coefficients*. They weight the sine and cosine functions of different frequencies and yield information about preferably arising periodicities. When we substitute the time t by the position x and the frequency ω by the space frequency $q = 1/d$ we get

$$f(x) = A_0 + \sum_{k=1}^{n} A_k \cdot \cos(2\pi \cdot k \cdot q \cdot x) + \sum_{k=1}^{n} B_k \cdot \sin(2\pi \cdot k \cdot q \cdot x), \quad (7.23)$$

i.e. the Fourier coefficients indicate which periodic distances (denoted by multiples of a basic frequency) dominate.

In other words: The Fourier analysis (or transformation) reveals information about arising periodicities. Within an image these are brightness fluctuations, e.g. in form of parallel stripes or lattices.

For digitalised images a numerical method is often used to determine which sine- or cosine-shaped masks with different periodic lengths describe best the brightness modulation within the image. Firstly, we would like to explain this method using only one image row.

---

[2] Joseph Fourier, French mathematician, 1768–1830.

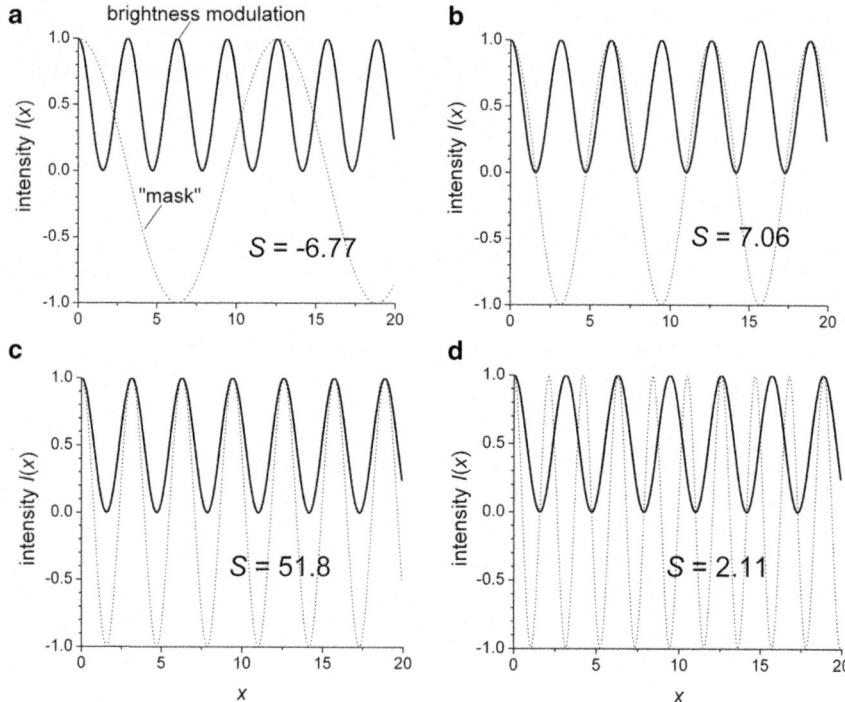

Fig. 7.11 Description of a periodic brightness modulation within an image row created by the square of a cosine function with frequency q_0. **a** Comparison with a mask created by the cosine function with frequency $q_0/2$. **b** With frequency q_0. **c** With frequency $2q_0$. **d** With frequency $3q_0$

In Fig. 7.11 the periodic brightness modulation within an image row is demonstrated by the square of a cosine function. This is overlaid by cosine functions with different periodic lengths as "masks". Obviously, the periodic length of the variant in Fig. 7.11c agrees very well with that one of the brightness modulation.

The "parameter of agreement" S for the space frequency q can be calculated by the relation

$$S(q) = \sum_k I(x_k) \cdot \cos(2\pi \cdot q \cdot x_k). \qquad (7.24)$$

The parameter S reaches its maximum for an optimum of agreement. In this way a numerical Fourier analysis is possible. The result is shown in a picture with S as brightness value (see Fig. 7.12). In this figure the single image row is expanded in vertical direction for better recognisability.

We recognise the inverse proportionality: the larger the periodic length of the brightness modulation the smaller the distance between the brightness maxima in the Fourier transform is.

7.4 Periodic Distribution of Brightness in Pictures: Fourier Analysis

Fig. 7.12 Image rows with brightness modulations of different periodic lengths and their corresponding Fourier transforms (*FT*)

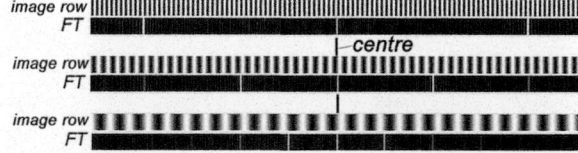

Fig. 7.13 Fourier transformation of a picture showing a quadratic lattice. **a** Original picture. **b** Fourier transformed image

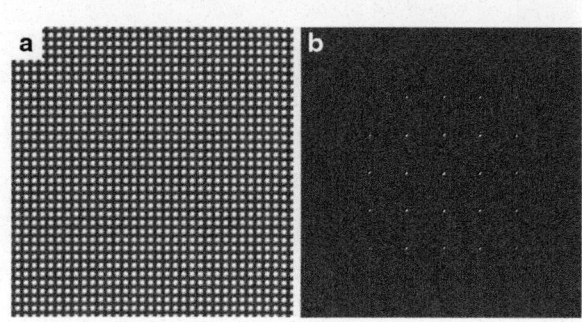

Let us transfer this method to a (two-dimensional) picture. The picture should have N_x pixels horizontally and N_y pixels vertically. The pixel brightness at position (x, y) is $I(x, y)$. Then, the brightness of the pixel (q_x, q_y) in the Fourier transformed image can be calculated by

$$I(q_x, q_y) = \sum_{x=0}^{N_x-1} \sum_{y=0}^{N_y-1} I(x, y) \cdot \cos(2\pi \cdot q_x \cdot x) \cdot \cos(2\pi \cdot q_y \cdot y). \quad (7.25)$$

q_x and q_y are space frequencies, i.e. reciprocal lengths. Instead of the cosine the sine function can also be used as well as a combination of both in form of a complex function

$$e^{2\pi \cdot i \cdot q \cdot x} = \cos(2\pi \cdot q \cdot x) + i \cdot \sin(2\pi \cdot q \cdot x). \quad (7.26)$$

In the latter case the results are complex numbers; their modulus is commonly displayed in an image. Such an image is also called *power spectrum*.

Figure 7.13 shows the result of the Fourier transformation of a lattice picture with quadratic meshes calculated by use of Eq. (7.25).

Back to the transmission electron microscope: Periodic atomic arrangements in crystalline samples generate distinct diffraction patterns in the back-focal plane of the objective lens. Following Bragg's law at small diffraction angles the distance between the position of a diffraction reflection and the centre is inversely proportional to the lattice spacing, i.e. directly proportional to the space frequency. In this way, the diffraction pattern can be understood as an analysis of the periodic lattice spacings in the sample. The Fourier transform of an aberration-free electron microscopic image of the lattice planes of a crystalline sample yields a similar information as the diffraction pattern of this sample.

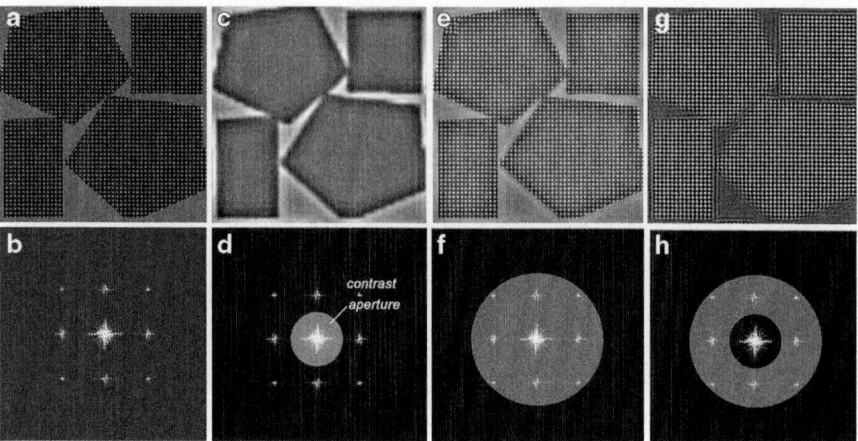

Fig. 7.14 Influence of the contrast aperture on the imaging properties (detailed explanations in the text). **a** Original picture. **b** Fourier transform of a). **c, e, g** Inverse Fourier transforms. **d, f, g** Different filters in the Fourier transforms. The Fourier transformation has been done using Gatan's software "Digital Micrograph", version 3.11.0, Gatan Inc. Pleasanton, USA

7.5 Mass Thickness and Phase Contrast

With the Fourier transformation into the reciprocal space and its inversion back into the real (position) space we have got a method for studying the influence of the contrast aperture on the electron microscopic image. The contrast aperture is positioned in the back-focal plane of the objective lens (cf. Sect. 6.2). In this plane the diffraction pattern is observable, too. We realise the analogy: A mathematical filter in the Fourier transform generates the same effect as a contrast aperture. After use of this filter and subsequent inverse transformation back into the position space we get the analogon to an electron microscopic micrograph obtained using a contrast aperture (Fig. 7.14).

Figure 7.14a represents the original picture, in electron microscope it would be the specimen. In this case there are four irregularly shaped crystallites on an amorphous support film. Within the "crystallites" the periodic atomic arrangement is modelled. The Fourier transform of this picture (Fig. 7.14b) shows the strong brightness periodicity within the "crystallites". After filtering in the Fourier space (also named reciprocal space) as shown in Fig. 7.14d and inverse transformation, in Fig. 7.14c we cannot see any atomic columns but only the blurred outer shape of the "crystallites". Transferred to the electron microscope this means that the waves diffracted at small details (atomic columns) into large angles are blocked by the contrast aperture, any interference with them becomes impossible. The information about the smallest structures in the specimen is lost. Only the smaller space frequencies are transferred, mass thickness contrast is visible.

Using a larger contrast aperture diffracted waves can also transmit the aperture hole, their interference is possible and the shape of the "crystallites" as well as the

7.5 Mass Thickness and Phase Contrast

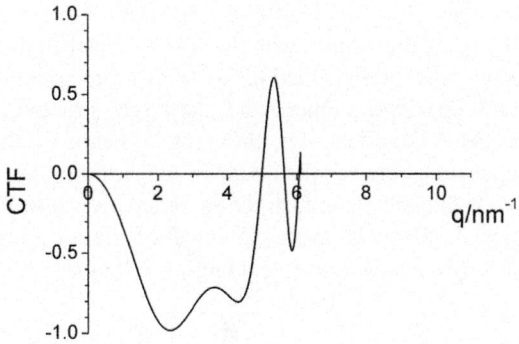

Fig. 7.15 Contrast transfer function CTF for $E_0=300$ keV, $\Delta E=0.7$ eV (Schottky field emission gun), $C_S=1.2$ mm, $C_C=1.5$ mm, $\Delta f=58$ nm (Scherzer focus) as well as a contrast aperture with a diameter of 35 μm

atomic columns within them are visible in the micrograph (Fig. 7.14e, f). We see the consequence: To get an atomic resolution a minimum size of the contrast aperture is needed dependent on the size of the structures which are to be imaged.

Finally, we would like to discuss a variant that cannot be easily performed in an electron microscope but (and this is an advantage of our mathematical filter) contributes to understanding of the problem: We imagine a ring-shaped aperture that blocks just small space frequencies (Fig. 7.14g, h). We observe the atomic columns with strong contrast while the edges of the "crystallites" can be only seen with the aid of the starting positions of the atomic columns.

The contrast transfer function shows the influence of the contrast aperture by a truncation of higher space frequencies. "Truncation" means that these space frequencies are not transferred into the image; the corresponding structure sizes are not visible in the electron microscopic image. An example is drawn in Fig. 7.15. The highest transferred space frequency follows from the radius of the contrast aperture. In the example is to $r_{CA}=17.5$ μm. Assuming that the focal length of the objective lens is equal to $f=1.5$ mm, the (half) aperture angle created by the contrast aperture is given by

$$\theta_{CA} = \frac{r_{CA}}{f} = 0.012 \text{ rad} = 12 \text{ mrad}. \tag{7.27}$$

Hence, the highest transferred space frequency is

$$q_{CA} = \frac{\theta_{CA}}{\lambda} = \frac{0.012}{1.97 \cdot 10^{-3} \text{ nm}} = 6.1 \text{ nm}^{-1}. \tag{7.28}$$

As disadvantage of the truncation the information limit cannot be reached, on the other hand the contrast transfer function oscillates to a lesser extend, i.e. under certain conditions the image interpretation becomes easier.

We learned that the contrast aperture needs to be large enough to obtain a high resolution. Nevertheless, it is beneficial for the imaging to insert an aperture to remove electrons diffracted to very high angles (i.e. which are meaning less for imaging).

Since the waves diffracted at small structures, e.g. lattice planes of a crystal, include larger angles with the optical axis than those which were diffracted at larger structures the first kind of waves is more strongly influenced by the spherical aberration. On the other hand, the larger structures are visible because of the mass thickness contrast. The different influence of the spherical aberration leads to a mismatch between the image of the lattice planes and the micrograph generated by mass thickness contrast, the so-called *delocalisation* (Sect. 10.16). This delocalisation δ_D depends on the spherical aberration (coefficient C_S), the defocus Δf, the electron wavelength λ, and the space frequency q characterising the lattice spacing:

$$\delta_D = \lambda \cdot q \cdot (C_S \cdot \lambda^2 \cdot q^2 + \Delta f). \tag{7.29}$$

δ_D can extend up to several nanometres and can make an accurate determination of the boundary between amorphous and crystalline sample areas difficult in high resolution electron microscopic images.

7.6 Contrast of Amorphous Samples

In Sect. 5.6 we described that the atoms within amorphous materials are not arranged lattice-like as known from crystals. There is no long-range order and no anisotropy in direction connected with a long-range order. Nevertheless, the atoms within an amorphous material are not randomly arranged. A so-called short-range order exists in form of most probable neighbour distances due to the preferred bond lengths. From Sect. 5.6 we also know that this short-range order can be determined by deconvolution of scattering curves measured in the diffraction mode of the transmission electron microscope.

In a reckless manner one could assume that these preferable distances can be imaged in the transmission electron microscope with high resolution at high magnification. But this assumption is absolutely wrong! We would like to discuss the reason for this.

Let us have a look at Fig. 7.16. In both parts of this figure the same sample area is imaged. Anyhow, the appearance is completely different: In the Fig. 7.16b the structures seem to be larger than in Fig. 7.16a on the left side, although between both images only the focus was changed. Obviously, we do not observe any sample structures but the property of the microscope imaging system. For a better understanding we think about the images of amorphous structures we have to expect. Therefore, we remember the contrast transfer by the objective lens.

If we look through a thin amorphous layer in a gedankenexperiment we will observe only the projection of the three-dimensional atomic positions onto the image plane. In the real-space the preferable distances cause periodic fluctuations of the atomic density, in the projection these density fluctuations are lost by averaging. This means that a wide variety of space frequencies is provided to be transferred by the objective lens in different ways (Fig. 7.17).

7.6 Contrast of Amorphous Samples

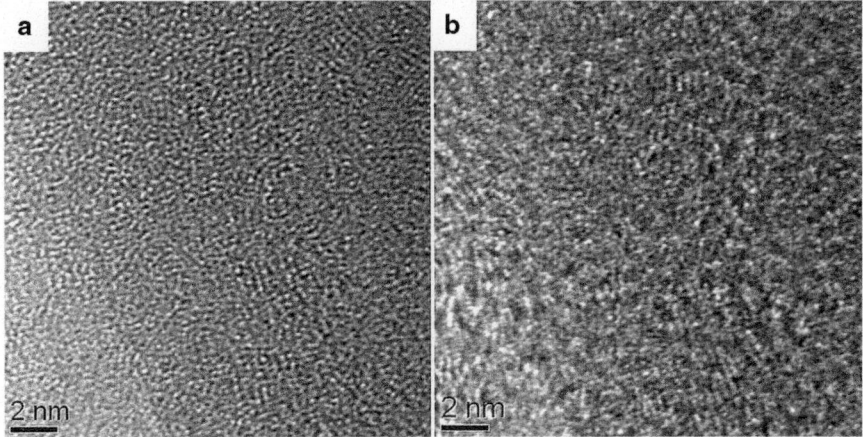

Fig. 7.16 Electron microscopic micrographs of a thin amorphous silicon film (parameters: $E_0=300$ keV, $C_S=1.2$ mm, $C_C=1.2$ mm, $\Delta E \approx 1$ eV (Schottky field emission gun). **a** Focussed. **b** Defocussed

The picture series in Fig. 7.17 was generated without any electron microscope. By means of an image processing program a picture with statistical ("white") noise was created (Fig. 7.17a). This picture does not show any contrast in agreement with our expectation on an electron microscopic micrograph of such an amorphous layer. Since there is no periodic contrast in this picture there are also no significant brightness maxima in the Fourier transform (Fig. 7.17b). In Fig. 7.17c the contrast transfer function using the microscope parameters described in Fig. 7.16 and the Scherzer focus is drawn. In this function the first three space frequency ranges are highlighted which are transferred with the same (negative) contrast. In the Fourier transform the distance to the centre is proportional to the space frequency. The oscillation of the contrast transfer function can be modelled by a series of mathematical annular filters in the Fourier transform. In Figs. 7.17d (for $\Delta f = 58$ nm) and 7.17g (for $\Delta f = 20$ nm) three of these annular filters are used. The inverse transformation back into the position space delivers the electron microscopic image which has to be expected.

In fact, we see that the defocussed image in Fig. 7.17h shows coarser (quasi) structures than the picture of Fig. 7.17e calculated for the Scherzer focus. The reason for it is the special kind of filtering by the contrast transfer function. Similar to the noisy starting picture within an amorphous sample all atomic distances exist in the projected plane. The objective lens filters some of this variety and images that with high contrast. In principle, the contrast transfer function describes this behaviour. In highly magnified electron microscopic images of amorphous samples we do not see any "amorphous structures" but a reflection of the contrast transfer function.

Often electron microscopic high resolution images are "Fourier-filtered" to reduce the noise. By doing so, one has to make sure that untypical details are not emphasised to be typical.

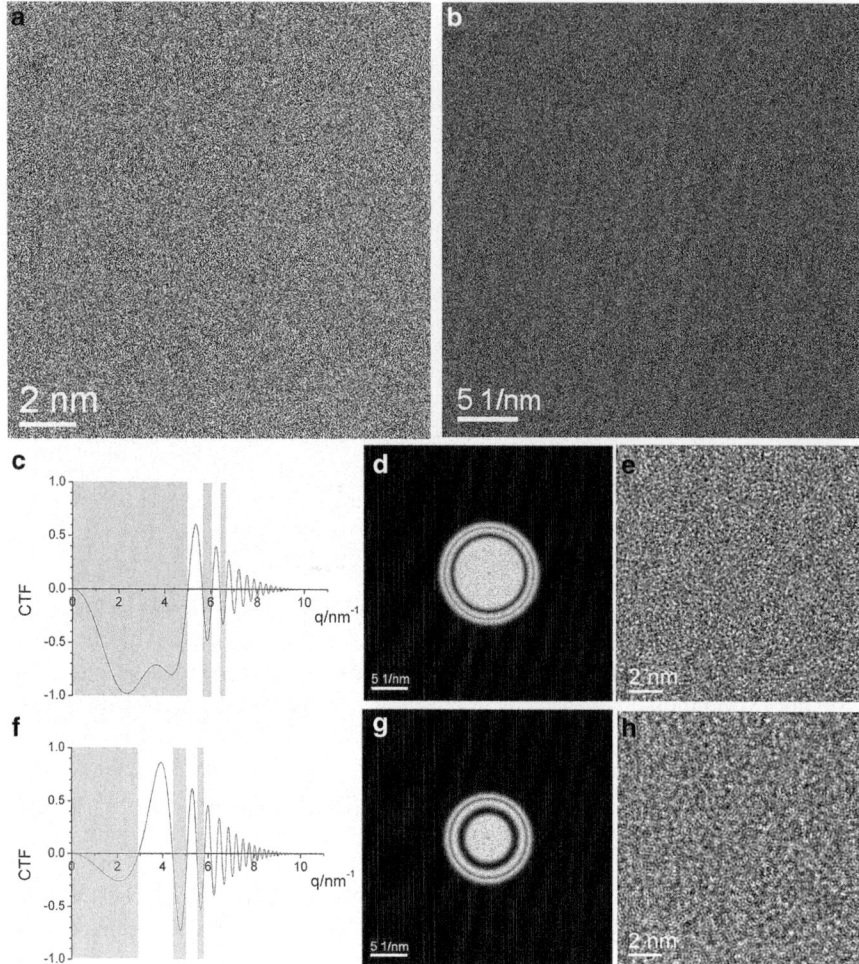

Fig. 7.17 Simulation of highly magnified micrographs of thin amorphous films for the demonstration of the influence of the contrast transfer function. **a** Computer-created, noisy picture. **b** Fourier transform. **c** Contrast transfer function for the parameter given in Fig. 7.16 at Scherzer focus ($\Delta f = 58$ nm). **d** Use of the greyly underlined windows in the contrast transfer function ("filter") in the Fourier-transformed image. **e** Inverse transformation of the filtered Fourier transform back into the position space. **f–h** As before, but for defocus $\Delta f = 20$ nm

7.7 Correction of Astigmatism

Following the explanations of the previous paragraph the Fourier transformation of an electron microscopic micrograph of a thin amorphous film reflects the contrast transfer function. We would like to analyse this fact more precisely by means of Fig. 7.18. Because of the oscillations of the contrast transfer function space frequencies

7.7 Correction of Astigmatism

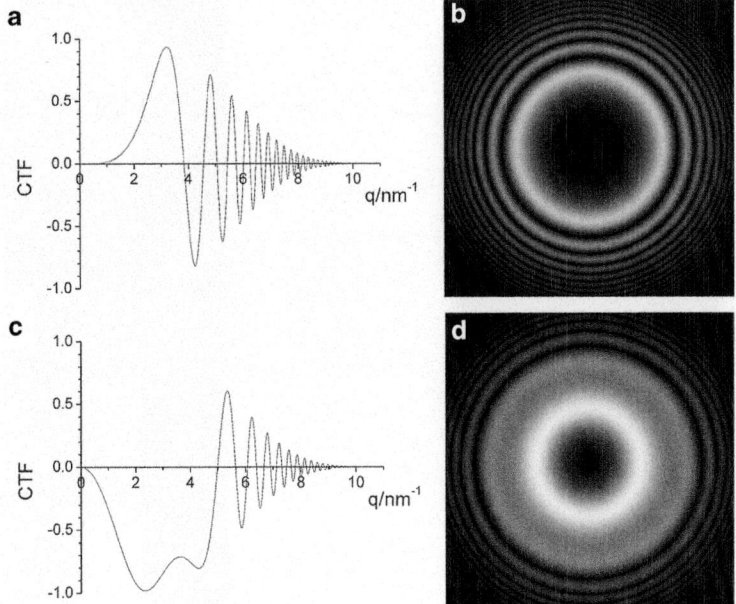

Fig. 7.18 Contrast transfer function and power spectra for defocus $\Delta f = 0$ (**a** and **b**) as well as $\Delta f = 58$ nm (Scherzer focus at $U_0 = 300$ kV and $C_S = 1.2$ mm—**c** and **d**)

are transferred with different efficiencies. The image intensity can be interpreted as the quadratic of the contrast transfer function.

Furthermore, let us rotate the function "around the optical axis", i.e. the power spectrum contains additionally information about the contrast transfer function at different azimuthal angles.

But, what happens for instance if the contrast transfer function is different in vertical (azimuth 90°) and horizontal (azimuth 0°) direction since there is a focal length difference between both sectional planes? We remember: This is the indication of astigmatism (cf. Sect. 2.3). In this case the distances of the brightness maxima from the centre are different in dependence on the azimuth within the power spectrum, the rings are distorted. Bizarrely shaped patterns are possible, at suitable foci nearly ellipses are generated (Fig. 7.19).

To correct the astigmatism a thin amorphous film is used as specimen. Often a small amorphous edge or area as part of the investigated sample is suitable for the correction.

By means of a CCD camera a highly magnified picture has to be recorded and synchronously Fourier-transformed. After setting of a proper defocus (some rings must be observable within the Fourier transform) we are able to recognise any astigmatism looking at the shape of the rings.

The (objective) stigmator (corrective for the image astigmatism) must be adjusted in a way that the rings become circular (Fig. 7.19e). Experimentally, one starts with a strong underfocus (more rings are visible) and refines while going closer to

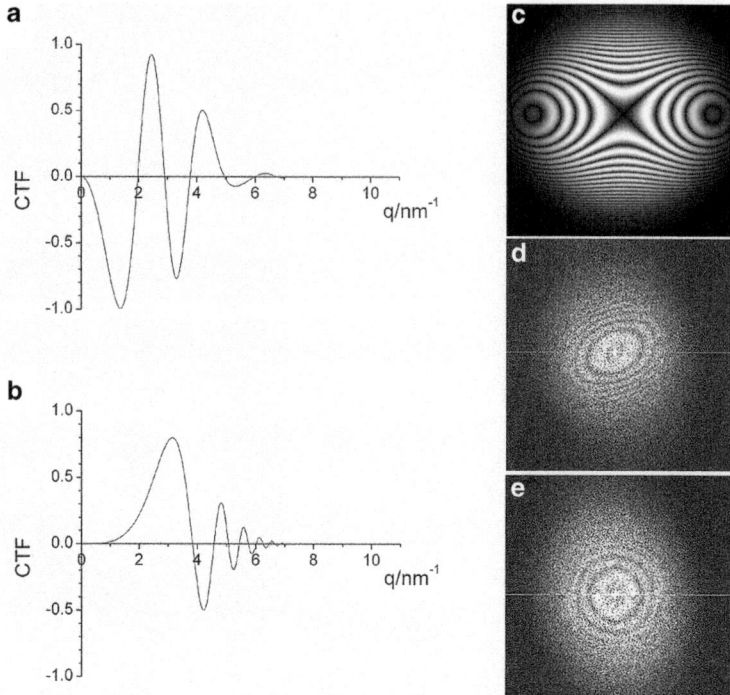

Fig. 7.19 Contrast transfer functions and power spectra with astigmatism. **a** Contrast transfer function in the horizontal sectional plane ($\Delta f = 0$ nm). **b** Contrast transfer function in the vertical sectional plane ($\Delta f = 140$ nm). **c** Resulting calculated power spectrum. **d** Fourier-transformed measured image of an amorphous Si film with astigmatism. **e** As **d** but after correction of astigmatism

the focus (cf. e.g. [2]). A multipole is the most important electron optical element of the stigmator. Its principle and its operation are described in Sect. 10.17.

7.8 Measurement of the Resolution Limit

Until the 1970s special specimens had been used to measure the resolution limit of a transmission electron microscope. These specimens covered heavy metal particles with sizes in the sub-nanometre range to deliver a mass thickness contrast sufficiently high for photographic recording. Since the electron-optical magnification of transmission electron microscopes was limited to about 100,000 in that time the pictures had to be photographically magnified up to 10 times. To exclude that the "dark points" arise from the grains of the photo plate two pictures on two different plates were taken under the same conditions (defocus). Small distances between "dark points" were only classified to be significant for the resolution limit if they could be seen in exactly the same arrangement on both plates. Additionally, the distances had to be proven in different directions to exclude the influence of astigmatism.

Fig. 7.20 Young's interference pattern (Fourier transform of an amorphous edge picture (thinned Si sample) demonstrating the measurement of infomation limit. The circle marks a space frequency of 6.7 nm^{-1}. The Si(220) reflex amounts to 5.2 nm^{-1}

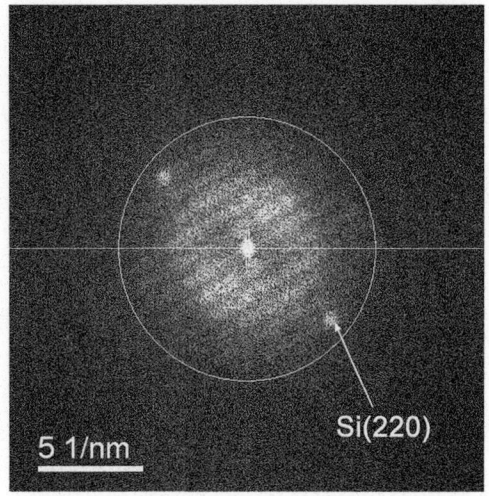

Later on, lattice fringe images, e.g. of small gold crystals, had been used to determine the resolution limit. Therewith only discrete values were detectable, just the selected lattice spacings. Furthermore, there is the danger to misinterpret double diffraction as imaging resolution.

Nowadays the resolution limit (more precisely: the information limit) is experimentally determined by the analysis of "Young[3] fringes". A thin amorphous film preferably of comparably heavy atoms serves as specimen. While the image is being recorded the specimen jumps by only a few nanometres within the object plane. This is normally done by a proper deflection of the electron beam ("image shift"). Alternatively, it is also possible to record two pictures with an image shift between them and subsequent overlay (addition) these pictures.

The next step is a Fourier transformation of this resulting image. We observe stripes with decreasing contrast towards the edge (Fig. 7.20).

The image shift generates an additional periodic brightness modulation in the direction of the shift which is displayed by the stripes within the Fourier transform. All distances given in the recorded image are shifted; however, their contrast decreases towards smaller distances (i.e. larger space frequencies) because of the damping of the contrast transfer function. Distances which are not imaged (they are smaller than the information limit) do not cause any brightness modulations, for the space frequencies connected to them no Young's interference pattern is observable. The fringes improve the visibility and facilitate a good distinction between background noise and information.

The reciprocal value of the largest space frequency detectable by the Young fringes is classified to be the information limit. It is marked by the white circle in Fig. 7.20. In this measurement the information limit is equal to 0.15 nm.

---

[3] Thomas Young, English polymath, 1773–1829.

This method is simple in its application and is used to certificate the specifications of an electron microscope, for instance. The impossibility to distinguish between different influences on the information limit is one point of criticism of this method. Besides the damping by the chromatic aberration, e.g. specimen drift, specimen thickness, and electronic instabilities influence the contrast of the Young fringes, too [3]. Furthermore, also this method is not free of influences of double diffraction.

7.9 Correction of Spherical and Chromatic Aberration

Astigmatism, spherical and chromatic aberration share one property: Electrons running off-axis through the lens field are wrongly deflected. At the two-fold astigmatism (cf. Sect. 2.3), whose correction we discussed in Sect. 7.7, this concerns the electrons in two perpendicular sectional planes. The correction can be in principle done by use of a quadrupole that drags the electrons in one sectional plane (that with the shorter focal length) to the outside, and squeezes them to the inside in the plane with the longer focal length. To tilt and to centre the field within the corrector without any mechanical movement an octupole can be used (Sect. 10.17).

The spherical aberration is expressed by the fact that in rotational-symmetric lens fields the electrons transmitting off-axis areas are deflected more than necessary (cf. Sect. 2.3). For correction a multipole is needed that, similar to the astigmatism correction in the plane of short focal length, drags the electrons to the outside. Nevertheless, this must happen in all azimuthal planes. Following H. Rose [4] a hexapole can be used for that. Unfortunately, the hexapole creates astigmatic intermediate images; an additional hexapole is needed to correct this. Basically, correctors for the spherical aberration ("C_S correctors") consist of pairs of multipoles and transfer lenses which ensure that the intermediate image planes are created "at the right positions".

The dependence of the refractivity on the velocity, i.e. on the wavelength of the electrons is the reason for the chromatic aberration. To understand the possibility of this correction we use a strongly simplified model of a magnetic electron lens: We think about the trajectory of an electron having the velocity v_{z0} within a homogenous magnetic field of the induction B_y (Fig. 7.21a).

The electron incident in z direction into the magnetic field is exposed to the Lorentz force

$$\mathbf{F_L} = -e \cdot \mathbf{v} \times \mathbf{B} \qquad (7.30)$$

and, with consideration of the directions indicated in Fig. 7.21,

$$\mathbf{F_L} = -e \cdot v_{z0} \cdot B_y \cdot \mathbf{e}_z \times \mathbf{e}_y = e \cdot v_{z0} \cdot B_y \cdot \mathbf{e}_x = F_{xL} \cdot \mathbf{e}_x. \qquad (7.31)$$

Inserted into the equation of motion it follows for the x component:

7.9 Correction of Spherical and Chromatic Aberration

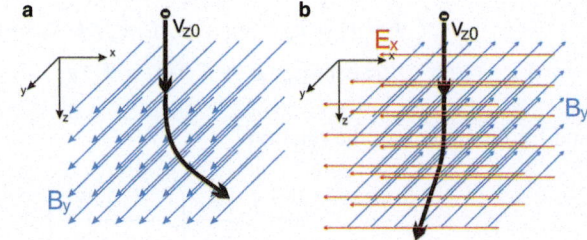

Fig. 7.21 Trajectory of electrons within (homogenous) fields. **a** Magnetic field. **b** Crossed magnetic (B_y) and electric (E_x) fields. (Wien(Wilhelm Wien, German physicist, 1864–1928, Nobel prize in physics in 1911) filter)

$$a_x = \frac{F_{xL}}{m_0} = \frac{e}{m_0} \cdot v_{z0} \cdot B_y$$

$$v_x = \int a_x \cdot dt = \frac{e}{m_0} \cdot v_{z0} \cdot B_y \cdot t + v_{x0} = \frac{e}{m_0} \cdot v_{z0} \cdot B_y \cdot t \quad (7.32)$$

$$x = \int v_x \cdot dt = \frac{1}{2} \cdot \frac{e}{m_0} \cdot v_{z0} \cdot B_y \cdot t^2 + x_0 = \frac{1}{2} \cdot \frac{e}{m_0} \cdot v_{z0} \cdot B_y \cdot t^2$$

(e/m_0: specific charge of the electron). For the z component it is:

$$\begin{aligned} a_z &= 0 \\ v_z &= v_{z0} \\ z &= v_{z0} \cdot t. \end{aligned} \quad (7.33)$$

Inserted into Eqs. (7.32) it delivers the wanted equation $x(z)$ of the trajectory:

$$x = \frac{1}{2} \cdot \frac{e}{m_0} \cdot \frac{B_y}{v_{z0}} \cdot z^2 \quad (7.34)$$

and, consequently, for the deflection angle:

$$\tan \alpha = \frac{dx}{dz} = \frac{e}{m_0} \cdot \frac{B_y}{v_{z0}} \cdot z, \quad (7.35)$$

i.e. slower electrons (longer wavelength) are more strongly deflected than faster electrons.

To correct this speed-dependent deflection we need a second magnetic field with reverse direction. However, as a consequence each deflection would be compensated. The goal is to compensate only the "wrong deflection" of the electrons with another velocity than v_{z0}. What we need is a counterforce to the Lorentz force. This counterforce is generated by an electric field E_x (Fig. 7.21b). The total force experienced by the electron is given by

$$\begin{aligned}\mathbf{F} = \mathbf{F_L} + \mathbf{F_E} &= -e \cdot v_{z0} \cdot B_y \cdot \mathbf{e_z} \times (-\mathbf{e_y}) + e \cdot E_x \cdot \mathbf{e_x} \\ &= e \cdot (-v_{z0} \cdot B_y + E_x) \cdot \mathbf{e_x} = F_x \cdot \mathbf{e_x}\end{aligned} \quad (7.36)$$

and for the trajectory the relation

$$x = \frac{1}{2} \cdot \frac{e}{m_0} \cdot \left(-\frac{B_y}{v_{z0}} + \frac{E_x}{v_{z0}^2} \right) \cdot z^2 \quad (7.37)$$

is derived and for the deflection angle

$$\tan \alpha = \frac{e}{m_0} \cdot \left(-\frac{B_y}{v_{z0}} + \frac{E_x}{v_{z0}^2} \right) \cdot z, \quad (7.38)$$

respectively. Electrons with the velocity v_{z0} are not deflected by the crossed electric and magnetic field if

$$B_y = \frac{E_x}{v_{z0}}. \quad (7.39)$$

If the electrons are faster ($v_z > v_{z0}$), we get

$$\tan \alpha = \frac{e}{m_0} \cdot \left(-\frac{E_x}{v_{z0} \cdot v_z} + \frac{E_x}{v_z^2} \right) \cdot z < 0, \quad (7.40)$$

are they slower ($v_z < v_{z0}$)

$$\tan \alpha > 0. \quad (7.41)$$

Using crossed electric and magnetic fields (Wien filter) it is in principle possible to correct the different refraction of electrons with different wavelengths. In reality, the construction of a corrector for the chromatic aberration is much more complicate. But the basic principle of the crossed fields preserves. They are realised in form of alternately arranged electric and magnetic dipoles within complex quadrupole and octupole systems [4]. The stability of the electronic systems has to satisfy extreme requirements [5].

From Sects. 7.3, 7.6, and 10.14 we know that the spherical aberration as well as the chromatic aberration both influence the contrast transfer function. In principle, this function is reflected in the power spectrum (i.e. after Fourier transformation) of highly resolved micrographs of thin amorphous films. Similar to the way for the correction of the astigmatism spherical and chromatic aberration can be measured by means of such power spectra. However, a single power spectrum is not enough for this purpose. Several images under different tilt and azimuthal angles have to be recorded (as at the conical darkfield imaging without contrast aperture—cf. Sect. 6.3) and Fourier transformed ("Zemlin tableau" [6, 7]). Consequently, the aberration coefficients can be calculated from the pattern in this tableau. Because of the remarkable calculation effort powerful computers are needed. This is one reason why such correctors could not be commercially obtained before about 2000.

Our discussion of imaging aberrations included just two-fold astigmatism and spherical aberration. However, there are also higher order aberrations which become important after correcting the low order ones. To really improve the imaging quality also some of the higher order aberrations have to be corrected or at least controlled. This can be done by modern correctors, too.

7.10 Interpretation of High Resolution TEM Images

What can we do for understanding and correct interpretation of high resolution images containing space frequencies in the oscillating range of the contrast transfer function without the help of an aberration corrected microscope shifting this critical range to higher space frequencies?

There are three possibilities:

First Starting point is an assumed atomic arrangement resulting from thermodynamics or other fundamentals. This arrangement is transferred into a "super cell" with periodic junctions at its borders (i.e. the atomic arrangements at the edges of the super cell agree) and the expected transmission electron microscopic image is calculated considering the contrast transfer function. Unfortunately, there are at least two parameters commonly unknown with the necessary accuracy: specimen thickness and defocus. Therefore, these two parameters are varied and a matrix of images is calculated [8]. The comparison with the measured image shows if the assumed atomic arrangement is proper to explain the electron microscopic image. Maybe, the super cell has to be changed to model the experiment. This "trial-and-error" method can be very time-consuming.

Second Using the electron microscope an image series at stepwise changed defocus is acquired, thereout the contrast transfer function and the object exit wave, respectively, can be reconstructed ("focal-series reconstruction" or "exit wave reconstruction" [9, 10]). Commonly, the problem is an insufficient stability of the specimen. Its mechanical drift in the object plane can be computationally corrected, but not its possible deformation. An uncontrolled movement of the specimen in z direction (along the optical axis) changes the defocus value, and consequently the defocus step size is not known with sufficient accuracy.

Third The operator calculates "backwards", i.e. he reconstructs the exit wave from the micrograph of the specimen. The influence of the aberrations can be computationally corrected. The wave is defined by its amplitude and phase. In the image brightness only the amplitude is recorded, therefore we need additional information about the phase. That can be obtained after interference of the object exit wave with a reference wave and record of the interference pattern. The reference wave does not penetrate the specimen (in case of thinned samples it transmits the hole or at FIB lamellas an area beside the edge of them). In the plane of the aperture for selected area diffraction a positively charged wire is positioned deflecting object and reference wave to each other, and both of them are overlaid ("holography" [11].) By means of the holography not only high resolution TEM images can be interpreted but also electric and magnetic fields within or at the edge of the specimen can be measured, for instance.

References

1. Stadelmann, P.: EMS—a software package for electron diffraction analysis and HREM image simulation in materials science. Ultramicroscopy **21**, 131–146 (1987). (used Java-EMS-Version: 6.6201U2011)
2. Möbus, G., Phillipp, F., Gemming, T., Schweinfest, R., Rühle, M.: Quantitative diffractometry at 0.1 nm resolution for testing lenses and recording media of a high-voltage atomic resolution microscope. J. Electron Microsc. **46**(5), 381–395 (1997)
3. Barthel, J., Thust, A.: Quantification of the information limit of transmission electron microscopes. Phys. Rev. Lett. **101** 200801-1–200801-4 (2008)
4. Rose, H.: Correction of aberrations, a promising means for improving the spatial and energy resolution of energy-filtering electron microscopes. Ultramicroscopy **56**, 11–25 (1994)
5. Haider, M., Müller, H., Uhlemann, S., Zach, J., Loebau, U., Hoeschen, R.: Prerequisites for a Cc/Cs-corrected ultrahigh-resolution TEM. Ultramicroscopy **108**, 167–178 (2008)
6. Haider, M., Uhlemann, S., Schwan, E., Rose, H., Kabius, B., Urban, K.: Electron microscopy image enhanced, Nature **392**, 768–769 (1998)
7. Zemlin, F., Weiss, K., Schiske, P., Kunath, W., Herrmann, K.-H.: Coma-free alignment of high resolution electron microscopes with the aid of optical diffractograms. Ultramicroscopy **3**, 49–60 (1978)
8. Möbus, G., Schweinfest, R., Gemming, T., Wagner, T., Rühle, M.: Iterative structure retrieval techniques in HREM: a comparative study and a modular program package, Journ. of Microscopy **190**(1/2), 109–130 (1998)
9. Thust, A., Lentzen, M., Urban, K.: Non-linear reconstruction of the exit plane wave function from periodic high-resolution electron microscopy images, Ultramicroscopy **53**, 101–120 (1994)
10. Thust, A., Coene, W. M. J., Op de Beek, M., Van Dyck, D.: Focal-series reconstruction in HR-TEM: simulation studies on non-periodic objects. Ultramicroscopy **64**, 211–230 (1996)
11. Lichte, H., Lehmann, M.: Electron holography—basics and applications. Rep. Prog. Phys. **71**, 1–46 (2008)

Chapter 8
Let Us Switch to Scanning Transmission Electron Microscopy

Abstract For analytical investigations in the transmission electron microscope (TEM) a nanoscaled illuminated (i.e. excited) area is wanted to maintain a high spatial resolution. The electron optics of the illumination system within a TEM (condenser system and a part of the objective field in front of the specimen) is able to focus very small electron probes with sizes down to the 0.1 nm range on the specimen plane. Deflection units allow to scan this small electron probe on the sample analogous to the method known from the conventional scanning electron microscope. Similar to the name "transmission electron microscope (or microscopy)—TEM" an abbreviation of the method is common practice: *STEM*. It stands for "*scanning transmission electron microscope*" or "*scanning transmission electron microscopy*".

Distinctions from the conventional scanning electron microscopy are on one hand the achievable smallness of the electron probe: Because of the advantageous electron optical conditions in the TEM (very small working distance) very small probe diameters can be reached. On the other hand, because of the electron-transparent specimen the contrast mechanism is based on the same principles as in conventional ("fixed beam") transmission electron microscopy: Weakly scattered electrons reach the detector, strongly scattered electrons do not reach it ("STEM brightfield image") or vice versa ("STEM darkfield image"). The electron-transparent specimens have a special advantage: Contrary to bulk samples routinely used in conventional scanning microscopy a widespread excited area cannot be generated within the thin film. This allows to reach a resolution limit of better than 0.1 nm in the STEM mode, too[1].

8.1 What Happens Electron-Optically?

At first sight it seems to be simple to focus the electron beam onto the sample and to scan it across the specimen row by row: We set the focal length of the condenser lens 2 (in case of a double condenser system) to convergent illumination as known

---

[1] The fundamental idea of the scanning electron microscope was published by M. von Ardenne in 1938 [1]. The STEM variant including the field emission gun was suggested by A. Crewe in 1968 [2]. Overview about STEM developments, see S. J. Pennycook [3].

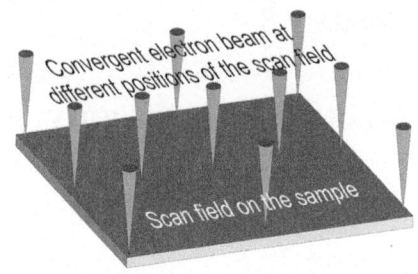

Fig. 8.1 Parallel incident direction of the convergent electron beam as a precondition for the STEM operating of the transmission electron microscope

from the convergent beam electron diffraction (cf. Sects. 2.7.1 and 5.3). For scanning a deflection system pivots the beam in two perpendicular directions. Using this approach two problems arise:

1. The condenser 2 lens is a lens with a long focal length. Consequently, its spherical aberration is comparably large (cf. Sect. 2.3). As we will see later (Sect. 8.2), it is impossible to reach a probe diameter of less than some nanometres due to the spherical aberration in this case.
2. The convergent electron beam must have equal incident directions independent on its position on the sample (Fig. 8.1). A simple deflection in two perpendicular directions does not completely satisfy this requirement.

To explain the scanning in more detail we have to define our model of the objective lens more precisely. Up to this point we have assumed that only one coil generates the magnetic field of the objective lens. On the other hand, we know that the specimen is positioned within the pole piece, i.e. we can split the magnetic field into two parts: a part in front of the specimen ("pre-field") and a part behind the specimen ("post-field"). Using more than one coil the pre-field and the post-field can be adjusted independently on each other. Following our refined model the objective lens consists of two parts: the objective pre-field and the objective post-field. The specimen is positioned between both, i.e. close to the pre-field. By means of this concept both problems mentioned above can be overcome.

We use the pre-field as an additional condenser lens, this procedure is called "*nanoprobe mode*" in the electron microscopic practice. Because of the very short working distance (distance between the last lens in front of the specimen and the specimen itself) the pre-field has a short focal length and consequently a small spherical aberration.

The parallel shift of the convergent electron beam during the scan on the specimen can be reached if the pivot point of the deflection unit is exactly positioned in the front focal point of the objective pre-field (Fig. 8.2). We remember: *Rays coming from the front focal point precede parallel to the optical axis in the image space.*

Since the contrast is mainly generated by different scattering of electrons within the sample it must be ensured that weakly and strongly scattered electrons are spatially divided after passing the specimen and the lenses. This allows their selection using an electron-sensitive detector with limited detection area. The spatial division

8.2 Resolution or: What is the Smallest Diameter of the Electron Probe?

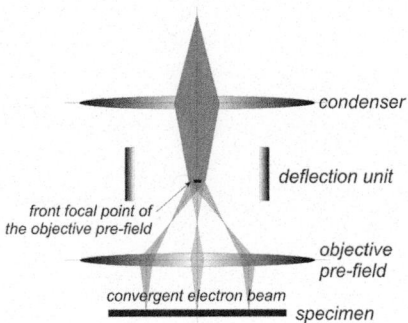

Fig. 8.2 Sketch of the ray path within the illumination system (including objective pre-field) in the STEM mode

of the different directions happens in the back-focal plane of the objective lens, following our model of the two-tier lens field in the back-focal plane of the post-field (cf. Sect. 2.7.2). The detector for the STEM signal is positioned close to the plane of the observation screen. To image the back-focal plane of the objective to the observation plane the diffraction mode is needed. Summarising, the electron-optical preconditions for high resolution STEM are *nanoprobe mode* for the illumination system and *diffraction mode* for the projective settings.

Since scanning transmission electron microscopic imaging is not a "true" optical imaging but an electronic signal processing the signal-to-noise ratio plays an important role. Therefore, during the preparation of the microscope for STEM an excellent alignment of the electron gun is a crucial precondition (cf. Sect. 4.3). It can be advantageous to check the alignment state in the nanoprobe mode before switching into the STEM mode and to improve this state if necessary.

8.2 Resolution or: What is the Smallest Diameter of the Electron Probe?

A small electron probe needed for the STEM method is generated by multistage demagnification of the cross over created by the electron gun. Independent on the specific kind and arrangement of the demagnification lenses some fundamentals have to be considered in this paragraph. The resolution in the STEM is in the range of the probe diameter in the sample plane.

Let the demagnified diameter of the cross over in the sample plane be d_{co}. Because of the conservation of the richtstrahlwert

$$R = \frac{j}{\Omega} \qquad (8.1)$$

(j: current density, Ω: solid angle of electron emission) along the ray path without apertures (cf. Sect. 2.6) in the probe plane one gets

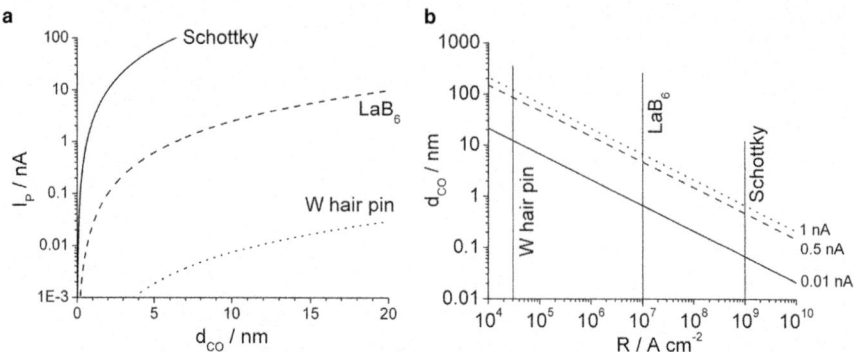

Fig. 8.3 Effects of the richtstrahlwert on beam current and minimum probe diameter in the STEM mode calculated for $\alpha = 10$ mrad. **a** Beam current and probe diameter for three richtstrahlwerts which are typical for the denoted cathodes. **b** Richtstrahlwert and probe diameter for three different beam currents

$$R = \frac{4 \cdot I_P}{\pi \cdot d_{co}^2} \cdot \frac{1}{\pi \cdot \alpha^2} \tag{8.2}$$

(α: beam aperture, i.e. half angle of the beam width) with the assumption of a circular probe cross section and beam current I_P. Hence, it follows for the probe diameter:

$$d_{co} = \sqrt{\frac{4 \cdot I_P}{R \cdot \pi^2 \cdot \alpha^2}} = \frac{2}{\pi \cdot \alpha} \cdot \sqrt{\frac{I_P}{R}}. \tag{8.3}$$

This equation is of broad importance: It is a relation between the probe diameter and the beam current, i.e. the smaller the probe is the smaller the beam current will be (Fig. 8.3a). On the other hand, to get a sufficient signal-to-noise ratio a minimum beam current is needed often restricting the smallness of the probe (Fig. 8.3b).

We realise the dramatic differences between the three kinds of cathodes. Assumed for the STEM imaging a beam current of 0.01 nA is necessary, then by use of a Schottky field emission cathode a probe diameter of about 0.1 nm is reachable, for the LaB_6 cathode ca. 1 nm can be achieved, and by use of a simple tungsten hairpin 10 nm. Analytical measurements require higher beam currents. 0.5 nA are reached at a probe diameter of 0.5 nm with Schottky field emission cathode, 5 nm are possible with LaB_6 cathode, and 85 nm with tungsten hairpin cathode. Analytics on the nanometre scale (we mean "true" nanometre areas) is only possible by a transmission electron microscope equipped with field emission gun.

Similar to the resolution limit of the transmission electron microscope in the fixed beam mode the electron-optical demagnification is impaired by aberrations, too. We would like to discuss the influence of the aberration by diffraction, spherical and chromatic aberration as well as astigmatism (cf. also Sect. 2.3).

8.2 Resolution or: What is the Smallest Diameter of the Electron Probe?

The diameter of the aberration disk by diffraction is given by

$$d_D = 1.22 \cdot \frac{\lambda}{\alpha} \tag{8.4}$$

with the assumption of very small apertures α and with the electron wavelength λ. The diameter d_s of the spherical aberration disk can be calculated by

$$d_S = 2 \cdot C_S \cdot \alpha^3. \tag{8.5}$$

for focussing into the Gaussian image plane and with C_s as spherical aberration coefficient. However, the Gaussian image plane does not represent the optimal image plane distance. As shown in *Sect. 10.18* the radius in the plane of least confusion is given by

$$r_S = \frac{1}{4} \cdot C_S \cdot \alpha^3, \quad \text{i.e.} \quad d_S = \frac{1}{2} \cdot C_S \cdot \alpha^3. \tag{8.6}$$

Analogously, the radius of the aberration disk generated by the chromatic aberration is given by

$$r_C = \frac{1}{2} \cdot C_C \cdot \left(\frac{\Delta E}{E}\right) \cdot \alpha, \quad \text{i.e.} \quad d_C = C_C \cdot \left(\frac{\Delta E}{E}\right) \cdot \alpha \tag{8.7}$$

in the plane of least confusion (ΔE: energy spread of the electrons, E: energy of the primary electrons, C_C: chromatic aberration coefficient). For the astigmatism

$$r_A = \frac{1}{2} \cdot \Delta f_A \cdot \alpha, \quad \text{i.e.} \quad d_A = \Delta f_A \cdot \alpha \tag{8.8}$$

is valid with Δf_A as astigmatic difference of the focal lengths.

To approximate the reachable probe size d we add the squared influence terms. We ignore the fact that the planes of least confusions of the different aberrations can be located at different positions. We get:

$$d^2 = d_{co}^2 + d_B^2 + d_S^2 + d_C^2 + d_A^2 \tag{8.9}$$

and, with consideration of Eqs. (8.5)–(8.8):

$$d^2 = \frac{1}{\alpha^2} \cdot \left(\frac{4 \cdot I_P}{\pi^2 \cdot R} + 1.5 \cdot \lambda^2\right) + \frac{C_S^2 \cdot \alpha^6}{4} + \left[C_C^2 \cdot \left(\frac{\Delta E}{E}\right)^2 + (\Delta f_A)^2\right] \cdot \alpha^2. \tag{8.10}$$

In theory, without aberrations the probe diameter could be unlimitedly demagnified by increase of the aperture α. In practice this is excluded since the lens fields and the deflection units properly work only close to the optical axis. This is why the aperture is limited to 50 mrad ... 100 mrad.

Considering the three mentioned aberrations we recognise a contrary dependence of the terms on the aperture: The first term includes α in the denominator, the second and third ones in the numerator. The consequence is an optimal aperture generating a minimum probe diameter:

$$\alpha_{opt} = \sqrt[4]{\frac{2}{3 \cdot C_S} \left[\sqrt{\left(\frac{K_{CA}}{C_S}\right)^2 + \frac{12 \cdot I_P}{\pi^2 \cdot R} + 4.5 \cdot \lambda^2} - \frac{K_{CA}}{C_S} \right]} \quad (8.11)$$

with $K_{CA} = C_C^2 \cdot \left(\frac{\Delta E}{E}\right)^2 + (\Delta f_A)^2$

(mathematical procedure Sect. 10.18). Using a field emission cathode (small energy spread of the electrons) and after correction of the astigmatism the spherical aberration dominates. In this case Eq. (8.11) is simplified to

$$\alpha_{opt} = \sqrt[8]{\frac{16 \cdot I_P}{3 \cdot \pi^2 \cdot C_S^2 \cdot R} + \frac{2 \cdot \lambda^2}{C_S^2}} \quad (8.12)$$

and the minimum probe diameter amounts to

$$d = \sqrt[8]{C_S^2 \cdot \left(\frac{16 \cdot I_P}{3 \cdot \pi^2 \cdot R} + 2 \cdot \lambda^2\right)^3}. \quad (8.13)$$

At modern instruments a correction of the spherical aberration even of the condenser system is possible by multipoles (cf. Sect. 7.9). For dominating chromatic aberration and astigmatism the term with the power of six is cancelled in equation (8.10). In this case the optimal aperture is given by

$$\alpha_{opt} = \sqrt[4]{\frac{1}{K_{CA}} \cdot \left(\frac{4 \cdot I_P}{\pi^2 \cdot R} + 1.5 \cdot \lambda^2\right)} \quad (8.14)$$

and the minimum probe diameter (neglecting higher order aberrations) amounts to

$$d = \sqrt[4]{4 \cdot K_{CA} \cdot \left(\frac{4 \cdot I_P}{\pi^2 \cdot R} + 1.5 \cdot \lambda^2\right)}. \quad (8.15)$$

These relations are illustrated at an example in Fig. 8.4. Since nano-analytics is only possible by use of a field emission gun we confine ourselves to the Schottky cathode. We recognise the advantage of the spherical aberration correction even for

8.2 Resolution or: What is the Smallest Diameter of the Electron Probe?

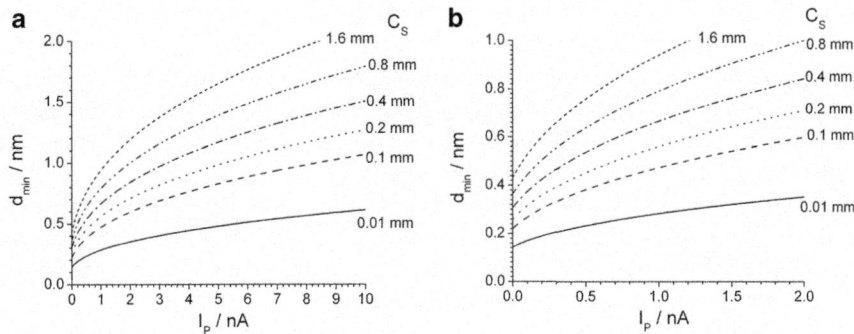

Fig. 8.4 Relationship between minimum probe diameter d_{min} and beam current IP at different coefficients C_S of spherical aberration. Parameters: $E=300$ keV, $\Delta E=0.7$ eV, $C_C=1.5$ mm, $R=10^9$ A/cm^2, $\Delta f_A=0$. **a** Overview. **b** Detail interesting for high resolution imaging and nano-analytics

analytics: For an assumed minimum beam current of 0.5 nA the minimum probe diameter amounts to 0.76 nm at a spherical aberration constant of 1.6 mm and after correction of the spherical aberration (here: $C_S=10$ μm $=0.01$ mm) to 0.23 nm, however[2].

The demagnification of the cross over is determined by the excitation of the condenser 1 lens ("spot size"). Its selection is firstly determined by the size of the structures which are to be imaged. For atomic resolution the probe must be smaller than the distances between the atomic columns (provided the aberrations are small enough to fulfil this demand). Meanwhile we know that the beam current decreases with decreasing probe diameter and, therefore, the signal-to-noise ratio impairs. Insofar the selection of the spot size equals the search for a good compromise considering the smallness of the electron probe and a sufficient signal-to-noise ratio. Of course, the signal-to-noise ratio can also be improved by a longer measuring time, i.e. by a longer dwell time of the probe on one scan pixel. As consequence the scan for the complete image needs more time. In case of nanostructures this can be a problem because of the specimen drift. Consequently, we have to consider the stability of the specimen in our compromise, too.

The goal of focussing in the STEM mode is to minimise the probe diameter in the object plane by a suitable setting of the focal lengths of condenser lenses and objective pre-field, i.e. the cross over has to be imaged into the object plane. Commonly, the quality of this setting can be observed by the sharpness of the scanned picture.

The magnification of the scanning method is equal to the ratio of the size of the scan field on the monitor and that on the specimen. Changing the magnification requires only another amplitude of the electron probe deflection on the specimen without any electron-optical influence on the probe size. Therefore, it is advantageous

---

[2] O.L. Krivanek et al. introduced a C_S-corrector especially for a "dedicated STEM" without any TEM part [4].

to focus the image at magnifications as high as possible. But be careful, please: Let us remember the useful magnification (see Sect. 1.4). At a (by spot size) adjusted probe size of 1 nm the useful magnification amounts to about 100,000. At higher magnifications it is impossible to get a sharp image without diminishment of the probe size. Please note additionally that the indicated STEM magnifications are often only relative to an unknown reference.

To correct the astigmatism the focal length of the objective pre-field is periodically changed. Any astigmatism can be seen as preferred direction in the scanned image which is turned by 90° during the periodic change of the focal length (in case of the normally dominating two-fold astigmatism).

At coherent illumination, i.e. using a field emission gun, there is an additional criterion to assess the sharpness: the *Ronchigram*[3]. This is the magnified diffraction disk of zero order at convergent illumination. While the fixed electron probe is penetrating an amorphous sample area the Ronchigram looks like the power spectrum of a highly magnified TEM image of an amorphous film. The rings visible in the Ronchigram reach their maximal diameters at optimal focussing; in case of astigmatism they are elliptically distorted.

Finally, we have to consider that the resolution of a digitalised picture is also determined by the number of pixels, i.e. by the number of points dwelled by the probe during the scan. The optimal number of pixels is given when the step width of the probe on the specimen is equal to the probe diameter.

In summary: The resolution of a STEM image is an interplay of the three components spot size, sampling range, and number of dwelled points.

8.3 Contrast in the Scanning Transmission Electron Microscopic Image

As for transmission electron microscopic fixed beam imaging the interaction between beam (primary) electrons and specimen atoms is the basis of the image contrast in the STEM mode, too. Different to conventional scanning microscopy our specimen is electron-transparent.

In Sect. 6.1 we introduced the mean free path Λ_{el} for elastic scattering as statistical quantity describing the scattering ability of a material characterised by density ρ and atomic number Z in dependence on the primary electron energy E_0 and the scattering angle α. For small angles $\alpha << 1$ we get

$$\Lambda_{el} = \frac{16 \cdot \pi \cdot \varepsilon_0^2}{e^4} \cdot \frac{E_0^2 \cdot \alpha^2}{Z^2 \cdot N_A \cdot \rho} \tag{8.16}$$

(ε_0: permittivity of vacuum, e: elementary electric charge, N_A: Avogadro's constant).

---

[3] Vasco Ronchi, Italian physicist (optics), 1897–1988.

8.3 Contrast in the Scanning Transmission Electron Microscopic Image 171

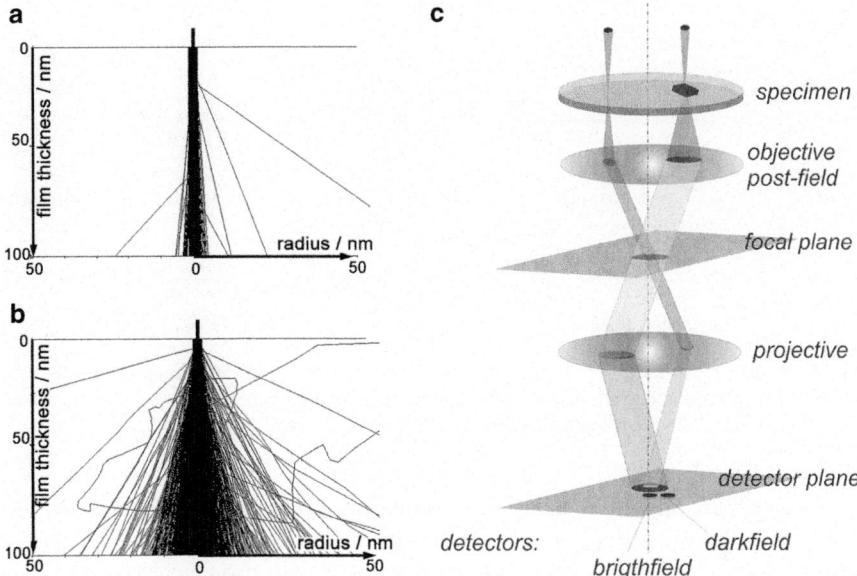

Fig. 8.5 Contrast in a STEM image. **a** Scattering of 200 keV electrons in a 100 nm thick carbon film. **b** Same but in a gold film (equal abscissa and ordinate scales as in part a). **c** Schematic illustration of the ray path below the specimen

In Fig. 8.5 the results of Monte-Carlo simulations following a model by D.C. Joy [5] are shown, demonstrating the different scattering abilities of carbon (Fig. 8.5a) and gold (Fig. 8.5b).

The consequence for the scanning transmission electron microscopic imaging is illustrated in Fig. 8.5c: The left part of the specimen consists of material with low atomic number, e.g. carbon. The electron scattering is weak; the electron beam is only weakly widened. A small intensity disk is generated in the centre of the back-focal plane of the objective post-field. This disk is imaged into the centre of the final image plane by the diffraction mode. The electron-sensitive (brightfield) detector is positioned in the centre of the last plane; the entrance aperture of this detector has approximately the same size as the intensity disk. The detector records the complete intensity of the weakly scattered electron beam and generates bright pixels on the monitor. The size of the "brightfield disk" is determined by the diameter of the condenser 2 aperture and the camera length. These three values must be synchronised to get a "true" brightfield STEM image.

Next to or around the brightfield detector a similar darkfield detector is arranged that receives only a small signal in the described case. This is changed if the electron beam meets material of a high atomic number, e.g. gold as shown on the right side of the specimen in Fig. 8.5c. The scattering is more strong and the electron beam is significantly more widened. The size of the intensity disk generated in the centre of the back-focal plane and in the final image plane is larger than for scattering in the lighter material.

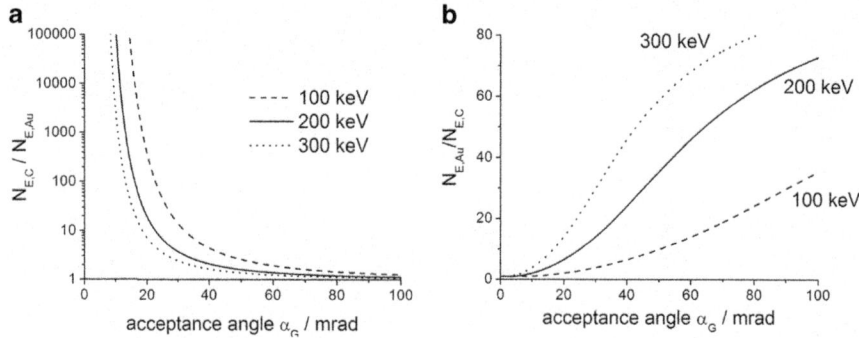

Fig. 8.6 Ratio of the numbers of electrons inside (**a**) and outside (**b**) an acceptance angle α_G at scattering in carbon and gold, respectively. Parameters: specimen thickness $s=100$ nm, carbon: $Z=6$, $M_r=12$, $\rho=2.26$ g/cm^3; gold: $Z=79$, $M_r=197$, $\rho=19.3$ g/cm^3

However, the brightfield detector acquires only a part of the intensity, now. The pixels from this sample area are darker than those from the weakly scattered part of the specimen. On the other hand, the larger intensity disk also reaches the darkfield detector; it delivers a higher signal than before.

We would like to confirm this circumstance quantitatively. Following the definition of the mean free path Λ_{el} for elastic scattering the ratio of the electrons deflected into angle α_G at specimen thickness s (cf. Eq. (6.4)) and the materials 1 and 2 is given by

$$\frac{N_{E,1}}{N_{E,2}} = \frac{e^{-s/\Lambda_{el,1}}}{e^{-s/\Lambda_{el,2}}} = e^{s\left(\frac{1}{\Lambda_{el,2}}-\frac{1}{\Lambda_{el,1}}\right)} \tag{8.17}$$

and, with consideration of (8.16),

$$\frac{N_{E,1}}{N_{E,2}} = e^{\frac{s \cdot e^4}{16 \cdot \pi \cdot \varepsilon_0^2} \cdot \frac{1}{E_0^2 \cdot \alpha^2}\left(Z_2^2 \cdot N_{A2} \cdot \rho_2 - Z_1^2 \cdot N_{A1} \cdot \rho_1\right)} \tag{8.18}$$

From this one derives the tailored quantity equation

$$\frac{N_{E,1}}{N_{E,2}} = e^{0.00392 \frac{s/\text{nm}}{\alpha_G^2 \cdot (E_0/\text{keV})^2}\left(Z_2^2 \cdot \frac{\rho_2/(\text{g}\cdot\text{cm}^{-3})}{M_{r,2}} - Z_1^2 \cdot \frac{\rho_1/(\text{g}\cdot\text{cm}^{-3})}{M_{r,1}}\right)} \tag{8.19}$$

In consequence the ratio depends on the following parameters: specimen thickness s, acceptance angle α_G as well as materials properties represented by atomic number Z, density ρ, and atomic or molecular weight M_r.

Regarding our already discussed case carbon (material 1) and gold (material 2) the result is shown in Fig. 8.6. Figure 8.6a represents the use of the brightfield

8.4 Speciality: High Angle Annular Darkfield Detector

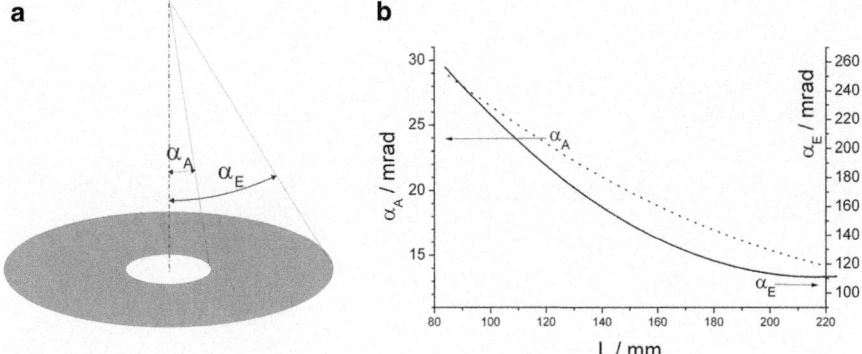

Fig. 8.7 High angle annular darkfield (HAADF) detector. **a** Sketch defining the angles. **b** Example of the dependence of the angle range acquired by the detector on the camera length L

detector. The ratio of the number of electrons scattered at carbon to those scattered at gold increases with decreasing acceptance angle α_G, i.e. the contrast becomes the better the smaller the entrance aperture of the detector is. Simultaneously the intensity decreases limiting the reduction of the aperture size. In practice the acceptance angle of the brightfield detector amounts to 5 mrad … 10 mrad.

For the darkfield imaging the strongly scattered electrons are used as brightness signal. The ratio of the electrons scattered at material 1 and 2, respectively, with a scattering angle greater than α_G is given by

$$\frac{N_{E,2}}{N_{E,1}} = \frac{1-e^{-s/\Lambda_{el,2}}}{1-e^{-s/\Lambda_{el,1}}}. \tag{8.20}$$

To obtain a contrast value greater than 1 we use the reciprocal ratio (gold is brighter than carbon in the darkfield image). In this case the contrast increases with increasing acceptance angle (Fig. 8.6b).

8.4 Speciality: High Angle Annular Darkfield Detector

The darkfield detector described up to now has a serious disadvantage: It acquires only a small azimuthal part of the strongly scattered electrons. The better alternative is an annular detector normally positioned above the observation screen. Often this detector is named "*HAADF detector*" as abbreviation for "*High Angle Annular Dark Field detector*". Using this, strongly scattered electrons can be recorded with high efficiency. The acquired range of scattering angles is influenced by the camera length (cf. Sect. 5.3; Fig. 8.7).

Figure 8.7 shows that scattering angles larger than 200 mrad can be captured by means of the HAADF detector advantageous for a high STEM contrast. For a

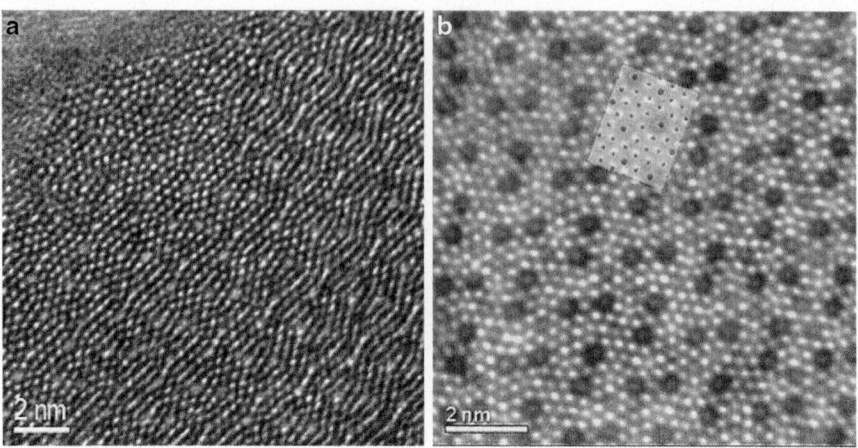

Fig. 8.8 Electron microscopic imaging of Cs(Nb, W)$_5$O$_{14}$ in [001] direction. **a** High resolution fixed beam image, i.e. conventional TEM imaging. **b** STEM-HAADF micrograph with inserted unit cell (in this unit cell the atomic positions are marked by small dark circles)

proper STEM darkfield image the good adjustment of camera length and illumination aperture (condenser 2 aperture) is absolutely necessary. The intensity disk of the weakly scattered electrons should be small enough that these electrons do not reach the annular detector. Otherwise we get a mixture of brightfield and darkfield image with a dominating brightfield signal.

Figure 8.8b shows an example for a STEM image recorded by an HAADF detector. We recognise the single atomic columns.

The comparison with the conventional fixed beam TEM image (Fig. 8.8a) illustrates the advantage of the STEM imaging: The excellent agreement with the inserted unit cell projected in [100] direction shows that the atomic columns can directly be localised at the position of the bright points. The dependence of the contrast on the specimen thickness and the focus is drastically reduced, i.e. the image interpretation becomes easier.

Furthermore, the brightness of the atomic columns in an HAADF-STEM image recorded at a reasonable camera length (often 100 mm ... 150 mm) is proportional to the squared atomic number, i.e. to Z^2. Hence, sometimes the term Z^2-imaging is used. However, often the brightness is proportional to $Z^{1.7}$ in fact.

References

1. von Ardenne, M.: Das Elektronen-Rastermikroskop. Theoretische Grundlagen. Z. Physik. **109**(9–10), 553–572 (1938)
2. Crewe, A.V., Wall, J., Welter, L.M.: A high-resolution scanning transmission electron microscope. J. Appl. Phys. **39**(13), 5861–5868 (1968)

3. Pennycook, S.J.: Seeing the atoms more clearly: STEM imaging from the Crewe era to today. Ultramicroscopy **123**, 28–37 (2012)
4. Krivanek, O.L., Dellby, N., Lupimi, A.R.: Towards sub-Å electron beams. Ultramicroscopy **78**, 1–11 (1999)
5. Joy, D. C.: Monte Carlo Modeling for Electron Microscopy and Microanalysis. Oxford University Press, New York (1995)

Chapter 9
Let us Use the Analytical Possibilities

Abstract In the description of the various imaging methods and contrast phenomena in the previous chapters the elastic scattering of electrons in a solid, i.e. the interaction without energy loss of the electrons, was used. Inelastic scattering processes with energy losses of the beam electrons were undesirable. They lead to changes of the electron wavelength and, consequently, to a loss of resolution by chromatic aberration and to a higher background in the diffraction patterns.

On the other hand, inelastic interactions are often element-specific, i.e. their results depend on the element to which the interacting atom belongs. Hence, a chemical analysis on the nanometre scale seems to be possible using this inelastic interaction. In this chapter we explain the inelastic scattering process, we describe the spectrometers and the methods needed to utilise this process in practice, and finally, we demonstrate these methods with the help of some examples.

9.1 Analytical Signals by Inelastic Interaction

Up to now we have only considered the Coulomb interaction between the atomic nuclei and the beam (primary) electrons without any energy transfer from the primary electrons to the atoms (elastic scattering). To understand which possibilities of energy transfer to the electron shell exist we need a model describing the properties of the atomic electron shell. The consequence for the primary electron is obvious: It loses energy. The atomic electron shell changes its energetic state and can e.g. emit characteristic X-rays. Electron energy losses and the energy of the X-ray radiation are measured in analytical transmission electron microscopy and are used for the chemical composition analysis. Therefore we confine ourselves to these two topics. Our plausible explanations are based on different atomic models. To describe the X-ray emission Rutherford-Bohr's[1] model is sufficient. Discussing the energy losses this model needs to be enhanced by binding-specific details.

---

[1] Niels Bohr, Danish physicist, 1885–1962, Nobel prize in physics in 1922.

9.1.1 Emission of X-rays

From the electrodynamics (Maxwell's equations) we know that accelerated electric charges generate fluctuating magnetic fields which generate fluctuating electric fields again. These periodically fluctuating fields are well-known as electromagnetic waves created during the acceleration of electrons. We remember the motion of electrons within a Coulomb force field of an atomic nucleus (Sect. 10.11): The electrons are deflected, i.e. their velocity direction changes. Each change of the velocity needs an acceleration; the same is true for a change of direction, too. It means that the deflection of electrons causes emission of electromagnetic radiation expanding with the velocity of light c. Its energy E determines the wavelength λ (Sect. 10.2):

$$E = h \cdot \frac{\omega}{2 \cdot \pi} = h \cdot \frac{c}{\lambda} \quad (9.1)$$

(h: Planck's constant). Electromagnetic radiation with wavelengths between 0.05 nm and 10 nm is named X-radiation (or—wave, respectively). The energy present in this radiation comes from the deflected electron, i.e. the electron is decelerated and the radiation is named "*bremsstrahlung*"[2].

The wavelengths of the bremsstrahlung establish a continuum; the energy cannot be higher than the primary energy E_0 of the beam electrons. Because of the inverse proportionality of energy and wavelength there is a lower short-wave limit λ_{min} of the bremsstrahlung:

$$\lambda_{min} = \frac{h \cdot c}{E_0} \quad (9.2)$$

The probability (cross section) to generate bremsstrahlung depends on its energy. Its intensity can be approximately described by Kramers'[3] equation

$$I_{Br}(E) = C \cdot Z \cdot \left(\frac{E_0}{E} - 1 \right) \quad (9.3)$$

([1], C: constant including, among others, the current of the primary electrons and the specimen thickness, Z: atomic number). We realise that the intensity of the bremsstrahlung increases with the atomic number and hyperbolically decreases with increasing energy. In measurements the efficiency of the detector has to be additionally considered. It is also energy-dependent and increases for increasing X-ray energies in the low energy range.

The bremsstrahlung does not contain element-specific signals but it influences the background of the X-ray spectrum (this is the intensity of the X-radiation in

---

[2] from the German words "bremsen" (to decelerate) and "Strahlung" (radiation).
[3] Hendrik Anthony Kramers, Dutch physicist, 1894–1952.

9.1 Analytical Signals by Inelastic Interaction

dependence on its energy or wavelength, respectively) and is able to complicate the quantitative interpretation of the spectra especially in the low-energy range.

However, the inelastic interaction between beam electrons and atoms can have another consequence: changes within the atomic electron shell. To understand this we have to deal more precisely with this shell and use Rutherford-Bohr's model. Starting point is the concept that the (one-fold) negatively charged electrons orbit the positively charged nucleus where the number of electrons agrees with the number of the (one-fold) positively charged protons within the nucleus and is equal to the atomic number in the periodic table of elements. In sum, the atom is electrically neutral. The Coulomb force acts as a radial force and holds the electrons on their orbit. This concept includes two problems:

1. Why are the positively charged protons within the nucleus not disassembled by the Coulomb force?

2. At the beginning of this chapter we pointed out that accelerated charge carriers emit electromagnetic radiation. Electrons on an orbit are accelerated charge carriers. They should be decelerated and finally fall into the nucleus.

The first problem is solved by the concept that the atomic core does not only include positive charges but also "gluons", i.e. a kind of cement holding the protons together. The electric repulsive force increases with the atomic number, more gluons are needed than can be delivered only by the protons. This dilemma is solved by insertion of additional particles without charge but with gluons, the neutrons, into the nucleus.

Solving the second problem we remember the wave character of the electrons. We assume that the electron moves on a circular orbit around the nucleus. Because of its wave character the perimeter of the orbit must be equal to a whole-number multiple of the electron wavelength, otherwise the wave and even also the electron itself cannot exist due to destructive interference:

$$2\pi \cdot r = n \cdot \lambda \quad (9.4)$$

(r: radius of the orbit, n: integer number). The radial force is equal to the Coulomb force (v: velocity, m_0: mass of the electron, e: elementary electric charge, ε_0: permittivity of vacuum, Z: atomic number):

$$m_0 \cdot \frac{v^2}{r} = \frac{1}{4\pi \cdot \varepsilon_0} \cdot \frac{Z \cdot e^2}{r^2} \quad (9.5)$$

With

$$E = \frac{m_0}{2} \cdot v^2 \quad (9.6)$$

as (kinetic) electron energy it follows:

$$E = \frac{1}{8\pi \cdot \varepsilon_0} \cdot \frac{Z \cdot e^2}{r} \quad (9.7)$$

and with consideration of de Broglie's Eq. (1.10) in non-relativistic approximation

$$\lambda = \frac{h}{\sqrt{2 \cdot m_0 \cdot E}} \tag{9.8}$$

(h: Planck's constant) as well as with (9.4):

$$E_n = \frac{1}{n^2} \cdot \frac{m_0}{8} \cdot \left(\frac{Z \cdot e^2}{\varepsilon_0 \cdot h}\right)^2 \tag{9.9}$$

Since n is an integer number the energy values given by Eq. (9.9) are discrete. We add the index n to point to this circumstance.

To avoid misunderstandings we would like to emphasise that E_n represents the *kinetic energy* of the shell electron. This is the higher the smaller the orbit radius is. The *potential energy* (the "potential") is given by

$$W_P(r_e) = -\frac{Z \cdot e^2}{4 \cdot \pi \cdot \varepsilon_0} \cdot \int_{r_e}^{\infty} \frac{dr}{r^2} = -\frac{Z \cdot e^2}{8 \cdot \pi \cdot \varepsilon_0} \cdot \frac{1}{r_e} \tag{9.10}$$

within an electrostatic force field, i.e. because of the negative sign the potential (or the energy level, respectively) becomes higher with increasing radius r_e. Later on, we will discuss single energy levels or energy states. These refer to the potential energy.

The wave character of electrons has an additional consequence: Two identical electrons, i.e. two electrons with exactly equal wave functions would interfere and interact. There are possibilities to conceive unequal but similar orbits (or wave functions): The orbits can be elliptical, the electrons can run in different directions, and the electrons can spin clockwise or counter-clockwise ("*spin*"). Similar energy levels are combined in shells. The different possibilities are described by four *quantum numbers*: The *principal quantum number* n denotes the orbit radius (or the longer half-axis of an elliptical orbit, respectively) and therefore the shell (n=1: K-shell, n=2: L-shell, n=3: M-shell etc.); the *orbital quantum number* l denotes the orbital angular momentum ($l=0,...,$ n−1; $l=0$: s-electrons, $l=1$: p-electrons, $l=2$: d-electrons, $l=3$: f-electrons etc.); the *magnetic quantum number* m denotes the z component of the orbital angular momentum (m=$-l$ ... $+l$), and the *spin quantum number* s the intrinsic angular momentum (s = $\pm 1/2$). The electrons must be different at least in one of these quantum numbers (Pauli's principle). Interestingly, a strong similarity with the planetary system can be seen, however on a completely different length scale.

From the quantum number classification the maximum number N_n of electrons in the shells can be derived (see also Table 9.1):

$$N_n = 2 \cdot n^2 \tag{9.11}$$

9.1 Analytical Signals by Inelastic Interaction

Table 9.1 Quantum numbers for the occupancy of the K- and the L-shell

Name of the shell	n	l	Name of the electron	m	s
K-shell (2 electrons)	1	0	1s	0	+1/2
	1	0	1s	0	−1/2
L-shell (8 electrons)	2	0	2s	0	+1/2
	2	0	2s	0	−1/2
	2	1	2p	−1	+1/2
	2	1	2p	−1	−1/2
	2	1	2p	0	+1/2
	2	1	2p	0	−1/2
	2	1	2p	1	+1/2
	2	1	2p	1	−1/2

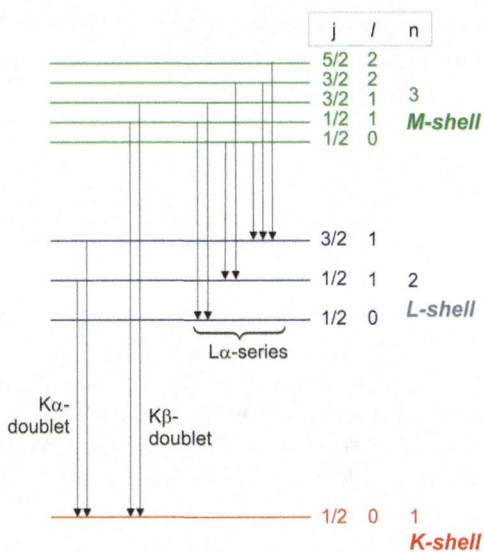

Fig. 9.1 Energy levels (schematic) of the first three shells K, L, and M including permitted transitions with emission of radiation

The total momentum quantum number is $j = l + s$. Transitions between different energy levels connected with the emission of radiation are quantum-mechanically only permitted for $\Delta l = \pm 1$ and $\Delta j = 0, \pm 1$.

The possible electron energy states of the inner three shells including their permitted transitions connected with emission of radiation are shown in Fig. 9.1. Because of Pauli's principle an ionised inner energy level is needed at the beginning which can then be occupied again by the mentioned transition as a second step. This is the point to come back to the inelastic interaction between beam electrons and atom: The beam electron can provide the "free inner energy level" by suitable energy transfer, it is able to ionise a core shell.

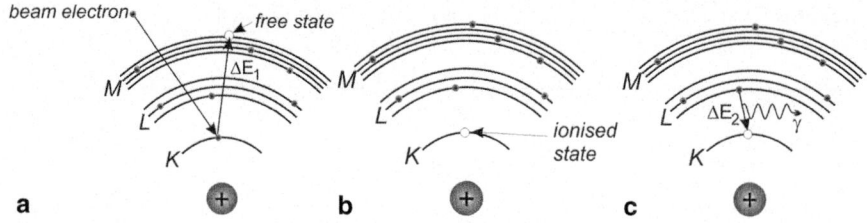

Fig. 9.2 Progress of the emission of characteristic X-radiation. **a** The beam electron (primary electron) ionises the K-shell. The electron positioned at this level before is shifted at least to the first (lowest) free state (energy difference ΔE_1). **b** The K-shell is ionised, i.e. one state is free. **c** An electron of the L-shell occupies the ionised state in the K-shell (transition with emission of radiation of energy ΔE_2)

Assuming one place on the K-shell (n=1) becomes free and is occupied again by a subsequent transition of an electron from the L-shell (n=2) with emission of radiation. Simplifying the procedure, we neglect the small differences between the energy levels within one shell and use Eq. (9.9) to evaluate the energy difference between the two shells K and L:

$$\Delta E_{n1,n2} = \frac{m_0}{8} \cdot \left(\frac{Z \cdot e^2}{\varepsilon_0 \cdot h}\right)^2 \cdot \left(\frac{1}{n_1^2} - \frac{1}{n_2^2}\right) \qquad (9.12)$$

For instance in case of chromium (Z=24) we get an energy difference of $9.413 \cdot 10^{-16}$ Nm or 5876 eV. Following Eq. (9.2) this corresponds to a wavelength of 0.211 nm, i.e. the emitted radiation is X-radiation. Since the differences of the energy levels depend on the element, the energy of the X-rays generated in this manner is element-specific, too. Contrary to the bremsstrahlung this element-specific X-radiation is named *characteristic X-radiation*.

9.1.2 Electron Energy Losses

After discussion of the characteristic X-ray emission as a result of changes within the electron shell in consequence of the interaction between primary electrons and specimen atoms we would like to deal with the change of the beam electron energy. At first view it seems to be simple: The beam electron loses exactly this energy which is transferred to the X-ray quants. In fact, this is correct regarding the bremsstrahlung. For emission of characteristic X-radiation the statement is wrong. Effectively, only a part of the electron energy loss is transferred to the X-ray quants within the two-step procedure (Fig. 9.2).

We assume that a primary electron loses the energy ΔE_1. This energy is used to ionise the K-shell, i.e. to shift one electron into a higher energy level or, perhaps,

9.1 Analytical Signals by Inelastic Interaction

to remove this electron completely from the atomic force field and giving it an additional kinetic energy (*secondary electron*). Compared with this the emitted X-ray quant has only the energy ΔE_2 adequate to the difference of an already occupied state (e.g. within the L-shell) and the K-shell energy which is less than ΔE_1. For instance, to ionise the K-shell of oxygen at least the energy of 532 eV is needed, the O-K X-ray line, however, implies only an energy of 523 eV.

The energy losses of the beam electrons depend on the energy levels of the electron shells. There are more possibilities than described up to now. So far, we have only considered single atoms. But, within the specimen there are plenty of atoms, in crystalline materials they are arranged periodically and closely together. Such an arrangement has consequences for the electron shells, i.e. we have to refine our atomic model. In this context the question for the fundamentals of the chemical binding arises. What does hold the atoms together? Actually, we have to expect a Coulomb repulsive force, when the atomic electron shells approach each other, i.e. we expect that the atoms drift apart.

However, this is avoided if the overlap of the electron shells leads to an energy reduction, i.e. the total energy of the binding state is less than the sum of the energies in the separated state. It is principally equal to a sledge on a sledding hill: Its potential energy in the valley is smaller than on the hill and the downhill force drives the sledge downwards. The stable state means that the sledge stands in the valley and the total energy becomes a minimum.

All the possibilities of binding share the overlap of the electron shells. One variant presumes that the outer shell of an atom is only weakly occupied. A typical example is sodium with atomic number 11. Its K-shell contains two s-electrons (notation: $1s^2$), the L-shell two s-electrons and six p-electrons ($2s^2\ 2p^6$), and the M-shell one s-electron ($3s^1$ or 3s). Commonly, for labelling the notation $1s^2\ 2s^2\ 2p^6\ 3s$ is used. On the other hand, chlorine with atomic number 17 has the electron configuration $1s^2\ 2s^2\ 2p^6\ 3s^2\ 3p^5$. Only one 3p-electron is missing to reach the energetic favourable electron configuration of a noble gas. This electron is given by the sodium resulting in two ions with opposed charge attracting each other (*ionic bond*). It is essential that this process leads to a reduced total energy, i.e. it generates a changed energy level typical for this kind of binding.

In another possibility of overlap some of the electrons are members of two atomic electron shells, e.g. in the hydrogen molecule. One hydrogen atom has the electron configuration 1s, in the molecule it is $1s^2$ which is energetically more favourable than two times 1s (*covalent bond*).

Finally, we can conceive an electron configuration in which the outer electrons are no longer locally bound, i.e. they are not part of a single atomic pair but can freely move within the crystal lattice (*metallic bond*). A quantum-mechanical calculation (solution of Schrödinger's equation within a periodic crystal potential) shows that *energy bands* exist in this case. The electrons within these bands are not bound to single atomic nuclei.

For a graphic illustration we remember that the energy levels are described by quantum numbers represented by different orbit shapes in a simple manner.

Fig. 9.3 Graphic illustration of the s- and p-electrons. The sketched figures are interpreted as locations of high probability of presence and called orbitals

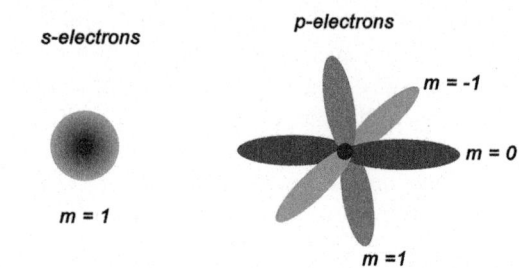

Fig. 9.4 Limiting cases of bondings. **a** s-s σ-bond. **b** p-p σ-bond. **c** p-p π-bond. (cf. also [2])

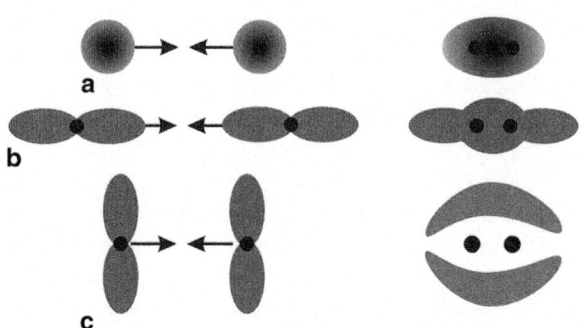

The orbital angular momentum $l=0$ (s-electrons) implies only one orbit shape (m=0) and is illustrated by a sphere, therefore, its orientation in the space does not matter. For $l=1$ (p-electrons) three different orbit shapes exist (m=−1, 0, 1), which are symbolised by ellipses stretching into the three directions in space (Fig. 9.3).

Using this kind of illustration three limiting cases can be designed corresponding to special kinds of bondings (Fig. 9.4).

There are also mixtures of these bondings called hybrid orbitals. All these bondings produce additional energy levels which can be occupied in consequence of inelastic collisions if they were not occupied before.

Within the energy bands the electrons can freely move, similar to a gas. Commonly, one speaks about the "*electron gas*". This gas oscillates after external excitation as it happens by inelastic interaction.

For a plausible explanation we conceive that the quasi-free electrons within the electron gas interact with a primary electron and shifted therefore. Because of the charge shifting an electric field E is generated and, connected with it, the restoring force

$$F = -e \cdot E \qquad (9.13)$$

(e: elementary electric charge). We conceive the charge shifting along an axis u and obtain from Maxwell's equations:

$$E = \frac{\rho}{\varepsilon \cdot \varepsilon_0} \cdot u \qquad (9.14)$$

(ρ: charge density, ε_0: permittivity of vacuum, ε: relative permittivity) and get with

$$\rho = \frac{Q}{V} = \frac{n \cdot e \cdot V}{V} = n \cdot e \qquad (9.15)$$

(Q: charge, n: number of electrons within the volume V) the equation of motion

$$\frac{d^2 u}{dt^2} + \frac{n \cdot e^2}{m \cdot \varepsilon \cdot \varepsilon_0} \cdot u = 0 \qquad (9.16)$$

(m: mass of the electron), i.e. the equation of harmonic oscillation with eigenfrequency

$$\omega_P = e \cdot \sqrt{\frac{n}{m \cdot \varepsilon \cdot \varepsilon_0}} \qquad (9.17)$$

which is equivalent to the energy

$$E_P = \frac{h}{2\pi} \cdot \omega_P \qquad (9.18)$$

(h: Planck's constant). These are additional possibilities of energy levels and therefore energy losses of the primary electrons. The energy losses affiliated to the eigenfrequencies of the electron gas are termed *plasmons*. The distribution of charge in the volume is different from that at the surface; hence it has to be distinguished between surface and volume plasmons.

Within the solid state model the bonds between the atoms are sometimes symbolised by mechanical springs. Despite the extremely different masses of electrons and atomic nuclei energy can be transferred from the beam electrons to the "construction of springs" and it starts to oscillate. The energies connected with the eigenfrequencies of these oscillations are termed *phonons* delivering additional very low energy levels in our binding model, too.

9.2 Energy Dispersive Spectroscopy of Characteristic X-rays ("EDXS")

From Sect. 9.1.1 we know that atoms in consequence of transitions with radiation emission send out X-radiation with an energy depending on the atomic number, i.e. on the chemical element. We would like to turn to practice and explain how the X-ray energy can be measured, how the spectra look like, which kinds

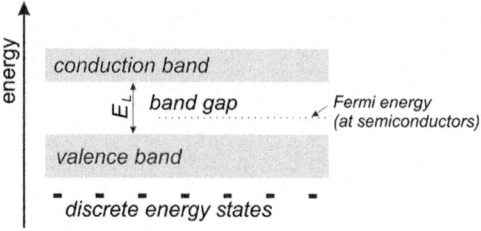

Fig. 9.5 Schematic illustration of the band structure of a semiconductor

of artefacts have to be considered, and which results can be concluded from the X-ray spectra.

9.2.1 X-ray Spectrometers and Spectra

Since energy and wavelength of X-radiation are inversely proportional (cf. Eq. (9.1)) the X-ray energy can be calculated by means of its wavelength. From Chap. 5 we know that the Bragg angle is determined by (among others) the wavelength. Using a crystal with well-known lattice constant it is possible to measure the wavelength by detection of the Bragg angle (*wavelength dispersive X-ray spectrometer* or *spectroscopy—WDXS*). However, the angle of incidence of the X-rays into the crystal and the angle of reflection (the sum of both yields the Bragg angle) must be equal during the measurement and therefore the crystal has to be moved on a circular orbit (Rowland[4] circle). It is geometrically very difficult to attach the needed device on a transmission electron microscope and its efficiency is also very small. For these reasons the wavelength dispersive spectrometry does not play any role in the practice of analytical transmission electron microscopy.

On the other hand, there is a further possibility to measure the energy of the X-rays. Understanding this we conceive that the ionisation energy of an electron shell comes from X-ray quants and not from beam electrons as assumed in Sect. 9.1.1. The X-ray quant shifts the electron at least to the first free energy state. In semiconductors this preferably happens between energy bands. The two upper ones are named valence and conduction band, respectively. In an undoped semiconductor the valence band is completely occupied at low temperatures, on the contrary the conduction band is unoccupied (Fig. 9.5). Therefore, the electrons cannot move freely and the electric resistance is high. The energy gap between valence and conduction band is comparably small in semiconductors, e.g. in silicon it amounts to about 1.2 eV.

At higher temperatures electrons can reach the conduction band hence the electric resistance decreases at increasing temperature typical for semiconductors.

The transfer of an electron into the conduction band can also be done by an X-ray quant of the energy E_{XQ} ("*inner photoelectric effect*"). For this purpose an energy E_L

---

[4] Henry Augustus Rowland, American physicist, 1848–1901.

9.2 Energy Dispersive Spectroscopy of Characteristic X-rays ("EDXS")

is needed per transferred electron. E_L must be at least equal to the size of the energy gap between the bands, i.e. the maximal number N_{max} of the "conduction electrons" amounts to

$$N_{max} = \frac{E_{XQ}}{E_L} \tag{9.19}$$

In the semiconductor crystal an electric field is generated by two electrodes. After impact of an X-ray quant a current is flowing for a short time; its intensity depends on the number of conduction electrons, i.e. on the energy of the X-ray quant. Using that it is directly possible to measure the X-ray energy (*energy dispersive X-ray spectroscopy—EDXS*).

However, the energy difference E_L of the band gap is only a minimum value. The energy needed per conduction electron can vary since different energy states within valence and conduction band are involved. The occupation is subject to a probability distribution, i.e. there are energy states which are preferably occupied. The number of conduction electrons and, connected with this, the measured X-ray energy follows this probability distribution, too. In the X-ray spectrum the inherent discrete X-ray energy is not shown as a line but as a curve similar to a Gaussian function ("*X-ray peak*").

The full width half maximum of this Gaussian curve is called *energy resolution* of the spectrometer. It is limited and depends on the X-ray energy. For an X-ray energy of 5.9 keV it amounts to about 120 eV ... 130 eV because of the statistical manner of emission and detection process.

The energy of 5.9 keV correlates to the Kα-transition of manganese. The selection of this reference energy is historic convenience. At the radioactive decay of the Fe55 isotope the line with the most intensity has just this energy. Therefore it is possible to test an EDX spectrometer without electron microscope by means of an Fe55-source. At lower X-ray energies the energy resolution improves and reaches about 60 eV for the Kα line of carbon (277 eV).

In Fig. 9.6 an EDX spectrum is shown. The names of the X-ray peaks are deduced from the sketch of the energy levels (Fig. 9.1):

The first capital (e.g. K) identifies the ionised core shell occupied again by an electronic transition connected with emission of X-radiation. The Greek letter marks the shell from which the electron comes refilling the ionised core shell. α means that the refilling electron comes from the energetically next higher shell, i.e. Kα means refilling from the L-shell, Kβ from the M-shell.

We observe that the background is higher in the low-energy range than in the high-energy range. This is the result of the bremsstrahlung.

Commonly, the thermal energy already at room temperature is high enough to shift some electrons into the conduction band leading to a low current between the detector electrodes connected with an increase of the background noise in the spectrum. Impurities of the semiconductor act as doping and increase the noise, too. Therefore, the detector crystal must be extremely pure and cooled by liquid nitrogen

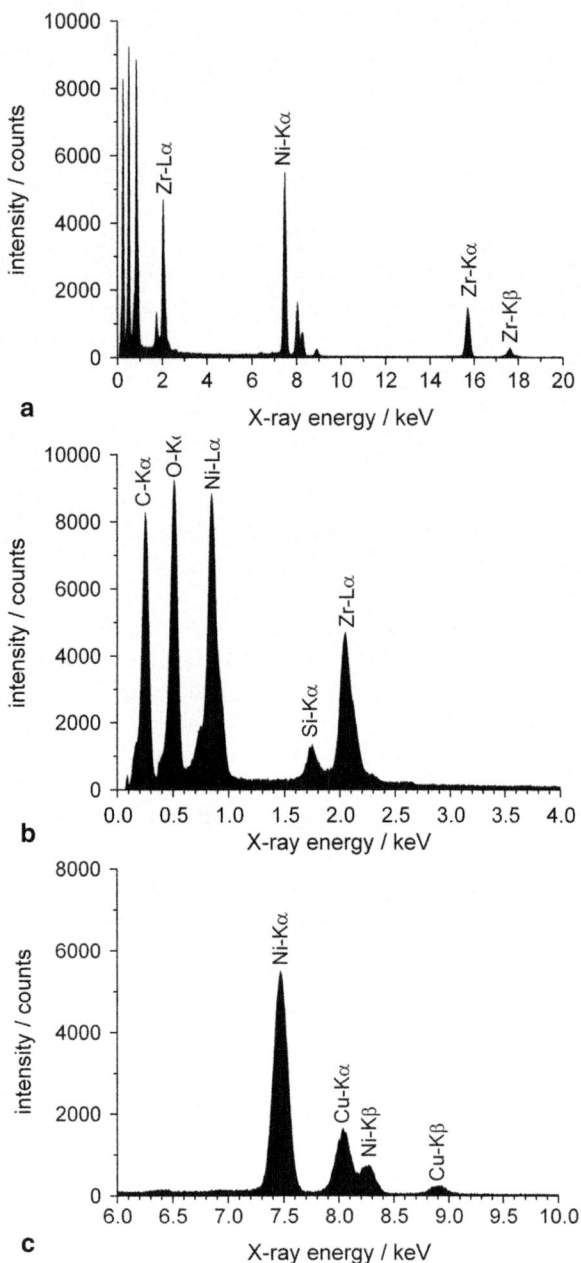

Fig. 9.6 Example of an energy dispersive X-ray spectrum (EDXS). The unit "counts" of the ordinate refers to the number of measured events within the measuring time and channel width. **a** Complete spectrum in the energy range 0–20 keV. **b** Detail of the low-energy range (0–4 keV). **c** Detail of the energy range 6–10 keV

or Peltier elements. The construction of an EDX detector is schematically illustrated in Fig. 9.7, its efficiency is explained in more detail in Sect. 10.20.

Lithium has been diffused into the semiconductor crystal (normally silicon) to compensate charge carriers which could be eventually generated by impurities. The goal is to create a zone free of charge carriers as large as possible. The reason is

9.2 Energy Dispersive Spectroscopy of Characteristic X-rays ("EDXS")

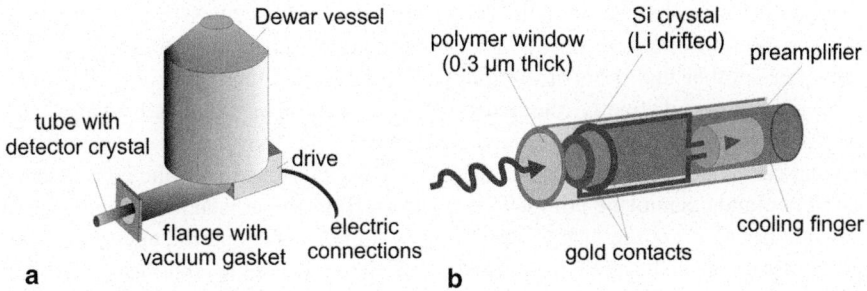

Fig. 9.7 Construction of an EDX detector with liquid nitrogen cooling. **a** Exterior view. **b** Detail: tube with detector crystal

that in this zone the freely movable charge carriers are only generated by invading X-ray quants. The electric field within the crystal is originated by electrodes, often made of gold. The Si crystal is about 3–5 mm thick with a cross sectional area of 10–30 mm². The detector unit is embedded in a tube pressure-tightly closed on the side of the X-ray source by an ultrathin polymer window and cooled including the preamplifier. During the X-ray analysis the tube with the crystal is inserted into the pole piece gap of the objective lens, before and after the analysis the tube must be retracted. Then, a small cap closes the end of the tube and avoids that high-energy backscattered electrons destroy the ultrathin polymer window in front of the crystal. It is not dangerous in the "normal" working mode of the electron microscope, i.e. with strongly excited objective lens with short focal length; the backscattered electrons are bundled by the lens field and cannot reach the detector. In this context it is important to know that at low magnifications ("low-mag mode") the objective is switched off or only weakly excited. To avoid the destruction of the polymer window the detector has to be retracted into its standby position before switching into the low-mag mode.

The vacua in the area surrounding the detector crystal and in the microscopic column, respectively, are separated by the polymer window. In this way it can be avoided that the cooled detector gets covered by ice if the column is vented. But be careful, please. The thin polymer film is supported by an aluminium grid on the detector side to be able to withstand a pressure difference of a bit more than 1 bar. A pressure of 1 bar in the microscopic column is permitted but vacuum in the column and a pressure of 1 bar or more in the surrounding of the detector can destroy the window! The vacuum in the detector tube is sustained by a sorbent cooled by liquid nitrogen in the Dewar vessel, too. After many years of operation the sorbent has been able to save so much gas that after warming-up the detector room a pressure of more than 1 bar can be reached. If the refill of the Dewar is forgotten in this time the polymer window can be destroyed. Therefore it is advantageous to perform a controlled warm-up of the detector and to evacuate it by a separate pump every few years as part of the maintenance.

A high-speed electronics ("pulse processor") makes sure that clouds of charge carriers generated by the X-ray quants are collected, separately determined, and, according to their intensity, sorted into energy channels. Within the time given to handle the charge carrier cloud the detector is blocked for new X-ray quants (*dwell or dead time*). It might be that at high X-ray intensities the quants follow so quickly one after another that it becomes impossible to separate them. The unusable dead time increases and, finally, the detector can be completely blocked. In this case the X-ray intensity has to be reduced or the electronics must be accelerated. In the second case the dwell time for the charge cloud is shortened with the consequence that the clouds cannot be completely removed from the crystal. The X-ray peaks become broader, i.e. the energy resolution becomes worse and the energy scale of the spectrum has to be newly calibrated. The pulse processor is computer-controlled and the dwell times are saved in calibration files together with the calibration of the energy scale.

Currently the described "classical" detectors are more and more substituted by *silicon drift detectors* (SDD—[3, 4]) where the silicon crystal is equipped with circular electrodes. Hence larger areas for the inflow of the X-rays become possible at the same time connected with a faster dissipation of the charge carrier clouds. These detectors are more sensitive and allow to handle higher X-ray intensities. Commonly a Peltier cooling is sufficient so that the liquid nitrogen handling is not necessary by the operator.

Especially for the detection of low-energetic X-radiation window-less detectors are available where the absorption of X-rays in the polymer window is cancelled. Because of the ditto nonexistent separation of the vacua in the detector room and microscopic column the handling is more complicated and needs more care. For the case of liquid nitrogen cooled devices such detectors are only sparsely available.

9.2.2 Qualitative Interpretation of X-ray Spectra

The energetic position of the characteristic X-ray lines broadened to the mentioned peaks are the topic of the qualitative interpretation of the spectra. They are overlaid by a background created by bremsstrahlung which is luckily significantly lower in the spectra recorded in transmission electron microscopes than in scanning electron microscopes due to the ultrathin specimen in the TEM.

The shape of the background curve in the measured spectrum is determined by the energy dependence of the bremsstrahlung (cf. Sect. 9.3) and by the detector efficiency (cf. Sect. 10.20 and Fig. 9.8). The intrinsic conduction of the semiconductor crystal depending on the purity of the crystal and its temperature generates an additional noisy background part.

The background by the bremsstrahlung reaches its maximum at X-ray energies of 0.8 keV … 1.5 keV and increases with increasing atomic number and sample thickness.

To identify the peaks the characteristic X-ray energies of the individual elements must be known. Figure 9.9 shows an overview of the X-ray energies. Today, they are part of a database included in the spectrometer software. Since the number of

9.2 Energy Dispersive Spectroscopy of Characteristic X-rays ("EDXS")

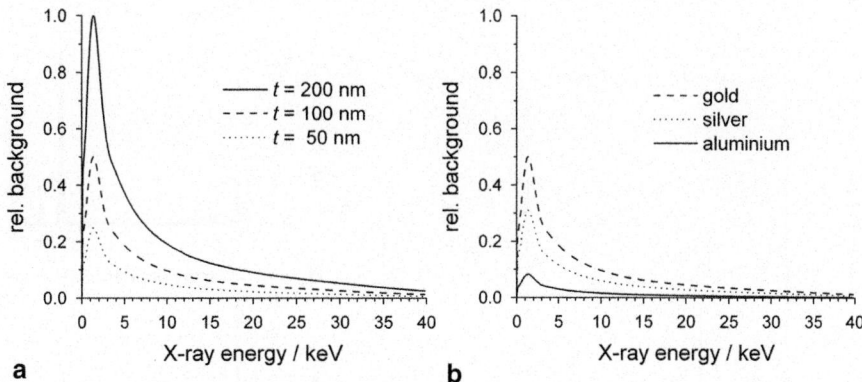

Fig. 9.8 Background by bremsstrahlung in the energy dispersive X-ray spectra at 200 keV primary electrons. **a** Gold with three different sample thicknesses. **b** For a sample thickness of 100 nm and three different elements. The values are related to the maximum value of gold with a thickness of 200 nm

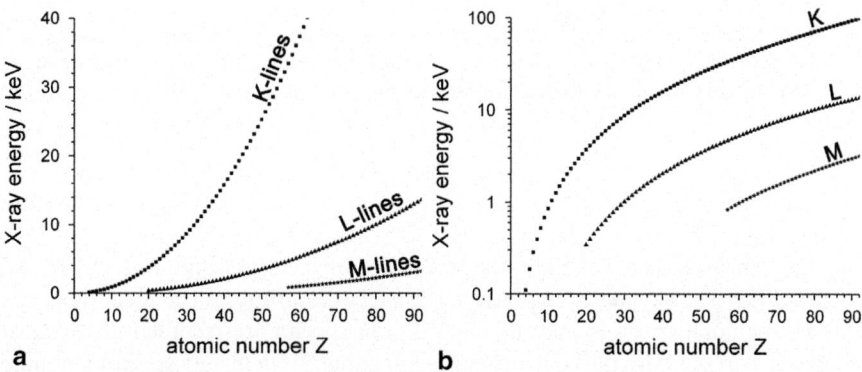

Fig. 9.9 Dependence of the characteristic X-ray energies (α-transitions) on the atomic number Z. **a** Linear representation. **b** Semi-logarithmic representation

shells with additional energy levels increases with higher atomic numbers the number of possible X-ray transitions becomes larger, too (cf. Sect. 9.1.1). Consequently, there are series with lines of similar energies. Figure 9.9 confines itself to the α-transitions.

We realise one problem from this figure: The possibility of overlap of X-ray peaks. This will be demonstrated in more detail at the example of chromium and oxygen (Fig. 9.10).

Chromium (Z=24) possesses the K-line at an energy of 5.41 keV and an L-line at 0.57 keV. The energy of the oxygen K-line amounts to 0.523 keV, i.e. it lies close to the Cr L-line.

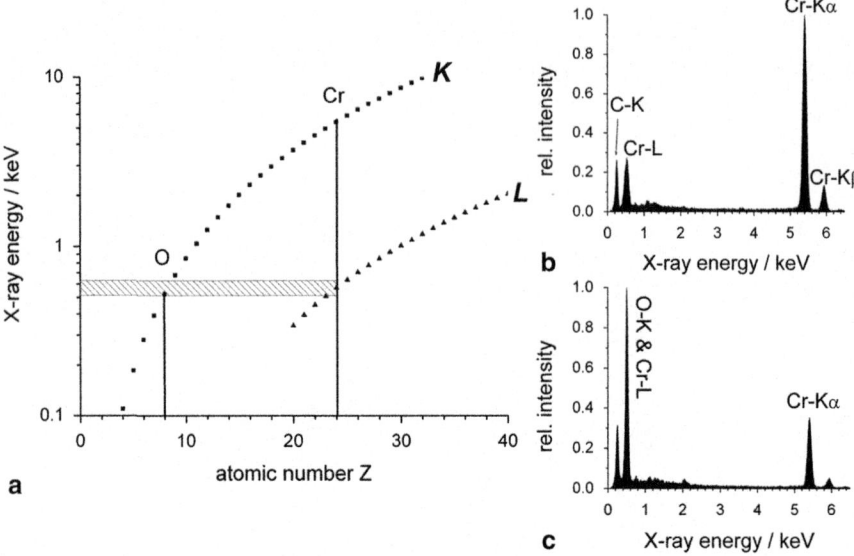

Fig. 9.10 Overlap of Cr L- und O K-Peak in energy dispersive X-ray spectra. **a** Accordance of both X-ray energies within a window given by the energy resolution limit of the spectrometer. **b** EDX spectrum of chromium. **c** EDX spectrum of chromium oxide

The energy resolution of the energy dispersive X-ray spectrometer is about 70–80 eV in this range, i.e. the L-peak of chromium and the K-peak of oxygen cannot be separated.

The comparison between the line intensities of the spectra in Figs. 9.10b and 9.10c shows that oxygen is involved in the spectrum of Fig. 9.10c. Unfortunately, this simple comparison is only possible if both spectra are recorded under equal conditions. In this case the conclusion is compelling. Without any spectrum of pure chromium one should be careful with some interpretations. The intensity ratio of the Cr K- and Cr L-peak depends among others on the energy-dependent spectrometer efficiency. It can be changed, e.g. by contamination of the detector window. Additionally, this ratio is also influenced by the specimen thickness as we will see in Sect. 9.2.3. Hence it is normally impossible to decide explicitly about the presence of a small content of chromium oxide without comparison measurements as mentioned above.

Additionally to peak overlaps there are two possible artefacts in the EDX spectra which can lead to wrong conclusions regarding the detected elements: escape and sum peaks.

To explain the *escape peak* we remember the events within the charge carrier free zone of the detector crystal: The incoming X-ray quant generates free charge carriers leading to the dismantling of its energy. But it is also possible that the X-ray quant ionises the silicon K-shell (assuming the detector crystal is made of silicon). Then, the electron from the K-shell becomes a free electron but the needed energy is

9.2 Energy Dispersive Spectroscopy of Characteristic X-rays ("EDXS")

equal to the energy of the absorption edge of silicon (1.84 keV), i.e. essentially larger than the band gap (1.2 eV). The refill of the free level in the K-shell is connected with the emission of an X-ray quant (energy: 1.74 keV) whose energy is normally dismantled by the generation of charge carriers. That is not the case if the X-ray quant *escapes* the detector crystal (therefore "*escape peak*"). Its energy is missing and an additional peak arises in the spectrum with an energy 1.74 keV smaller than that of the reference peak. For instance, the Cr Kα-peak (5.41 keV) is seconded by an escape peak at 3.67 keV. This is just the Kα energy of calcium. Without further thinking one could conclude that a low content of calcium is included in the sample but in fact, it is only the escape peak of chromium Kα. The probability for arising of such escape peaks depends, among others, on the size of the detector crystal but it should be considered during the interpretation of the X-ray spectra.

Sum peaks arise when two X-ray quants from the same element impinging into the detector crystal at the same time. "At the same time" means that the pulse processor is not able to separate them. In this case an additional peak with double energy of the reference peak arises in the spectrum. The probability of sum peaks increases with increasing intensity of the X-radiation. It is low in modern EDX spectrometers since their pulse processors work very fast.

We would like to point out a fact which is often underestimated. Despite the scanning transmission electron microscopic mode with a sub-nanoscaled electron probe the detector acquires X-rays coming from the surrounding area of the probe position. The size of this area can reach up to some micrometres, these distances can be travelled by elastically scattered electrons. If, for instance, copper is to be detected within the specimen the support grid or the holder of the TEM lamella must not consist of copper! Otherwise one cannot be sure: Is there really a copper content in the sample or is it only a stray radiation from the holder? It can be also a problem in case of layers on a substrate. Often radiation from the substrate is detectable even if the electron probe is not directly positioned on the substrate.

This is why the TEM specimen holder used for EDXS analyses is covered by a beryllium plate in front of the detector or the complete specimen surrounding is made of beryllium ("low-background holder"). The low-energetic beryllium radiation (110 eV) cannot be detected by a detector equipped with polymer window.

To get a "free view" of the detector to the analysed sample position it is advantageous to tilt the specimen (holder) about 10 ... 25° into the direction of the detector.

Finally, we would like to specify the elements which can be determined by EDXS. The efficiency of detectors with polymer windows decreases at X-ray energies below 0.15 keV to less than 20%, i.e. X-radiation of such a low energy cannot be practically detected. The Kα radiation of beryllium is at 0.11 keV, this of boron is at 0.18 keV. The critical border lies between them: beryllium is not detectable, boron, in principle, is (Fig. 9.11).

At X-ray energies higher than about 40 keV the detector efficiency significantly decreases but in this case one can use L- and M-lines, i.e. there is no limitation regarding the detection of high atomic number elements.

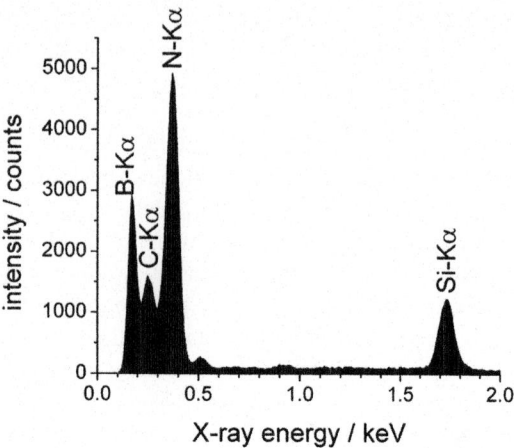

Fig. 9.11 Detection of boron in boron carbide (BN) by means of the EDX spectrometer. Sample: BN as fine powder on carbon support film containing silicon

Obviously, the minimum detection limit of energy dispersive X-ray spectroscopy in the transmission electron microscope is also of interest. It depends on the surroundings of the element which is to be determined, on the efficiency of the characteristic X-ray emission by the element, and on the detector efficiency for this radiation. Medium-heavy elements provide the best conditions, the detection limit amounts to about 2 atom-% in this case. The opposed extreme is given by boron carbide. In that case differences in the energy resolution of only a few electron volts already decide whether the boron content can be detected or not.

Please be careful if using the possibility of automatic peak identification in the spectra included in the spectrometer software. Often a range of "rarely used" elements is excluded from the search and result together with peak overlaps in a completely wrong determination. "What cannot recognised by the naked eye is not detected!" For significant detection the signal should be at least three times higher than the noise (Rose criterion [5]).

9.2.3 Quantifying X-ray Spectra

The number of free charge carriers generated in the detector crystal reflects the energy of the X-ray quants and is recorded as the height of a current pulse. The intensity of the X-radiation depends on the number of excited atoms and determines the number of current pulses ("counts") during the measuring time and, consequently, the peak size, i.e. the area of the peak. The ratio of the peak sizes I_A and I_B of two elements A and B is proportional to the ratio of the number of atoms N_A and N_B, which are excited. Let the proportionality factor be k_{AB}:

$$\frac{I_A}{I_B} = k_{AB} \cdot \frac{N_A}{N_B}. \tag{9.20}$$

9.2 Energy Dispersive Spectroscopy of Characteristic X-rays ("EDXS")

The elemental concentration c_A is equal to the quotient of the number of atoms of sort A and the total number of atoms. The same holds for sort B:

$$c_A = \frac{N_A}{N_A + N_B}, \quad c_B = \frac{N_B}{N_A + N_B}. \tag{9.21}$$

Commonly, concentrations are given in percent. In the case mentioned above these are atomic percents. The information in weight or mass percent, respectively, is common practice, too. For this purpose the number of atoms has to be multiplied by the atomic mass M. Therewith the concentration ratios are given by:

$$\frac{c_A}{c_B} = \frac{N_A}{N_B} \quad \text{and} \quad \frac{c_{A,M}}{c_{B,M}} = \frac{N_A \cdot M_A}{N_B \cdot M_B} \tag{9.22}$$

($M_{A,B}$: atomic weights of the elements A and B). Hence, we calculate the ratio of the concentrations of two elements within the sample from the ratio of the intensities of the X-ray peaks which are relevant for these elements:

$$\frac{c_A}{c_B} = \frac{c_{A,M}}{c_{B,M}} \cdot \frac{M_B}{M_A} = \frac{1}{k_{AB}} \cdot \frac{I_A}{I_B}. \tag{9.23}$$

This equation is known as "Cliff-Lorimer equation", the proportionality factors as "*Cliff-Lorimer factors*" [6]. Using the normalisation

$$c_A + c_B = 1 \tag{9.24}$$

the individual concentrations can be calculated from the concentration ratio:

$$c_A = \frac{c_A / c_B}{1 + c_A / c_B} \quad \text{and} \quad c_B = \frac{1}{1 + c_A / c_B}. \tag{9.25}$$

If not only two but n elements participate one reference element B is selected and the concentration ratios are calculated using Eq. (9.23):

$$\frac{c_1}{c_B}, \frac{c_2}{c_B}, \frac{c_3}{c_B}, \ldots, \frac{c_{n-1}}{c_B}. \tag{9.26}$$

Now, the condition for the normalisation is

$$c_B + \sum_{i=1}^{n-1} c_i = 1. \tag{9.27}$$

Using this, firstly the concentration of the reference element B has to be calculated:

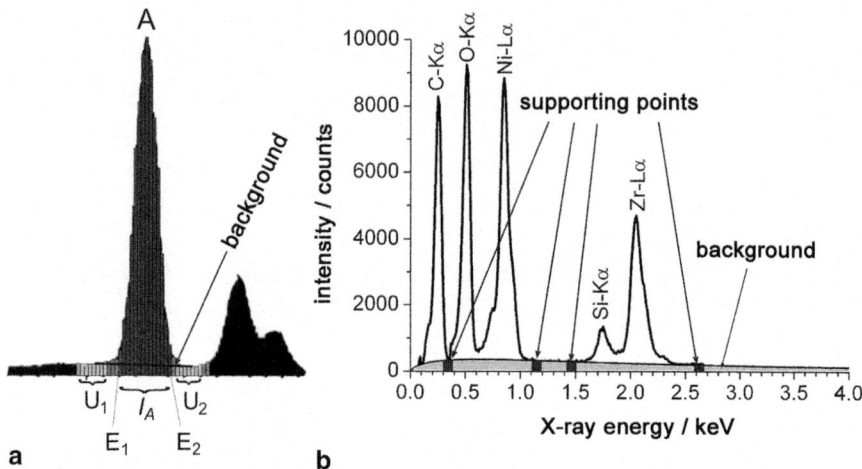

Fig. 9.12 Determination of the peak areas (sizes) and background approximation in EDX spectra. **a** Summation of the channel contents and background curve in the high-energy part of the spectrum. **b** Background in the low-energy part of the spectrum

$$c_B = \frac{1}{1+\sum_{i=1}^{n-1} c_i / c_B}. \tag{9.28}$$

The determination of the other concentrations is straightforward:

$$c_i = \frac{c_i}{c_B} \cdot c_B. \tag{9.29}$$

We recognise that there are two fundamental tasks for the quantification of X-ray spectra: the determination of the peak intensities and of the proportionality factors $1/k_{AB}$.

Determination of the peak intensities: The solution of this task seems to be simple: We determine the peak areas. But we have to subtract the energy-dependent background $U(E)$ from these values. As already mentioned (cf. Sect. 9.2.1) the background is generated by bremsstrahlung and noise as a consequence of intrinsic conduction within the detector crystal. Let the name of the spectrum function be $S(E)$. This function is given in the form of a look-up table, i.e. as numbers (counts) within an energy interval from $E - \Delta E/2$ to $E + \Delta E/2$ called energy E ("channel of energy E"—Fig. 9.12). The borders i_A and i_E denote the channels at the begin and at the end of the peak originated by element A:

$$I_A = \sum_{i=i_A}^{i_E} (S_i - U_i). \tag{9.30}$$

9.2 Energy Dispersive Spectroscopy of Characteristic X-rays ("EDXS")

With consideration of very thin specimens in transmission electron microscopy the background by bremsstrahlung changes at higher energies something close to linearly with the energy (Fig. 9.8). The background curve at higher energies can be approximated by a straight line whose equation can be determined with the help of the mean intensities U_1 and U_2 within two supporting points outside of the X-ray peaks (Fig. 9.12a).

The background approximation in the low-energy spectrum range is more problematical. In this range bremsstrahlung as well as spectrometer efficiency drastically change with the X-ray energy (Fig. 9.8). Neglecting the effects of absorption edges in the spectrometer efficiency (cf. Sect. 10.20) the background curve can be approximated by a function of the form

$$U(E) = c_1 \cdot \left(\frac{E_0}{E} - 1\right) \cdot (1 - e^{-c_2 \cdot E}) \tag{9.31}$$

(c_1 and c_2: parameters for the fit procedure).

Up to now we have assumed that the X-ray peaks are well separated. This is not always the case as shown in Fig. 9.10. At peak overlaps a simple summation of the channel intensities fails. The assignment of the intensities to the involved peaks (i.e. elements) can be performed by modelling of the measured peaks by Gaussian (or similar shaped) functions. The maxima of the Gaussian curves are set to the characteristic energies of the participating element peaks, their full width half maximum is given by the energy resolution of the spectrometer. The total area enclosed by the Gaussian curves is fitted to the channel intensities in the spectrum by variation of the maxima of both Gaussian functions. After ending the fit procedure the separated peak intensities are given by the areas under the Gaussian curves. Sometimes the spectrometer software needs an experimental determination of the peak shape which is well suitable to fit the peaks. In this case exemplary spectra of standards have to be measured during the installation to find the best mathematical peak-form model.

Determination of the proportionality factors $1/k_{AB}$: The measurement of the proportionality factors by means of standards with well-known composition represents the best method. The elemental ratio c_A/c_B for the standard is known, the ratio of the peak intensities I_A/I_B is determined from the measured spectrum, and the proportionality factor can be calculated using Eq. (9.23). This method has the advantage that all individual features of the spectrometer and the data analysis are considered. But it has serious disadvantages, too. We need standards and we have to prepare electron-transparent specimens for transmission electron microscopy. This effort is only worthwhile if a specific material system is to be investigated for a longer time.

Therefore another method is preferred for a "fast" quantification of X-ray spectra which does not need any standards: the calculation of the proportionality factors. The name of this method is *standardless quantification*.

The basis of the calculation is a model that multiplies the probabilities of the several steps of X-ray emission and detection. These several steps are:
- the ionisation of a core shell (cross section for ionisation Q),
- the refill of this shell with emission of an X-ray quant (fluorescence yield ω), and
- the detection of the X-ray quant by the detector (detector efficiency D_{eff}).

To get a better overview it is useful to write the proportionality factor in the form

$$\frac{1}{k_{AB}} = \frac{k_A}{k_B}. \tag{9.32}$$

With consideration of the steps mentioned above and the reciprocal we get

$$\frac{k_A}{k_B} = \frac{Q_B \cdot \omega_B \cdot D_{eff,B}}{Q_A \cdot \omega_A \cdot D_{eff,A}}. \tag{9.33}$$

In the literature these factors are commonly referred to mass percents. In this case the ratio of the atomic weights M_B and M_A has to be additionally considered:

$$\left(\frac{k_A}{k_B}\right)_M = \frac{Q_B \cdot \omega_B \cdot D_{eff,B} \cdot M_B}{Q_A \cdot \omega_A \cdot D_{eff,A} \cdot M_A}. \tag{9.34}$$

The X-radiation is isotropically emitted into the total space. The part recorded by the detector depends on the solid angle including the entrance window of the detector, i.e. it depends on the geometry of the detector and its distance to the sample. The angle between the sample surface and the detector axis plays a role, too. The geometric influences are summarised in the "take-off angle". Its influence is stronger at low X-ray energies and rough surfaces. Therefore the geometric conditions should be kept at least similar.

The detector efficiency is closer described in Sect. 10.20. The cross section for core shell ionisations can be calculated using different models on the basis of different approximations (cf. Sect. 10.21).

Insofar the standardless quantification is burdened with systematic errors depending on the quality of the model. Often the models are valid only for a limited range of X-ray energies. For the quantification of medium-heavy elements (atomic numbers between 20 and 40) using K-lines the smallest errors can be expected (relative about 5 %). The errors are significantly larger for the comparison of K- with L- or M-lines, respectively. The k-factors shown in Fig. 9.13 are referred to mass percents and silicon Kα radiation as reference as usual in the literature.

Figure 9.13 also shows that the k-factors of the individual elements are very different. Therefore it would be wrong to imply the concentration ratios by a simple comparison of the peak heights!

The lower the efficiency of X-ray emission and detection is the larger the corresponding k-factor has to be to compensate these differences during the conversion

9.2 Energy Dispersive Spectroscopy of Characteristic X-rays ("EDXS")

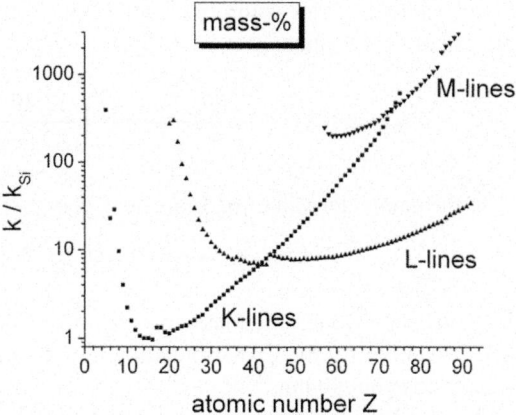

Fig. 9.13 Calculated Cliff-Lorimer-k-factors referred to the Si Kα-line for 300 keV-electrons (calculation Sect. 10.21)

of intensity ratios into concentration ratios. This also means: the larger the k-factor the weaker the measured signal is. Looking at Fig. 9.13 we recognise that elements with atomic numbers up to about 40 can be most efficiently measured using the K-lines. For elements with atomic numbers larger than 40 the L-lines should be preferably used.

As a rule, at larger differences of the k-factors used for the quantification larger errors have to be expected.

Absorption correction: The strong absorption of low-energetic X-radiation even within the thin specimen can additionally falsify the result of the quantification (Sect. 10.22). The specimen thickness t for which the systematic error originated by different absorption remains less than 10% can be calculated by

$$t < \frac{0.2 \cdot \sin \delta}{\rho \cdot \left|{}^{EA}(\mu/\rho) - {}^{EB}(\mu/\rho)\right|} \tag{9.35}$$

(δ: take-off angle of the X-rays, ρ: specimen density, ${}^{EA}(\mu/\rho)$ and ${}^{EB}(\mu/\rho)$, respectively: mass absorption (also: attenuation) coefficient of the sample material for X-rays with energies of EA and EB, respectively—cf. also Eq. (10.396)). We would like to calculate this limitation of the thickness for two materials: $Cr_{25}Si_{75}$ and CrO.

The concentrations (in atomic percent) are for $Cr_{25}Si_{75}$: $c_{Cr}=0.25$ and $c_{Si}=0.75$. We approximately assume that the density and the mass absorption coefficients can be calculated from the values of chromium and silicon with concentrations as weighting factors:

$$\rho = c_{Cr} \cdot \rho_{Cr} + c_{Si} \cdot \rho_{Si}$$
$$= 0.25 \cdot 7.19 \text{ g/cm}^3 + 0.75 \cdot 2.33 \text{ g/cm}^3 = 3.54 \text{ g/cm}^3 \tag{9.36}$$

To calculate the mass absorption coefficients we need a mean atomic number Z for the use of the equation

$$\frac{^E(\mu/\rho)_Z}{\text{cm}^2 \cdot \text{g}^{-1}} = C \cdot \left(\frac{12.396}{E/\text{keV}}\right)^\alpha \cdot Z^\beta \tag{9.37}$$

with the parameter values of Table 10.4. Z is given by

$$Z = c_{Cr} \cdot Z_{Cr} + c_{Si} \cdot Z_{Si} = 0.25 \cdot 24 + 0.75 \cdot 14 = 16.5. \tag{9.38}$$

By doing this we get the mass absorption coefficients 5,360 cm²/g for the Si-Kα- and 290 cm²/g for the Cr-Kα-radiation. For a take-off angle of 20° the thickness is limited to about 40 nm.

For CrO we get after an analogous calculation as before c_{Cr}=0.5 and c_O=0.5 with ρ=3.58 g/cm³, Z=16, $^{OK}(\mu/\rho) \approx 101{,}000$ cm²/g, $^{CrK}(\mu/\rho) \approx 270$ cm²/g $^{CrK}(\mu/\rho) \approx$ a thickness limit of less than 5 nm.

Obviously, at the given sample composition for specimen thicknesses of about 100 nm as usual in transmission electron microscopy the absorption of the X-radiation must be considered if the error is to be less than 10%.

How is it possible to correct this absorption influence? Certainly by a calculation as described in Sect. 10.22. However, sample thickness, sample density, and mass absorption coefficients must be known with sufficient accuracy. In our examples described above we had used approximations for density and mass absorption coefficients. The measurement of the absolute specimen thickness is often problematic, too. We would like to point to a method by Z. Horita [7] for an absorption correction working without any knowledge of the values mentioned above.

Precondition for the use of this method is, firstly, a specimen with continuously varying thickness as normally given at the wedge-shaped edge of the hole in conventional ion milled samples. Secondly, we need a material that emits at least one high-energy X-ray line with negligible absorption within the specimen. During the measurement several spectra are recorded at positions with different specimen thickness under identical conditions (beam current, take-off angle, measuring time) and quantified without consideration of absorption. The results are drawn in a diagram as ordinate values against the intensities of the high-energetic X-ray line (which should depend linearly on the specimen thickness). The intersection of the line of best fit with the ordinate (it represents the specimen thickness 0) yields the absorption-corrected result of the quantification (Fig. 9.14).

This figure shows the correction for a sputtered silicon-chromium layer. The sputter target consists of 74 atom-% silicon and 26 atom-% chromium. The spectra were recorded in three sample areas being differently thick (about 50 nm, 100 nm, and 150 nm), yielding three groups of c_{Si}/c_{Cr}-ratios: For Cr-Kα intensities of about 1,500 counts from the thinnest area with a c_{Si}/c_{Cr} mean value of 2.9, those of about 3,000 counts with the mean value of 2.65, and those from the thickest area with 4,000–5,000 counts and a mean value of 2.5. The extrapolation up to the ordinate

Fig. 9.14 Absorption correction for an EDXS measurement of the silicon-chromium ratio using the method by Horita [7]

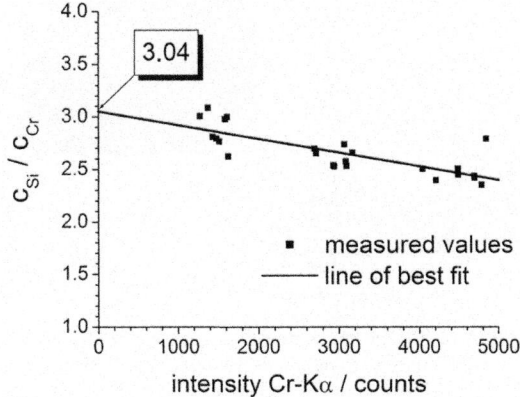

delivers the absorption-corrected c_{Si}/c_{Cr} ratio of 3.04 giving a silicon content of 75.2 atom-% which is close to the value of the target as expected in this case.

At the measurement shown in Fig. 9.14 a sample thickness of ca. 40 nm (calculated as thickness resulting in an absorption error of 10% as mentioned above) would produce a signal of about 1,300 counts within the measuring time. At this abscissa the error by unconsidered absorption differences between Si- and Cr-Kα-radiation should amount to about 5%. The agreement of the error orders indicates that our model is suitable to quantitatively explain the absorption effects.

Random errors: Finally, we would like to point to some errors at EDXS measurements which can be influenced by the operator. At first, we have to consider that the X-ray emission as well as its detection are statistical processes. At a number N of statistical events the relative error is

$$1/\sqrt{N} \tag{9.39}$$

or in other words: If the statistical error at the analysis of two X-ray peaks with the (detected) counts $N_1=N$ and $N_2=a \cdot N$ is supposed to be less than 1% the number N of counts in the reference peak must be

$$N > \frac{10000}{a} \cdot \left(1+\sqrt{a}\right)^2, \tag{9.40}$$

at two equal peaks just 40,000 counts for each peak.

Barriers on the path of the X-rays between specimen and detector can be an additional problem ("shading"). The crosspiece of a support grid, an overhanging sample detail as well as the edge of the specimen holder can be such a barrier. In the EDX spectrum such an effect is visible by means of missing or at least reduced signal at low energies including a missing background by bremsstrahlung.

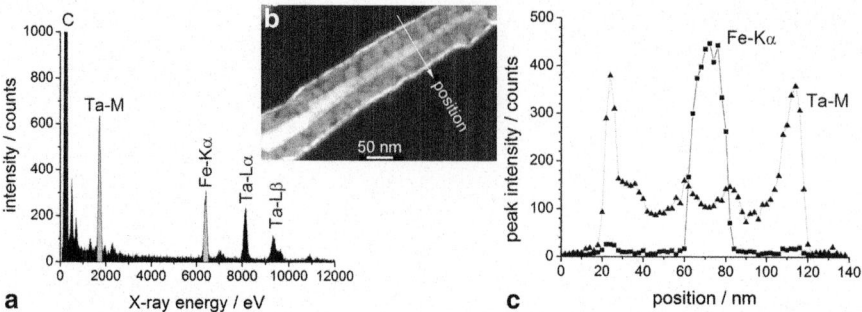

Fig. 9.15 EDXS line scan measurement at iron-filled and tantalum (*ALD* Atomic Layer Deposition) ALD-coated carbon nanotubes. **a** EDX spectrum as sum of all the spectra. The energy windows are grey colourised and cover the Ta-M peak and the Fe-Kα peak. **b** STEM-HAADF image with marked line (position) along which 70 spectra were recorded. **c** Profile of the X-ray intensities of the Ta-M and the Fe-Kα peak along the position line. The small Ta peaks at the edges of the Fe profile point to a Ta shell of the iron filling. (Sample by courtesy of S. Menzel, IFW Dresden)

Sometimes the situation can be improved by a large tilt of the specimen towards the detector.

Already during the insertion of the specimen into the holder one should pay attention to a "free path" of the X-rays to the detector. For thin films on support grids the film should be on the top (attention, in the inserting position under the light optical microscope the holders are often tilted by 180° around their axis against the working position in the TEM). Girders with welded FIB lamellas (cf. Sect. 3.4) should be inserted in a way that the open side of the girder points to the detector. In the case of thin layer stacks in the specimen it should be inserted such that the layers do not overlap if the specimen is tilted towards the detector. It is worth to think about these things before inserting the specimen into the holder since each specimen insertion and disassembling can damage the specimen.

9.2.4 Line Profiles and Elemental Mappings

In the scanning transmission electron microscopic mode it is possible to guide the electron probe step by step along a given line and to acquire information (e.g. an EDX spectrum) at each stopping point. This method is called "*line scan*". For the interpretation a characteristic X-ray peak has to be selected, a window with a width of 100 eV … 200 eV is laid over the selected peak, the intensities within the window are extracted in all spectra and plotted against the position of the electron probe while the spectra had been recorded. The result is a proportion profile of the element corresponding to the selected X-ray peak along the given line (Fig. 9.15).

9.2 Energy Dispersive Spectroscopy of Characteristic X-rays ("EDXS")

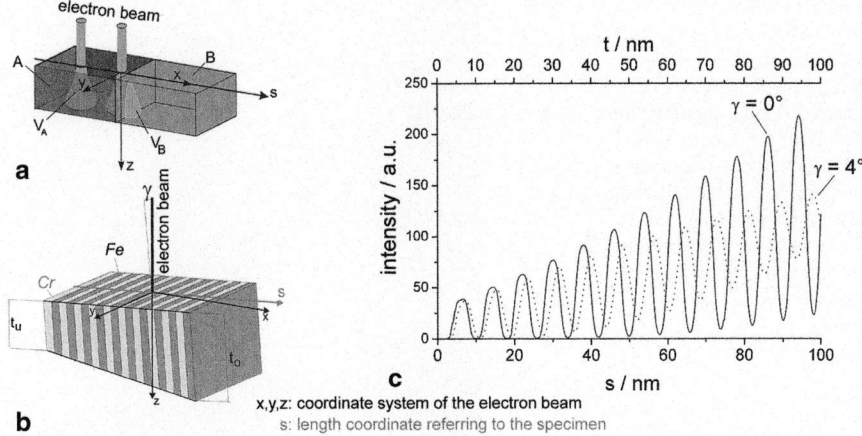

Fig. 9.16 Spatial resolution of the line scan method. **a** Sketch of a broadened electron probe moving across an interface between the materials A and B. **b** Schematic illustration of a layer stack from Fe/Cr multilayers with single layer thicknesses of 4 nm and a wedge-shaped cross-section specimen starting with thickness t at t_U and ending with t_O. The interfaces are tilted against the electron incident direction by an angle γ. **c** Calculated intensity profile of iron at tilt angles γ of 0° and 4° (assumed probe diameter at the top of the sample: 1 nm). At an assumed wedge angle of 45° the s- and t-scale agree (t_U=0, t_O=100 nm)

The intensity profile of the Ta-M line in Fig. 9.15c shows a typical curve of a tube-coating [8].

Using this line scan method it is possible to measure diffusion profiles on cross-sections of specimens (cf. Sects. 3.3 and 3.4). In this context the spatial resolution of the method is of interest. From Fig. 8.5 we already know that the electron beam is broadened by elastic scattering within the sample. This is schematically shown in Fig. 9.16a. The broadening b of the probe diameter can be evaluated by an approximation equation by J. I. Goldstein [9]:

$$\frac{b}{\text{nm}} = \frac{0.126 \cdot Z}{E_0 / \text{keV}} \sqrt{\frac{\rho/(\text{g} \cdot \text{cm}^{-3})}{M} \cdot (t/\text{nm})^3} \qquad (9.41)$$

(Z: atomic number, E_0: energy of the primary electrons, ρ: density, M: atomic weight, t: specimen thickness). For a specimen thickness of 100 nm and a primary energy of 200 keV it follows a broadening of 5.6 nm in carbon and 15.6 nm in gold. Monte-Carlo simulations allow more accurate calculations. As shown in Fig. 9.16a the broadened electron probe gradually enters the interface between two materials. The excited volumes within materials A and B do not abruptly change, the intensity profiles are "smeared". They are additionally broadened if the interface is not exactly oriented parallel to the incident electron beam but tilted by an angle γ.

The problem can be very well demonstrated using the example of a multilayer stack investigation (Fig. 9.16b). We assume an iron-chromium layer stack with 4 nm layers and wedge-shaped increasing sample thickness starting at 0 and ending

Fig. 9.17 Elemental mappings of chromium and iron recorded by EDXS in the STEM mode, coloured (Cr: green, Fe: red), and overlaid. The layer stack starts with a 15 nm thick chromium layer on the left side, the following layer thicknesses amount to 4 nm ... 5 nm. The specimen thickness increases wedge-shaped from the left to the right

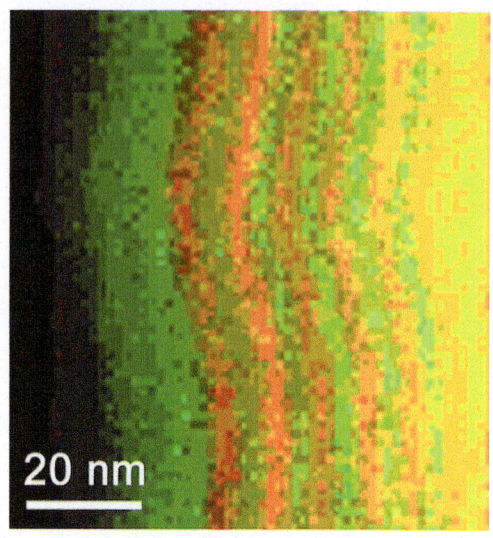

with 100 nm. The results of a calculation [10] for the exactly orientated ($\gamma = 0°$) and a layer stack whose interfaces are tilted by 4° against the incidence direction of the electron beam (Fig. 9.16c) show that for the exactly orientated interfaces the iron signal does not reach zero in the chromium area at wedge thicknesses t > 40 nm. Its modulation depth, however, furthermore increases. Contrary, for the tilted sample the modulation depth drastically decreases; after extrapolation of the curve it can be assumed that the layer stack is not detectable at a specimen thickness t ≈ 150 nm or larger.

If the electron probe is not guided along a line but scans a complete area we get a two-dimensional picture of the elemental distribution (*"elemental mapping"*).

Often such images are evaluated for different elements, differently coloured and overlaid. Figure 9.17 shows the result of the Cr/Fe multilayer mentioned above. The image agrees with the expectation: With increasing specimen thickness the colours smear more and more. On the right side referring to the thickest part of the imaged sample area only yellow as a mixture of green and red can be seen.

9.3 Electron Energy Loss Spectroscopy ("EELS")

After the discussion of electron shell transitions with emission of X-radiation and their measurement we would like to come back to the primary electrons. Their energy before entering the sample can be calculated from the acceleration voltage between cathode and anode. When we measure their energy after penetration of the sample we can derive the energy loss which the electrons suffer within the specimen. From Sect. 9.1.2 we know that these energy losses are

9.3 Electron Energy Loss Spectroscopy ("EELS")

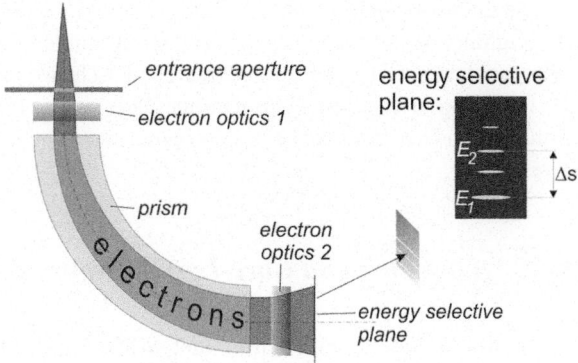

Fig. 9.18 Schematic illustration of an electron energy spectrometer

characteristic for the elements as well as for the bonding type between the atoms in the solid.

Now, we are going to explain topics of this *electron energy loss spectroscopy* (*EELS*) in practice: How can the electron energies be measured? What do the spectra look like and what can we learn from them? What are the possibilities in the context with transmission electron microscopy?

9.3.1 Electron Energy Spectrometer

It is the purpose of the spectrometer to separate the incoming electrons in terms of their energy, i.e. electrons with different energies shall leave the spectrometer device at different positions or under different path inclinations, respectively. In the light optics this is done by glass prisms. In Sect. 10.23 it is shown that magnetic or electric fields are able to perform this job for electrons. The magnetic or electric prism has to be combined with electron-optical components which, among others, arrange that the electrons enter the prism in a parallel manner and allow changing the dispersion (Fig. 9.18).

Together with the electron-optical units (multipoles) the prism creates an energy-selective plane; electrons with identical energy meet at the same position in this plane. The distance Δs between the positions of electrons with the energies E_1 and E_2 is determined by the dispersion (sensitivity)

$$D = \frac{\Delta s}{E_1 - E_2} \qquad (9.42)$$

of the spectrometer. The energy of the primary electrons influences the dispersion, too. The lower the primary energy the larger is the dispersion (Sect. 10.23). In principle it is possible to decelerate all electrons by immersion lenses (often called "drift tube") uniformly to increase the dispersion in this way.

The energy-selective plane is imaged onto a CCD camera (cf. Sect. 2.7.4) and the brightness values are read out. The spectrometer alignment includes the control with eventual correction of the perpendicular electron incidence into the prism and the focussing of electrons with identical energy into the energy-selective plane. As a rule, in modern spectrometers a computer with a suitable software undertakes this alignment.

9.3.2 Low-Loss and Core-Loss Regions of the Spectra

Let us have a look at an electron energy loss spectrum typical for the spectra acquired in a transmission electron microscope (Fig. 9.19).

The spectra in this figure are normalised, i.e. the highest intensity within the selected energy range is set equal to one. The energy loss, i.e. the difference of the primary energy determined by the acceleration voltage between cathode and anode and the electron energy measured by the spectrometer, is the abscissa.

Besides the energy loss the intensity also depends on the beam current and on the spectrometer dispersion. The pixel size of the CCD camera determines an energy window within which the electrons cannot be distinguished regarding their energy. The width of this window decreases with increasing dispersion, i.e. the number of registered electrons per pixel decreases, too. Therefore, it has to be expected that the spectra measured by a spectrometer with high dispersion (or high energy resolution) and unchanged further conditions (beam current, apertures etc.) are more noisy than those measured by a low-dispersion spectrometer.

At sufficiently thin specimens the far highest intensity ($>90\%$) is generated by elastically scattered electrons (Fig. 9.19a). Please remember: Elastic scattering happens without energy loss, thus the name of this part of the spectrum is *zero-loss peak*. Actually, this should be a single line and not a broad peak. The reasons of the broadening are, on the one hand, the energy differences of the electrons emitted from the cathode (energy spread—cf. Sect. 2.5) and, on the other hand, the limited energy resolution of the spectrometer (dispersion and pixel size). The spectrometer is adjusted such that the energy loss zero agrees with the maximum of the zero-loss peak. The intensity at the negative part of the abscissa is not a consequence of any acceleration of the electrons but it is caused by the finite width of this peak.

The peak of the elastically scattered electrons is followed by the *low-loss* region. Such low energy losses are created by excitation of transitions between energy levels being close together, e.g. transitions within the energy bands or oscillations of the complete electron gas (plasmons) or the network of solid atoms (phonons with energies less than 0.1 eV). For the EELS practice in the transmission electron microscope phonons do not play any role.

Higher energy losses are generated by ionisation of core shells (*core-loss* region). In this region the intensity is low so that the beam current and/or the apertures must be enlarged and the measuring time has to be prolonged to get sufficiently high signals. To avoid damage of the CCD camera the zero-loss peak and, if needed, also

9.3 Electron Energy Loss Spectroscopy ("EELS") 207

Fig. 9.19 Electron energy loss spectrum of nickel oxide. **a** Complete spectrum with labelling of the three essential ranges. **b** Core loss region, separately acquired at higher intensity and longer measuring time than (**a**). **c** Range of the O-K and the Ni-L edge of (**b**) after additional amplification and background subtraction

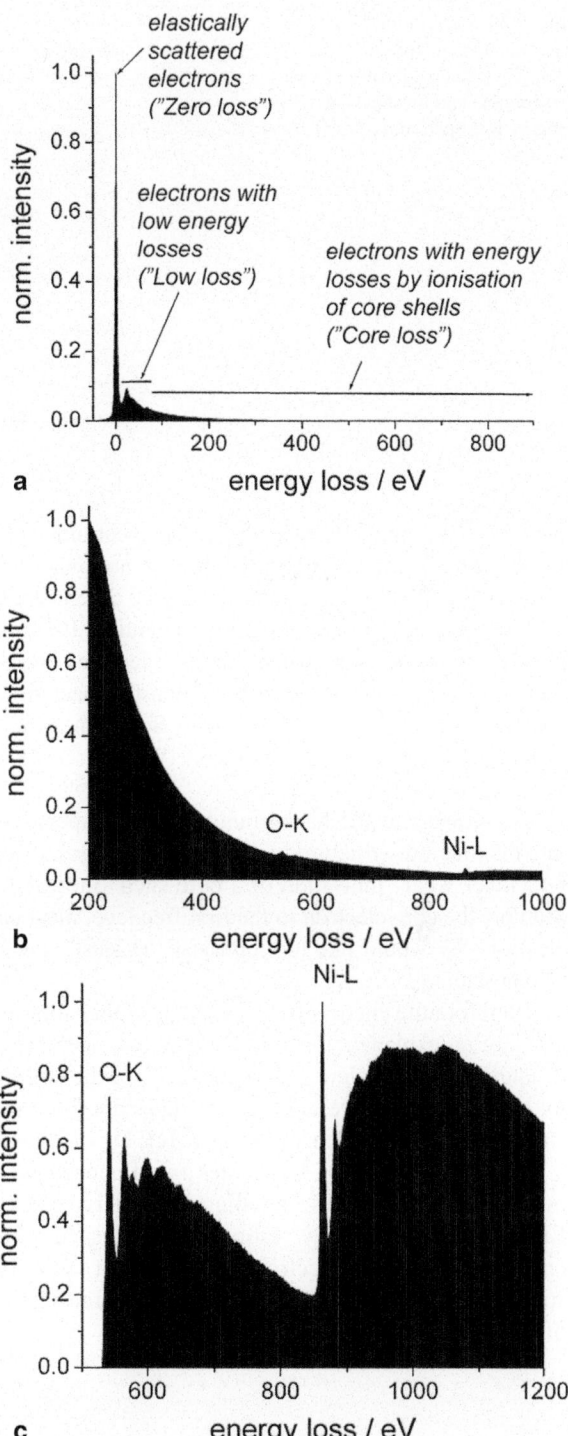

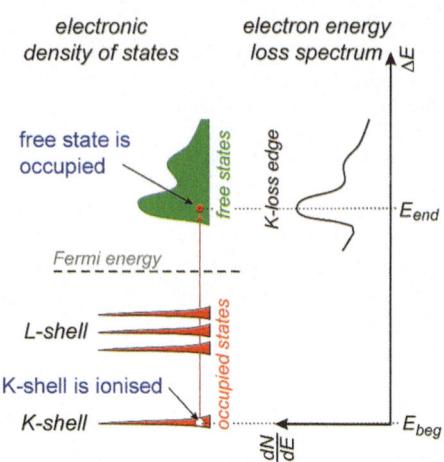

Fig. 9.20 Correlation between the electronic density of states (density of free states) and the electron energy loss spectrum

the low-loss range is truncated, i.e. the spectrum is shifted in the energy-selective plane (offset) such that these regions are no longer within the field of view of the camera (Fig. 9.19b). Especially within the core-loss region the high and hyperbolically decreasing background attracts attention. It is generated by non-characteristic inelastic interactions between primary electrons and the sample (e.g. excitation of bremsstrahlung—cf. Sect. 9.1.1). Often this background leads to problems in the identification of *characteristic edges*. Sometimes they can be discerned only after background subtraction (Fig. 9.19c). We will come back to this circumstance in Sect. 9.3.7.

We observe in the background-corrected spectrum that the edge curve is not monotonous but structured (*edge fine structure*). To understand this structure we remember what "ionisation of a core shell" means: At least the energy needed to promote the core electron to the first free level must be transferred to the core shell electron. We would like to emphasise "at least", i.e. higher energy transitions are also possible.

Therefore the energy-loss edge starts at this minimum (*onset*) and slopes gradually. For an explanation of the structure we need the knowledge of the free levels ("states") distribution or, more precisely, of the *density of states* ("DOS"). Since a lot of electrons are present the free states are filled step by step and the edge fine structure reflects the density of states (cf. Fig. 9.20 and Sect. 9.3.6).

Since the density of free states is influenced by, among others, the bonding between the atoms these bonding characteristics can be analysed by means of EELS, too. If the energy resolution of the spectrometer is high enough and the energy spread of the electrons emitted from the cathode is sufficiently small all solid properties reflected in the density of free states can be measured by means of EELS.

9.3 Electron Energy Loss Spectroscopy ("EELS")

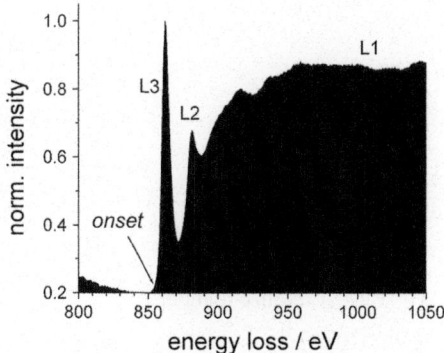

Fig. 9.21 Labels of the parts of the split L-edge of nickel. The energies of the absorption edges mark the start of the slope of the edge ("onset")

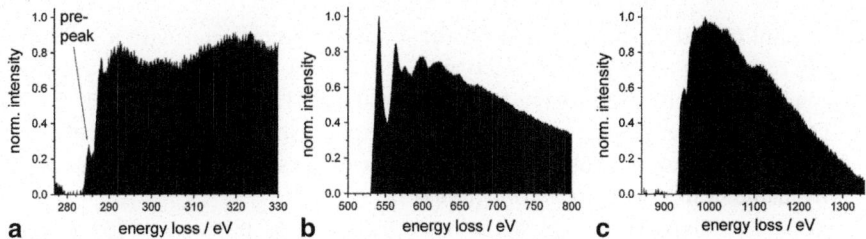

Fig. 9.22 Typical shapes of edges after background subtraction. **a** K-edge of graphitic carbon (*pre-peak*). **b** K-edge of oxygen in nickel oxide (*white line*). **c** L-edge of copper (*delayed edge*). Note the different abscissa scales

9.3.3 Qualitative Elemental Analysis

The determination of the elements involved in the energy losses is preferably performed in the core-loss region of the energy-loss spectrum. An accurately calibrated spectrometer and the knowledge of the ionisation energies of the elements are needed. In Fig. 9.19 we already assigned the element abbreviations of oxygen and nickel at the energy-loss edges visible in the spectrum.

Because of the increase of possible energy levels at the L-, M-, N- etc. shells a number of edges exists following each other immediately. They are differently shaped due to differences of the ionisation cross section. The sequence of these edges influences the edge fine structure and are named by the label of the ionised shell and a consecutive number, e.g. L1, L2, L3. For nickel the energy losses due to the ionisation of the L-shell amount to L1 = 1008 eV, L2 = 872 eV, and L3 = 855 eV (Fig. 9.21). Contrary to the X-ray spectra the energies of the absorption edges are given at the start of the slope in the EEL spectrum (called *onset*) and not at the peak maximum.

Since the shape of the edge is determined by the density of free states and this density is different for individual elements the shape of the edges is already different for the isolated atoms. In Fig. 9.22 three variants are shown with the help of

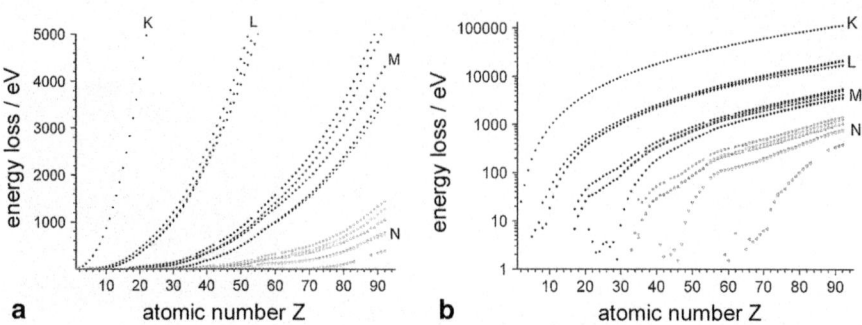

Fig. 9.23 Energy loss edges (onsets) of the K-, L-, M-, and N-series in linear (**a**) and semi-logarithmic (**b**) representation in dependence on the atomic number Z

the spectra of carbon, oxygen, and copper. Using a spectrometer with a sufficient energy resolution a pre-peak is observable at the onset of the K-edge of graphitic carbon, generated by π-bonds (cf. Sect. 9.1.2 and e.g. [11]).

The oxygen K-edge starts with a surged slope and a sharp peak resulting in a fine white line in the case of photographic registration of the spectrum. This is the reason for the name of this behaviour: *white line* (Fig. 9.22).

At the L-edge of copper the slope increases gradually, it is a so-called *delayed edge*.

Despite the more than 100 times better energy resolution of the EEL spectrometer compared with the EDX spectrometer overlays are possible similar to those of the X-ray spectra. This is caused by the broad edge structure which limits the unique interpretation of such an EEL spectrum.

The characteristic energy losses are almost equal to the energies of the absorption edges of the X-radiation (cf. Fig. 9.23 and Sect. 10.20).

Because of the comparatively poor energy resolution of the energy dispersive X-ray spectrometer the splitting of the energy levels within the shells in EDXS normally does not play any role and only the mean values of the absorption energies of several shells are listed in this case.

In principle the plasmon losses differ from element to element, too. However, the differences are small and depend sensitively on the chemical bonding of the element. Furthermore they lie in the low-loss region and are overlaid by a high background so that the qualitative element analysis with the help of plasmon losses is done only in exceptional cases.

9.3.4 Background and Multiple Scattering: Requirements to the Sample

In Fig. 9.19b we have seen that the element-specific edges only weakly rise above the hyperbolically decreasing background and they can often be identified only after background subtraction. Now, we would like to think about the reasons for this

9.3 Electron Energy Loss Spectroscopy ("EELS")

background, how it can be mathematically approximated, and what are the possibilities for reducing it.

Reasons for the background are non-characteristic energy losses of the primary electrons: the excitation of *bremsstrahlung*, electrons which are completely removed from the atomic shells are transferred into the vacuum with any kinetic energy (*secondary electrons*), and transitions between energy levels in valence and conduction band (*inter-band transitions*) are some of the possibilities. Furthermore it is possible that a primary electron suffers a characteristic energy loss and additionally non-characteristic ones before leaving the sample. In this way it contributes to the background, too.

Obviously, to get a high net edge signal an optimisation of the specimen thickness is possible: The probability of characteristic energy losses increases with the number of the participating atoms, i.e. with the specimen thickness. On the other hand, thereby the probability of a subsequent non-characteristic energy loss also increases and, consequently, such electrons do not contribute to the net edge intensity. The probability of non-characteristic energy losses is given by the curve of the energy loss spectrum itself. Besides the peak of the elastically scattered electrons the low-loss region offers the highest intensity (Fig. 9.19) and thereby the largest probability for a subsequent interaction. In the mathematical sense, each electron energy loss edge is convoluted with the zero- and low-loss region of the spectrum (Sect. 10.24). The removal of the multiple scattering parts from the spectrum can be effected by the deconvolution of the spectrum. Of course, a spectrum including the zero- and low-loss region must be measured at the same specimen position as the core-loss spectrum.

To quantitatively gather the background we have to find a formalism for the background curve $U(\Delta E)$. We do not want to discuss the physical fundamentals of this curve but confine ourselves to a purely mathematical approximation. At least in small energy intervals of 200 eV ... 300 eV it can be very well performed by the function

$$U(\Delta E) = U_1 \cdot \left(\frac{\Delta E}{\Delta E_1}\right)^{-a} \quad (9.43)$$

(U_1: background height at energy loss ΔE_1, a: approximation parameter). In Fig. 9.24 it is demonstrated for a nickel-manganese layer as an example.

For the determination of a we need at least two supporting points $U_1(\Delta E_1)$ and $U_2(\Delta E_2)$. Because of the noise in the measured curves small windows are used as supports and the background heights U_1 and U_2 are the mean values within these windows (Fig. 9.24a). Using these, the parameter a can be calculated by

$$a = \ln\left(\frac{U_1}{U_2}\right) \bigg/ \ln\left(\frac{\Delta E_2}{\Delta E_1}\right). \quad (9.44)$$

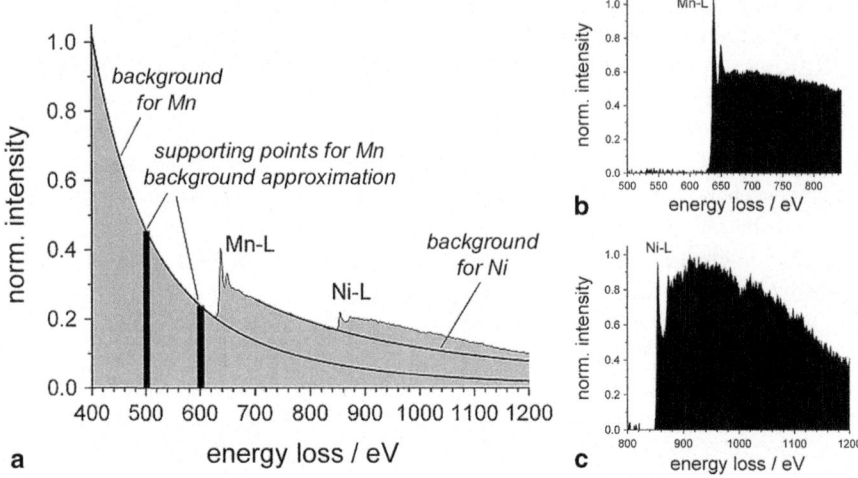

Fig. 9.24 Background approximation and subtraction in the EEL spectrum of a Mn-Ni layer. **a** Original spectrum with marked background for the Mn-L edge and the Ni-L edge. **b** Profile of the Mn-L edge after background subtraction. **c** Profile of the Ni-L edge after background subtraction

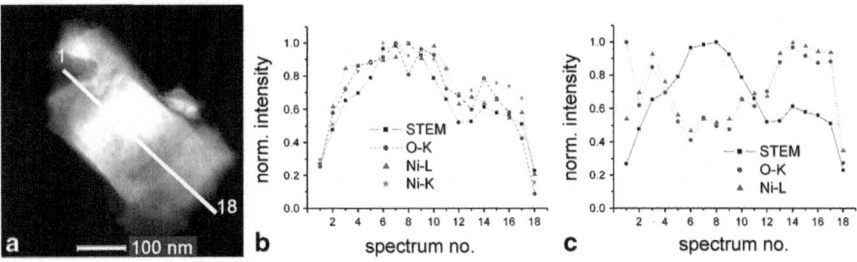

Fig. 9.25 Net intensities at varying specimen thickness. **a** STEM-HAADF image with marked line (positions of the electron beam at which the spectra are acquired). **b** EDXS signal. **c** EELS signal

Most analysis programs for EEL spectra use every measuring point within a selected energy range to fit the background parameters (i.e. they use a higher number of supporting points).

We realise that the background is well modelled for the respective energy loss edge. On the other hand, we also realise that the measured curve and the modelled background do not come together within the energy loss interval up to 1,200 eV, since the width of the energy loss window selected to determine the net edge intensity has an influence on the result (Fig. 9.24b, c).

Let us come back to the optimisation of the specimen thickness to get a high net edge signal. We would like to demonstrate it for the example of a nickel oxide crystal with varying thickness (Fig. 9.25).

In the STEM-HAADF image (Fig. 9.25a—cf. also Sect. 8.4) we mainly observe a mass thickness contrast, i.e. for constant chemical composition it is thickness contrast: the thicker the sample at the measured position the brighter is the image pixel

9.3 Electron Energy Loss Spectroscopy ("EELS") 213

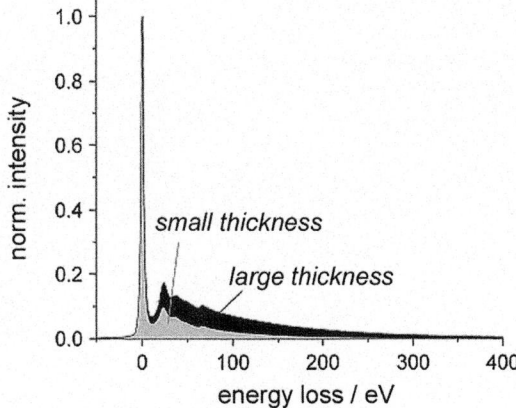

Fig. 9.26 Comparison of zero-loss and low-loss region of two EEL spectra acquired of a nickel oxide crystal at positions with different specimen thickness

at this position. In Fig. 9.25a a white line is drawn, along which eighteen EDX and EEL spectra and the brightness of the STEM-HAADF signal were acquired.

In Fig. 9.25b the brightness ("STEM") as well as the EDX intensities of the O-K peak, the Ni-L peak, and the Ni-K peak are drawn against the number of the spectrum. We realise in agreement with the expectation: The four curves correlate, the thicker the sample the higher is the EDX signal.

The situation is completely changed for the EELS signals (Fig. 9.25c): In the spectra recorded at the positions with the largest sample thickness the net edge intensities of O-K and Ni-L decrease because of the larger amount of multiple inelastic scattering and the limited width of the registration window for the determination of the net intensity. This indicates the existence of an optimal specimen thickness which is obviously exceeded at the positions with the largest specimen thickness. To quantify the optimal thickness we have to approach the measurement of the specimen thickness by electron energy loss spectroscopy in the next paragraph.

An additional point that is to mention here is the choice of the registration window. In principle, we would have to sum all the net intensities behind the onset of the edge. However, there are two problems:

1. There are other edges that overlap the signal and
2. the background extrapolation becomes worse for larger energy ranges.

Both factors limit the accuracy, therefore, a compromise for the registration width has to be found.

9.3.5 Measurement of the Specimen Thickness

For understanding a possibility to measure the specimen thickness by EELS let us have a look at Fig. 9.26. In this figure the peaks of the elastically scattered electrons together with the low-loss regions are compared for two nickel oxide spectra recorded at positions with different thicknesses of the nickel oxide crystal. The

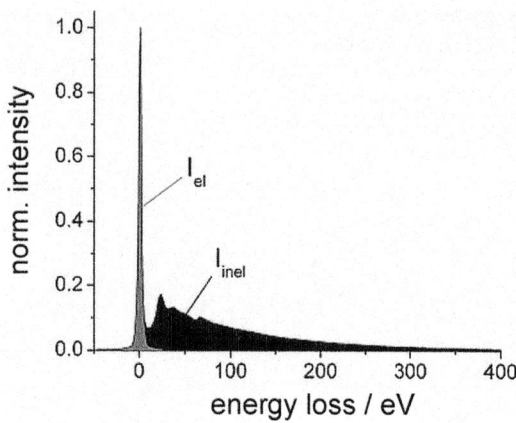

Fig. 9.27 Distinction of the intensity I_{el} within the elastic peak and the intensity I_{inel} of the inelastically scattered electrons

spectra are normalised, i.e. the intensity of the channel with energy loss zero is set to one.

We realise that the background within the low-loss region increases relative to the zero-loss peak at larger specimen thickness. This is expected since the probability of inelastic interactions increases with increasing specimen thickness.

This effect is used to measure the specimen thickness. An EEL spectrum is needed like that of Fig. 9.26, i.e. a spectrum including the peak of the elastically scattered electrons (Fig. 9.27). The portions of the elastically and the inelastically scattered electrons are mathematically separated (areas under the spectrum curve) as illustrated in Fig. 9.27.

Similar to the elastic scattering (cf. Sect. 6.1) we define a *mean free path of inelastic scattering* Λ_{inel}. Within this length the portion of the elastically scattered electrons in the total intensity I_0 is reduced to $1/e$ (e = 2.71828..., Euler's number):

$$I_{el} = I_0 \cdot e^{-t/\Lambda_{inel}} = (I_{el} + I_{inel}) \cdot e^{-t/\Lambda_{inel}} \tag{9.45}$$

and, respectively

$$\frac{t}{\Lambda_{inel}} = \ln\left(\frac{I_{el} + I_{inel}}{I_{el}}\right) \tag{9.46}$$

(*t*: specimen thickness). With it we get the specimen thickness as multiple of the mean free path of inelastic scattering. For the evaluation of Λ_{inel} approximations exist, e.g. an equation by R. F. Egerton [12]:

$$\frac{\Lambda_{inel}}{nm} \approx \frac{106 \cdot F \cdot E_0 / \text{keV}}{E_m \cdot \ln\left(\frac{2 \cdot \beta}{\text{mrad}} \cdot \frac{E_0 / \text{keV}}{E_m}\right)} \tag{9.47}$$

9.3 Electron Energy Loss Spectroscopy ("EELS")

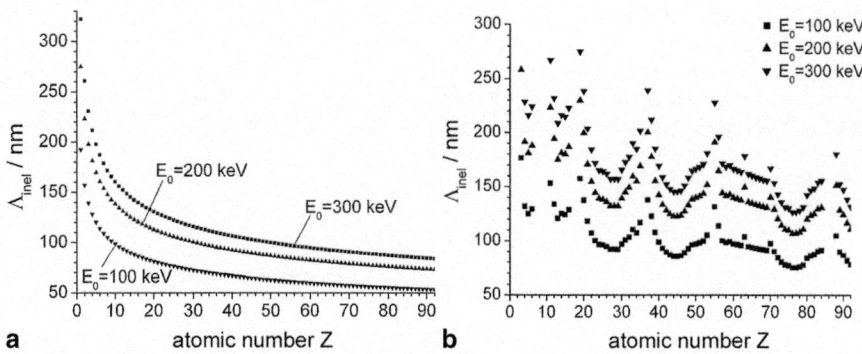

Fig. 9.28 Mean free path of inelastic scattering in dependence on the atomic number for three primary electron energies and an acceptance angle of 10 mrad. **a** After Eq. (9.49). **b** After Eq. (9.50)

(β: acceptance angle, i.e. maximum angle of inelastically scattered electrons detected by the spectrometer) with the relativistic correction factor

$$F = \frac{1+E_0/1022 \text{ keV}}{(1+E_0/511 \text{ keV})^2} \qquad (9.48)$$

and the atomic number-dependent quantity E_m [13]

$$E_m \approx 7{,}6 \cdot Z^{0{,}36} \qquad (9.49)$$

with Z as weighted atomic number of the involved elements. Other authors [14] suggest a (more plausible) dependence on the material's density ρ instead of that on the atomic number. Following this experimentally well-supported suggestion the (a bit simplified) formalism is given by

$$\frac{\Lambda_{\text{inel}}}{\text{nm}} \approx \frac{18{.}2 \cdot F \cdot E_0/\text{keV}}{\left(\rho/\text{g} \cdot \text{cm}^{-3}\right)^{0{,}3} \cdot \ln\left[\frac{1}{2} + \left(\frac{F \cdot E_0/\text{keV} \cdot \beta/\text{mrad}}{7{.}8 \cdot \left(\rho/\text{g} \cdot \text{cm}^{-3}\right)^{0{,}3}}\right)^2\right]} \qquad (9.50)$$

In Fig. 9.28 we realise the order of the mean free path of inelastic scattering: For the parameters commonly used in transmission electron microscopes it amounts to about 50 nm ... 300 nm.

Using Eq. (9.45) we are able to quantify the spectra which are the basis of Fig. 9.25 and, in this way, to find out an optimal thickness for EELS investigations (Fig. 9.29).

It is evident from Fig. 9.25 that the maximum net edge intensity is reached in the spectra 14–19 and its "wrong" reduction arises at the spectra 4–10. From Fig. 9.29

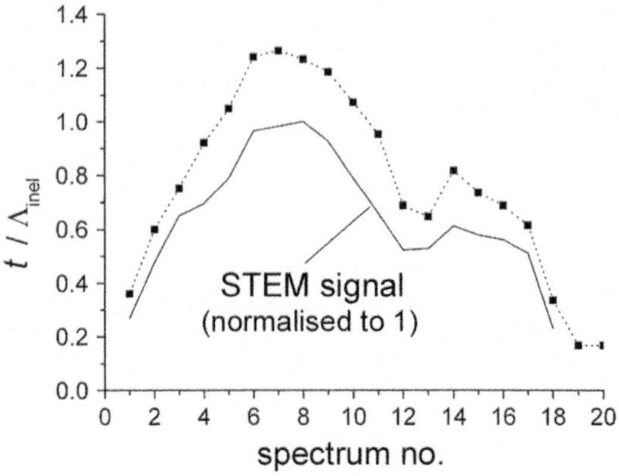

Fig. 9.29 Ratio of specimen thickness t and mean free path Λ_{inel} of inelastic scattering in the EEL spectra of NiO (cf. Fig. 9.25). By comparison the (normalised to 1) STEM signal of Fig. 9.25 is also drawn

Fig. 9.30 Regions and differences of the EELS edge fine structure

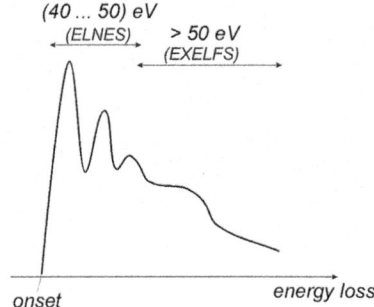

we retrieve a ratio $t/\Lambda_{inel} > 0.9$ for the spectra 4–10 and t/Λ_{inel} between 0.2 and 0.8 for the spectra 14–18. In practice it had been proved that the specimen thickness t/Λ_{inel} of (0.3 … 0.4) is optimal for EELS investigations.

9.3.6 Edge Fine Structure: Bonding Analysis

From Sect. 9.3.2 we know that the density of free (energy) states reflects the edge fine structure and, insofar, the kind of the chemical bond between the atoms. However, this only refers to the region of the edge fine structure close (i.e. in an energy distance of 40 eV … 50 eV) to the onset of the ionisation edge. The name of this region is "Energy Loss Near Edge Fine Structure" leading to the abbreviation ELNES for the method of bonding analysis by EELS.

ELNES is continued by an edge fine structure extending up to more than 100 eV from the onset ("Extended Energy Loss Fine Structure"—EXELFS, cf. Fig. 9.30). Interferences of the electron waves scattered at the atomic arrangement are responsible for this extended structure of the edge.

Firstly, we would like to discuss the near edge fine structure. The calculation of the band structure of solids and, connected with it, the density of states needs

9.3 Electron Energy Loss Spectroscopy ("EELS")

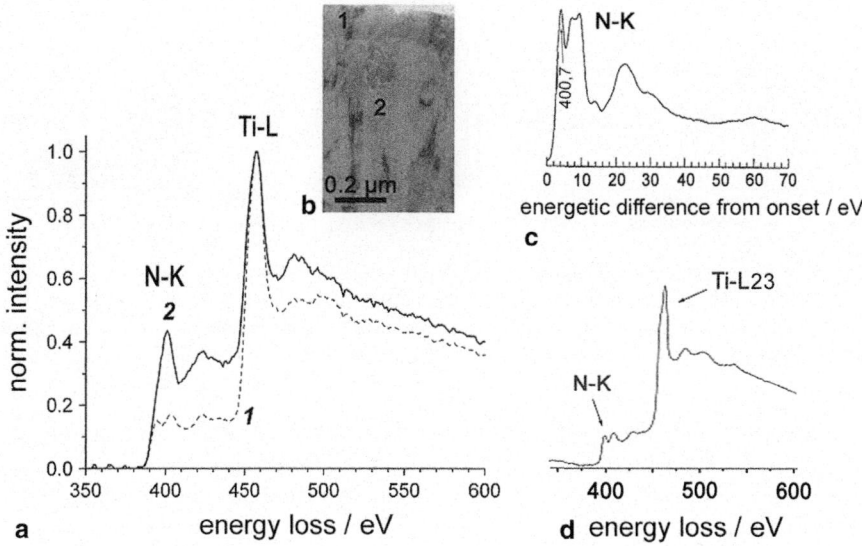

Fig. 9.31 Near edge fine structure of N-K. **a** Measurement of two EEL spectra at the positions 1 and 2. **b** Detail of a TEM brightfield image with marking the positions 1 and 2. **c** N-K edge of hexagonal AlN [19]. **d** N-K edge of TiN [20]. Obviously the nitrogen is preferably bound to Ti at position 1, however to Al at position 2

the solution of Schrödinger's or Dirac's[5] equation within the solid based on the potentials created by atomic nuclei and electrons. There are approximations for doing this, e.g., among others, the density functional theory [15–18]. Alternatively, multiple-scattering approaches may be employed which yield a correlation of the ELNES features with the local coordination shells around the scattering centre.

Commonly, the practitioner often eschews such calculations and uses a kind of "fingerprint", i.e. he compares the measured fine structure with that of well-known chemical compounds and identifies the measured kind of bonding in this way. Figure 9.31 shows this for the example of hexagonal aluminium nitride and titanium nitride.

For comparison of the spectra essential features have to be recognised which characterise the differences even between similar spectra. Such features can be:

- the exact energy loss at the onset position ("chemical shift"),
- monotone or non-monotone slope starting at the onset ("pre-peak" or not),
- the energetic differences of the maxima close to the onset (e.g. L3-L2 transitions) as well as their intensity ratios.

The quantification of such differences allows, for example, the determination of the ratio of diamond-like to graphitic carbon in Cr-doped carbon films [11] or the determination of the manganese oxidation numbers in lanthanum-strontium manganates [21, 22].

---

[5] Paul Dirac, British physicist, 1902–1984, Nobel prize in physics in 1933.

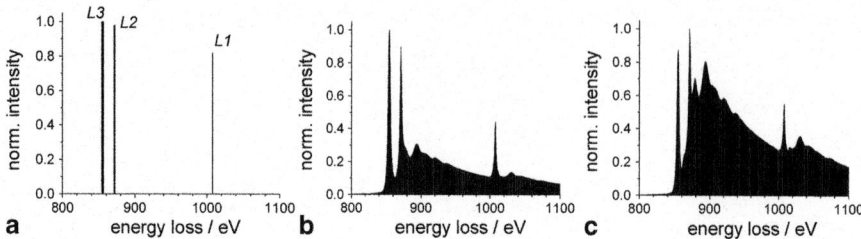

Fig. 9.32 Influence of the specimen thickness on the edge fine structure. **a** Model of the L1, L2, and L3 edge of nickel (intensities calculated via Eq. (10.374)). **b** Convolution of the model with the zero- and low-loss region measured at a specimen thickness of about a half of the mean free path of inelastic scattering. **c** As (**b**) but with a specimen thickness of about 1½ times of mean free path

Especially when own measurements are to be compared with references the influence of the energetic spectrometer resolution and the energy spread of the primary electrons have to be considered. They essentially determine how far details of the edge fine structure can be perceived, anyway. Beside the quality of the spectrometer itself (aberrations of electron optics, stability of supply voltages and currents) the energy spread of the electrons emitted by the cathode (cf. Sect. 2.5) essentially decides on the reachable energy resolution in energy loss spectra. For high resolution the electron microscope should be equipped by at least a Schottky field emission cathode (energy spread ≤ 1.5 eV in dependence on the operating parameters during the measurements). A cold field emission cathode decreases the energy spread to about 0.3 eV, it can be furthermore reduced by use of a monochromator (e.g. a Wien filter—cf. Sect. 7.9), nevertheless at cost of intensity.

In this context the specimen thickness is important, too. Because of multiple scattering the edge fine structure is convoluted with the shape of the low-loss region (Fig. 9.32).

The energy resolution of the complete system (electron gun and spectrometer) determines the width of the elastic (zero-loss) peak. It influences the spectrum together with the multiple scattering mainly reflected in the low-loss region. In principle it is possible to correct this mathematically by deconvolution of the spectrum with the zero- and low-loss region. However, this method is dangerous: The noise can be eventually amplified in a way that a fine structure results as an artefact.

To exploit the energy resolution a spectrometer dispersion (cf. Sect. 9.3.1) is recommended for which the full width half maximum of the zero-loss peak extends over five channels. Of course, another dispersion can also be selected, e.g. if a larger energy loss range is to be recorded in only one spectrum and the edge fine structure is not of interest in the measurement.

In comparison with the mentioned influence factors the *natural line width* (decay width) often plays only a secondary role. It is possible to clarify this natural line width by means of Heisenberg's[6] uncertainty principle

$$\Delta x \cdot \Delta p \geq h \tag{9.51}$$

---

[6] Werner Heisenberg, German physicist, 1901–1976, Nobel prize in physics in 1932.

9.3 Electron Energy Loss Spectroscopy ("EELS")

(Δx: uncertainty of the position, Δp: uncertainty of the momentum, h: Planck's constant). After substitution of the differences by differentials we get

$$dx \cdot dp \geq h. \tag{9.52}$$

In non-relativistic (classical) approximation we can write:

$$v \cdot dt \cdot m \cdot dv \geq h \tag{9.53}$$

(v: velocity, m: mass). With $v \cdot dv = d(v^2/2)$ it follows

$$d\left(\frac{m}{2}v^2\right) \cdot dt = dE \cdot dt \geq h, \tag{9.54}$$

i.e. the line width dE is determined by the life time dt of the states (energy levels) which are involved into electronic transitions. The order of life time evaluated e.g. by G. Wentzel [10.15] is $1/(10^9 \cdot Z^4)$ s (Z: atomic number). Using this approximation we obtain from Eq. (9.54) line widths between 0.003 eV and 3 eV for atomic numbers between 5 (boron) and 30 (zinc).

The extended edge fine structure is generated by the influence of the next and more distant neighbour atoms on the scattering process. The wave of the electron which has been excited from a core shell (ionisation) is scattered (reflected) at the neighbour atoms and interferes with the initial wave. The wavelength depends on the energy which was transferred from the primary electron during the ionisation process, i.e. it depends on the energy loss of the primary electron. On the other hand, the distance of the next neighbour atoms determines the path difference of the waves interfering with each other. With it, the distances of the neighbour atoms influence these energy losses at which constructive and destructive interference arises. Similar to the analysis of diffraction diagrams of amorphous materials it is possible to determine next neighbour distances by measuring the extended edge fine structure. Since the modulations in the energy loss spectrum arise behind an element-specific ionisation edge it is clear to what element the reference atom appertains. Unfortunately, often the modulations are weakly developed, so that noise can lead to wrong conclusions. Furthermore, other edges like e.g. L1-edges often overlap the EXELFS modulations leading to misinterpretations, too.

9.3.7 Quantifying Energy Loss Spectra

In principle quantifying EEL spectra is similar to that of EDX spectra (cf. Sect. 9.2.3): The ratio of two element concentrations c_A and c_B is proportional to the ratio of the net intensities I_A und I_B of the corresponding element-specific loss edges. The proportionality factor is equal to the reciprocal ratio of both ionisation cross sections (excitation probabilities of the loss edges A and B) Q_A and Q_B:

$$\frac{c_A}{c_B} = \frac{Q_B}{Q_A} \cdot \frac{I_A}{I_B} \tag{9.55}$$

However, there are four serious differences which make the quantifying of the EEL spectra more difficult:

1. Commonly the background in an EEL spectrum is much higher than that of an EDX spectrum and it varies strongly with the mass thickness and the vicinity of additional loss edges, i.e. the background approximation influences essentially the determination of the net edge intensities.

2. Generally the spectrum curve does not reach the background level even at larger energetic distance from the characteristic edge (Fig. 9.24) so that the width of the energy window for the measurement influences the net intensity, too.

3. The emission of X-rays occurs isotropically into the complete solid angle. Dependencies on the crystallographic orientation [23] are weak and do not play any role for the quantification in practice. In contrast, for the inelastically scattered electrons the scattering angle θ_{inel} and energy loss ΔE depend on each other. Following the classical laws of collisions (Sect. 10.19) there is the relation

$$\theta_{inel} = \frac{\Delta E}{2 \cdot E_0} \qquad (9.56)$$

(E_0: energy of the primary electrons) which connects those quantities. Therefore, in the ionisation cross sections of Eq. (9.55) beside the energy dependence the angle dependence must also be considered since only a small but variable angle range is collected:

$$Q = Q(\Delta E, \theta_{inel}). \qquad (9.57)$$

The probability for an electron losing an energy ΔE and being inelastically scattered into a certain angle is named *"generalised oscillator strength"* (*GOS*) or Bethe[7] surface, respectively [24, 25].

Therefore, the result of quantification also depends on the acceptance angle of the spectrometer, which is determined by the geometry, the lens settings of the transmission electron microscope (imaging or diffraction mode) as well as by the diameter of the contrast and spectrometer entrance aperture. Normally, one can find advice for the calculation of the acceptance angle in the operating instructions of microscope and spectrometer. The angle amounts to an order of 5 mrad ... 30 mrad. It should be larger than θ_{inel}. At convergent illumination (e.g. STEM mode) the acceptance angle should be larger than the angle of convergence of the illumination [26].

To calculate the ionisation cross sections (solution of Schrödinger's equation) different approximations exist as part of the quantification software (cf. e.g. [27]). Because of the variety of influence parameters a comparison with measurements on standards is especially important [28].

4. The edge profiles are convoluted with the zero- and low-loss region of the spectrum, i.e. deconvolution of the measured spectrum is desirable despite the problems connected with it (cf. Sect. 9.3.6).

---

[7] Hans Bethe, German/American physicist, 1906–2005, Nobel prize in physics in 1967.

9.4 Energy Filtered Imaging

As described for the energy dispersive X-ray spectroscopy (cf. Sect. 9.2.4) line scan profiles and two-dimensional distribution maps can be measured by EELS with extraction of the net edge intensities in STEM mode of the electron microscope. Doing this an EEL spectrum has to be recorded at each "pixel" position of the focussed electron probe. The analysis is a bit more sophisticated than in case of EDXS since the loss spectra have to be background-corrected, anyhow, to determine the net edge intensity. If the differences in the edge fine structure are sufficient, bonding-specific distribution maps are possible, too.

In respect of the electron energy loss spectroscopy there is another possibility to acquire elemental maps: the *energy filtered transmission electron microscopy (EFTEM)*. An "imaging energy filter" is the precondition for doing it. This is an electron-optical system which transfers a real intermediate image into another one; besides it includes an electron prism. In principle this unit can be integrated into the projective lens system ("in-column filter") or connected in series after the final image plane ("post-column filter"). The most important property of this imaging filter is the creation of an *energy-selective plane*, i.e. a plane in which all electrons with the same energy meet in one point (or in one line, respectively—cf. Sect. 9.3.1).

We have already learnt about such special planes (cf. Sect. 2.7.2): The back-focal plane of the objective lens represents an *angle-selective plane*. The diffraction pattern is generate in it. Using the contrast aperture positioned in this plane strongly scattered electrons are removed from the ray path. In this way mass thickness (scattering absorption) and diffraction contrast are arranged.

The real intermediate plane of the objective lens is a *position-selective plane*. With an aperture located in this plane we select a field of view and, with it, an area on the specimen which is the information source of the diffraction pattern (selected area electron diffraction—cf. Sect. 5.3).

The energy-selective plane makes possible to select electrons of a predefined energy, more precisely: of a predefined energy range (because of the finite aperture size). Only these electrons remain in the ray path and contribute to the image brightness. Because of the electron-optical properties of the electron prism the electrons with one and the same energy do not meet in a point but in a line. Therefore, a slit is used instead of a circular aperture.

The procedure will be explained by means of the example shown in Fig. 9.33. Transferred to an EEL spectrum the slit works like an energy window. By setting this energy window on an element-specific electron energy loss, e.g. on 532 eV (oxygen—Fig. 9.33b), all oxygen-rich sample areas should be bright in the image.

This is not the case in Fig. 9.33c. We know that in the brightfield image of Fig. 9.33a the tantalum grains appear dark because of the high atomic number (=73) followed by mass thickness contrast. In Fig. 9.33c, recorded "in the light" of electrons with energy losses between 532 eV and 557 eV, they appear brighter than the surrounding. The obvious assumption that these grains are from tantalum oxide is wrong! We did not consider the background. The tantalum grains more strongly

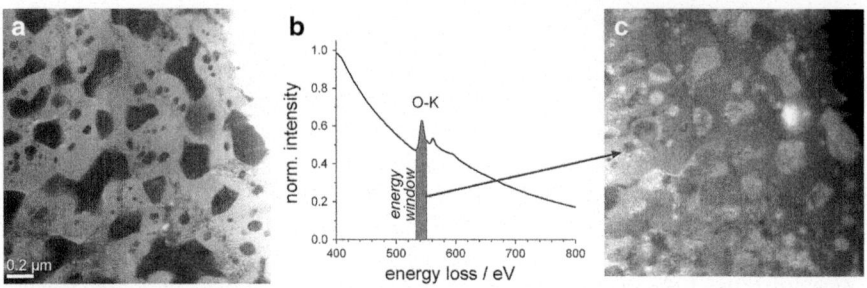

Fig. 9.33 Energy filtered imaging of a Ta-Mg-O layer without background subtraction. **a** TEM brightfield image (zero-loss peak with 10 eV energy window). **b** EEL spectrum of the sample with O-K edge and marked 25 eV wide energy window for recording the picture in (**c**)

scatter electrons resulting in a higher background which erroneously yields a higher oxygen signal at the tantalum positions. This example shows the importance of the background subtraction in EFTEM.

Mainly two methods for background treatment are used: the two- and the three-windows method. In the *two-windows method* beside the energy window already set in Fig. 9.33b on (or immediately after, respectively—*post-edge window*) the characteristic edge a second energy window immediately before the edge (*pre-edge window*) is used and an additional "pre-edge image" is recorded. The brightness values of the pre- and post-edge image are divided by each other pixel by pixel and the result is displayed as third image with the calculated ratios as brightness values (*jump-ratio method*). With it, these brightness values are a measure of how much the energy loss edge raises over the background determined immediately before the edge.

If both windows are set in the zero- and low-loss region (elastically and inelastically scattered electrons) the two-dimensional distribution of the specimen thickness can be determined in this way. A disadvantage of the two-window method is that only brightness ratios are indicated and not "true" net edge intensities.

The three-windows method avoids this disadvantage (Fig. 9.34 taken for the same sample as in Fig. 9.33). Using this method three images are recorded with different energy windows: two images before the energy loss edge (*pre-edge1* and *pre-edge2*) and one image immediately after the loss edge (*post-edge*). By use of the two pre-edge images the background brightness is approximated pixel by pixel (cf. Sect. 9.3.4, Eqs. (9.43), (9.44)) and this value is subtracted from the pixel brightness in the post-edge image. In this way the distribution map is calculated (Fig. 9.34f, map of oxygen as example). We realise that, contrary to the assumption of Fig. 9.33c, the tantalum grains do not include any oxygen.

The three images have to be taken from exactly the same sample area. This is not guaranteed because of the specimen drift and exposure times from some seconds to a few minutes per image. As a rule, the images have to be drift corrected before starting the background approximation and subtraction. It becomes an extra problem if the specimen deforms while the images are being recorded. Especially in the case of abrupt changes of the element concentration (e.g. layer boundaries) connected with an abrupt change of the background wrong elemental concentrations can be

9.4 Energy Filtered Imaging

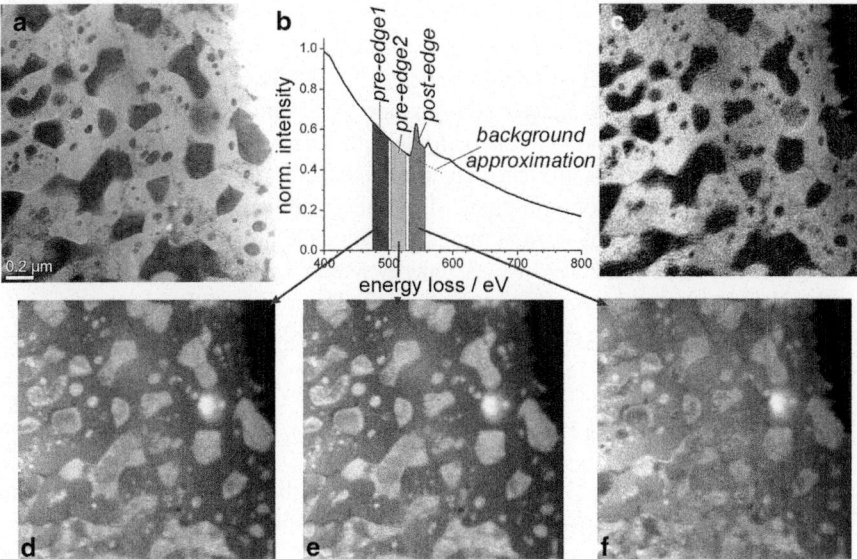

Fig. 9.34 Energy filtered imaging of a Ta-Mg-O layer using the three-window method. **a** TEM brightfield image (zero-loss peak with 10 eV energy window). **b** EEL spectrum with labelling of the energy windows used for the three images. **c** Pre-edge1 image (energy loss range 475–500 eV). **d** Pre-edge2 image (energy loss range 503–528 eV). **e** Post-edge image (energy loss range 532–557 eV). **f** Oxygen distribution map calculated using the images **c**, **d** and **e**

pretended in such areas. This is the more critical at smaller investigated structures. It is recommended to control the brightness values in the calculated maps. An accumulation of negative values indicates problems of the background approximation. It is helpful in these cases when all raw pictures are saved and not only the calculated maps as a result. Using the raw data one can try to improve the drift correction or to reach a better background approximation by averaging over a larger number of neighboured pixels.

The practitioner tries to mitigate this problem by short exposure times. This needs a setting of the electron gun which ensures a high beam current and energy windows as wide as possible. Unfortunately, a setting for a high beam current often implicates a larger energy spread of the electrons emitted by the gun and, as consequence, a broader zero-loss peak which is disadvantageous for high energy resolution in the spectrum.

Finally, we would like to discuss some electron-optical points of view in the energy filtered imaging. After selection of electrons with energy loss ΔE by a slit in the energy dispersive plane these electrons are also used for the imaging process. The image focussing happens in the "normal" TEM micrograph, i.e. mainly by use of elastically scattered electrons ($\Delta E = 0$). After switching to EFTEM the focus changes because of the changed electron energy. The problem is reduced when the slit is not shifted in the energy-selective plane but the acceleration voltage increases by $\Delta U_0 = \Delta E/e$ (e: elementary electric charge). After an energy loss ΔE in the sample the electrons have got the same energy as for the initial setting of focus and alignment units again. However, this is invalid for the condenser system. It is positioned in front of the sample and works mainly with electrons of higher energy, i.e. focal lengths and the alignment of the condenser system have to be readjusted after changing the energy. As a changed illumination varies the number of electrons per area this needs to be considered especially for quantitative analyses.

The spatial resolution is influenced by the same factors as for the "normal" transmission electron microscopic imaging. The energy spread of the elastically scattered electrons used for the normal imaging amounts to about 1 eV (Schottky field emission cathode). For EFTEM we use energy window widths of more than 10 eV, so that the chromatic aberration gains influence. The radius of the chromatic aberration disk depends on the energy spread δE (cf. Sect. 2.3):

$$\delta_C = \alpha \cdot C_C \cdot \frac{\delta E}{E_0} \tag{9.58}$$

(α: objective aperture, C_C: coefficient of chromatic aberration, E_0: energy of the primary electrons). Using this radius as spatial resolution we get, e.g. with aperture $\alpha = 10$ mrad, chromatic aberration coefficient $C_C = 1.5$ mm, and primary electron energy $E_0 = 200$ keV the relation

$$\delta_C / \text{nm} = 0.075 \cdot \delta E / \text{eV}, \tag{9.59}$$

i.e. a spatial resolution of 1.5 nm at a width of the energy window of 20 eV. An improvement is possible by reducing of the energy window, however, the intensity becomes smaller, too. To get a signal which is significant above the noise background the measuring time must be increased. This is connected with the problems mentioned above (drift etc.).

In practice the energy filtered TEM needs a compromise determined by the structure size to be imaged, the specimen thickness, the shape of the loss edge, the energetic position of this edge and other edges in its neighbourhood, and the richtstrahlwert of the electron gun.

9.5 Comparison Between EDXS and EELS

For the practitioner this comparison is mainly related to the question that he asks at the beginning of analytical measurements in the transmission electron microscope: "Wherewith do I start?"

Firstly, a lot of facts favour the energy dispersive X-ray spectroscopy: The spectra are clear with low background. The influence of the specimen thickness is as expected: A thicker specimen results in higher signals. The quantification of the spectra in the scope of the explained errors is possible by "pressing a button" using the software normally delivered together with the spectrometers.

On the other hand, the electron energy loss spectroscopy needs more background knowledge: The background in the spectra is high, sometimes the element-specific loss edges can be unambiguously identified only after background subtraction. The background is strongly influenced by the specimen thickness, the net edge intensity increases at first with increasing specimen thickness but decreases later on even up to the invisibility of an edge. The specimens have to be prepared in a suitable thickness. The shape of the energy loss edges is different (white lines, delayed edges), therewith the edges are differently observable above the background. An expected but not observable edge does not mean that the expected element is not really in the specimen. The quantification of the energy loss spectra needs the input of parameters like acceptance angle and width of the energy window for the background approximation and the determination of the net edge intensity. This can discourage an inexperienced operator.

However, there are cases where EDXS does not help. The energy resolution of the energy dispersive X-ray spectroscopy amounts to about 130 eV. Overlays of X-ray peaks are not uncommon, e.g. titanium and oxygen cannot be clearly distinguished in the X-ray spectrum. Contrary, the resolution of the energy loss spectrometers comes to about 1 eV or better. Titanium and oxygen can be distinguished and bonding relations can be analysed, i.e. it is possible to decide whether it is titanium oxide or not.

Commonly, the axis of the X-ray detector lies a few millimetres above the eucentric height. Barriers on the X-ray path to the detector especially influence the low-energetic radiation of light elements. Often, the specimen must be tilted into the direction of the detector. For the investigation of interfaces and layer stacks this reduces the possibilities to adjust the interfaces regarding the electron beam direction. Such disadvantages are avoided at EELS.

Differences are given in respect of the detection efficiency: At suitable specimen thicknesses light elements can be detected with higher efficiency by EELS. For elements in the mean range of atomic numbers, between about 11 and 30, there are no significant differences, whereas EDXS is better adapted for heavier elements.

Finally some words about the spatial resolution in the STEM mode which enables the acquisition of line scan profiles and elemental maps by recording of a multiplicity of spectra (cf. Sect. 9.2.4). The elastic scattering happens in larger angles than the inelastic scattering. X-rays are emitted from all specimen areas reached by electrons, i.e. from the complete range just hit by elastically scattered electrons. However, these elastically scattered electrons do not play any role in EELS. The angle range and the excitation zone are smaller than in EDXS and are additionally limited by the entrance aperture of the EEL spectrometer. Therefore, the spatial resolution in the STEM mode is better for EELS than for EDXS.

References

1. Kramers, H.A.: On the theory of X-ray absorption and of the continuous X-ray spectrum. Philos. Mag. **46**, 836–871 (1923)
2. Kreher, K.: Festkörperphysik (p. 25). Akademie-Verlag, Berlin (1976)
3. Gatti, E., Rehak, P.: Semiconductor drift chamber—an application of a novel charge transport scheme. Nucl. Instrum. Methods Phys Res. **225**, 608–614 (1984)

4. Lechner, P., Eckbauer, S., Hartmann, R., Krisch, S., Hauff, D., Richter, R., Soltau, H., Struder, L., Fiorini, C., Gatti, E., Longoni, A., Sampietro, M.: Silicon drift detectors for high resolution room temperature X-ray spectroscopy. Nucl. Instrum. Methods Phys. Res. A **377**, 346–351 (1996)
5. Rose, A.: The sensitivity performance of the human eye on an absolute scale. J. Opt. Soc. Am. **38**, 196–208 (1948)
6. Cliff, G., Lorimer, G.W.: The quantitative Analysis of thin specimens. J. Microsc. **103**, 203–207 (1975)
7. Horita, Z., Sano, T., Nemoto, M.: Determination of the absorption-free k_{ANi} factors for quantitative microanalysis of nickel base alloys. J. Electron Microsc. **35**, 324–334 (1986)
8. Thomas, J., Gemming, T.: Shells on nanowires detected by analytical TEM. Appl. Surf. Sci. **252**, 245–251 (2005)
9. Goldstein, J.I.: Principles of thin film X-ray microanalysis. In: Hren, J.J., Goldstein, J.I., Joy, D.C. (eds.) Introduction to Analytical Electron Microscopy, p. 101. Plenum Press, New York (1979)
10. Thomas, J., Rennekamp, R., van Loyen, L.: Characterization of multilayers by means of EDXS in the analytical TEM. Fresen. J. Anal. Chem. **361**, 633–636 (1998)
11. Fan, X., Dickey, E.C., Pennycook, S.J., Sunkara, M.K.: Z-contrast imaging and electron energy-loss spectroscopy analysis of chromium doped diamond-like carbon films. Appl. Phys. Lett. **75**, 2740–2742 (1999)
12. Egerton, R.F.: Electron Energy-Loss Spectroscopy in the Electron Microscope, 2nd edn, p. 305. Plenum Press, New York. (1996)
13. Malis, T., Cheng, S.C., Egerton, R.F.: EELS log-ratio technique for specimen-thickness measurement in the TEM. J. Electron Microsc. Tech. **8**, 193–200 (1988)
14. Iakoubovskii, K., Mitsuishi, K., Nakayama, Y., Furuya, K.: Thickness measurements with electron energy loss spectroscopy. Microsc. Res. Tech. **71**, 626–631 (2008)
15. Hohenberg, P., Kohn, W.: Inhomogeneous electron gas. Phys. Rev. **136**, B864–B8871 (1964)
16. Kohn, W., Sham, L.J.: Self-consistent equations including exchange and correlation effects. Phys. Rev. **140**, 1133–1138 (1965)
17. Rez, P., Alvarez, J.R., Pickard, C.: Calculation of near edge structure. Ultramicroscopy **78**, 175–183 (1999)
18. Hérbert, C., Luitz, J., Schattschneider, P.: Improvement of energy loss near edge structure calculation using Wien2k. Micron **34**, 219–225 (2003)
19. Serin, V., Colliex, C., Brydson, R., Matar, S., Boucher, F.: EELS investigation of the electron conduction-band states in wurtzite AlN and oxygen-doped AlN(O). Phys. Rev. B **58**, 5106–5115 (1998)
20. Contreras, O., Duarte-Moller, A., Hirata, G. A., Avalos-Borja, M.: EELS characterization of TiN by the DC sputtering technique, Journ. Electron Spec. and Rel. Phenom. **105**, 129-133 (1999)
21. Riedl, T., Gemming, T., Wetzig, K.: Extraction of EELS white-line intensities of manganese compounds: Methods, accuracy, and valence sensitivity, Ultramicroscopy **106**, 284-291 (2006)
22. Riedl, T., Gemming, T., Gruner, W., Acker, J., Wetzig, K.: Determination of manganese valency in $La_{1-x}Sr_xMnO_3$, Micron 38, 224-230 (2007)
23. Spence, J. C. H., Taftø, J: ALCHEMI: a new technique for locating atoms in small crystals, Journ. of Microscopy, **130**, 147-154 (1983)
24. Reimer, L. (Ed.): Energy-Filtering Transmission Electron Microscopy, Springer, Berlin, p. 9 (1995)
25. Hofer, F.: Inner-Shell Ionization, ibidem, p. 225
26. Egerton, R. F.: Electron Energy-Loss Spectroscopy in the Electron Microscope, 2nd Edition, Plenum Press, New York, London, p. 334 et sqq. (1996)
27. Rez, P.: Electron Ionisation Cross sections for Atomic Subshells, Microsc. Microanal. **9**, 42-53 (2003)
28. Hofer, F.: Determination of inner-shell cross-sections for EELS-quantification, Microsc. Microanal. Microstruct. **2**, 215-230 (1991)

Chapter 10
Basics Explained in More Detail (with a Bit More Mathematics)

Abstract Can the intensity fluctuations generated while a wave is impinging on an edge really be explained by Huygens' principle? Do rotational-symmetric magnetic fields really possess lens properties for electrons? How can the equations for the calculation of distances and angles between diffraction reflections be deduced in the case of a non-cubic lattice? How can the efficiency of an energy dispersive X-ray detector be calculated? How does an electron prism work? We would like to answer these and similar questions in this chapter. In doing so, it is necessary that the mathematics plays a much more important role than in the chapters before. In some cases we have written little computer programs to get quantitative results using our simplified models.

10.1 Diffraction at an Edge (Huygens' Principle)

Huygens' principle is adapted for the description of diffraction effects of the wave propagation. It implies that each point of a wave front is the origin of an elementary wave and the overlay of these elementary waves yields the new wave front.

Let us imagine that a plane (electron) wave meets an edge from the top as it happens when we image a thin film with a hole in a transmission electron microscope. In Fig. 10.1a it is schematically illustrated what happens according to Huygens' principle. However, this kind of illustration is only coarse and not suitable for quantitative observations. The elementary waves are not discrete circles but their intensity is cosine-like distributed circularly propagating in our plane-view (spherical waves in space). Additionally, we have to assume that the spherical waves are very close together ("*Each point of the wave front ...*"). Considering these preconditions and summing up all intensities of all waves we obtain Fig. 10.1b.

We find the well-known fact confirmed that the wave is diffracted around the edge and shows maxima and minima ("interference patterns") behind the edge. The next step is to determine the angular distribution of the intensity behind the edge. Let us have a look at Fig. 10.2 for this purpose.

We realise that the intensity maximum is not at $\alpha = 0°$ (imaginary extension of the edge in vertical direction) but it that is shifted by some degrees. The shift increases

Fig. 10.1 Diffraction of a plane wave at an edge (Huygens' principle). **a** Sketch. **b** Result of a calculation

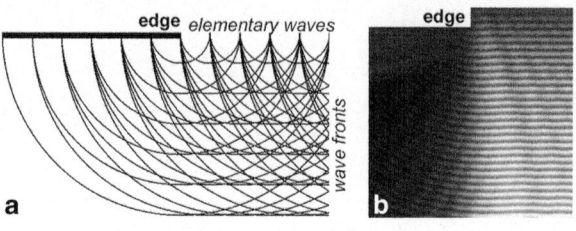

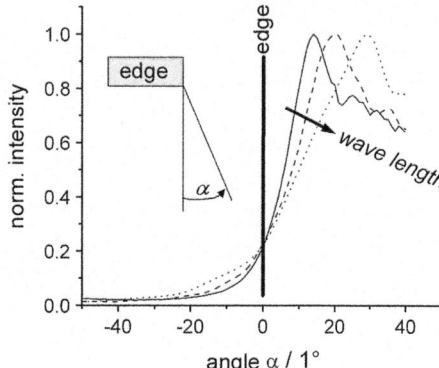

Fig. 10.2 Intensity distribution of a diffracted wave behind an edge in dependence on the angle α at different wavelengths

with increasing wavelength. This means that in observation planes which do not exactly agree with the edge plane bright stripes, so-called diffraction or Fresnel[1] fringes, appear.

10.2 Wave Function for Electrons

The word "electrons" implies that they are particles. Many results of observations can be very well explained by the particle model of the electrons: Electrons are accelerated within an electric field, they experience forces in electric and magnetic fields and are deflected, they generate defects in solids etc.

However, there are also phenomena which do not confirm with the particle model: Electrons show interferences which can only be explained using their wave character.

For a profound mathematical treatment of the phenomena affiliated to the wave character the knowledge of the position- and time-dependent wave function $\psi(\mathbf{r},t)$ is of fundamental importance and we would like to deduce this function in detail.

The quantity ψ tells us something about the conditions within the wave in dependence on position and time. The simplest way to demonstrate this is the example of a mechanical transverse wave (Fig. 10.3).

---

[1] Augustin Jean Fresnel, French physicist, 1788–1827.

10.2 Wave Function for Electrons

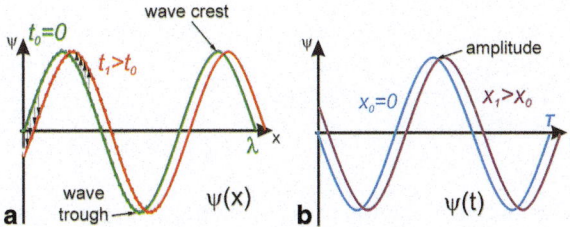

Fig. 10.3 Sketch of a mechanical transverse wave. **a** Snapshots at the times t_0 and $t_1 > t_0$. **b** Time function of two oscillators at the positions x_0 and $x_1 > x_0$

Precondition is an arrangement of oscillators (in Fig. 10.3 drawn in green and in red, respectively) coupled with each other. At the time $t=t_0=0$ let the oscillators elongate in a way that their connecting line results in the green curve.

Each oscillator moves according to a time function as illustrated in the right subframe (Fig. 10.3b), i.e. the first oscillators close to $x_0=0$ move a bit to the bottom, these after the first (green) wave crest a bit to the top etc. The connecting line of the oscillators at the time $t=t_1$ is drawn in red. Because of the motion of the oscillators the wave crest (or "phase") is shifted a bit to the right side. Please note: The oscillators do <u>not</u> move to the right side, only the phase (that is the information about the momentary elongation of the oscillators) shifts to the right!

The mathematical treatment starts with the equation of the harmonic oscillation:

$$\frac{d^2\psi(t)}{dt^2} + \omega^2 \cdot \psi(t) = 0 \tag{10.1}$$

with $\omega = 2\pi/T$ (*T*: oscillating period) and its solution

$$\psi(t) = A \cdot \cos(\omega \cdot t + \phi) \tag{10.2}$$

(*A*: amplitude=constant, ϕ: phase shift). For the oscillator positioned at x_0 with the time function drawn in blue in Fig. 10.3b we find $\phi = \pi/2$. We also realise that the time function of the oscillator at the position x_1 just requires another value of ϕ. Therefore, it is possible to integrate the position dependence into the wave function by a position-dependent phase shift $\phi(x)$:

$$\psi(x,t) = A \cdot \cos(\omega \cdot t + \phi(x) + \phi_0). \tag{10.3}$$

We suppose a constant motion velocity of the wave crest (more general: of the *phase*, i.e. a constant *phase velocity*). Obviously, the wave crest moves on by a wavelength λ while the oscillator is executing a complete oscillation with the oscillating period *T*. Therefore, the phase velocity is given by

$$c = \frac{\lambda}{T} = \lambda \cdot \nu \tag{10.4}$$

(*v*: oscillation frequency). For the wave crest

$\psi = \psi_0 = A$, i.e. $\cos(\omega \cdot t + \phi(x) + \phi_0) = 1$ and $\omega \cdot t + \phi(x) + \phi_0 = 0$ (10.5)

is valid and we get the relation

$$\phi(x) = -\omega \cdot t - \phi_0. \tag{10.6}$$

The motion of the wave (phase, respectively) in x-direction can be described by

$$x = c \cdot t \text{ and so } \phi(x) = -\omega \cdot \frac{x}{c} - \phi_0 = -\frac{2\pi}{T} \cdot \frac{T}{\lambda} \cdot x - \phi_0 = -2\pi \cdot \frac{x}{\lambda} - \phi_0. \tag{10.7}$$

The term

$$k = \frac{1}{\lambda} \tag{10.8}$$

is called wave number. Please note that also the definition $k = 2\pi/\lambda$ is common. For the wave function it follows:

$$\psi(x,t) = A \cdot \cos 2\pi (v \cdot t - k \cdot x). \tag{10.9}$$

In principle (but more difficult to understand) it is also possible to insert directly the position dependence of the wave function into the differential equation of the harmonic oscillation. In doing so, we consider the equivalence of the position ($-k \cdot x$) and the time dependence ($v \cdot t$) within the argument, change to partial derivatives and get:

$$\frac{1}{v^2} \cdot \frac{\partial^2 \psi(x,t)}{\partial t^2} - \frac{1}{k^2} \cdot \frac{\partial^2 \psi(x,t)}{\partial x^2} = 0 \tag{10.10}$$

and

$$\frac{1}{c^2} \cdot \frac{\partial^2 \psi(x,t)}{\partial t^2} - \frac{\partial^2 \psi(x,t)}{\partial x^2} = 0, \tag{10.11}$$

respectively. By two times partial differentiation and subsequent insertion of the results into the differential equation it can be shown that our wave function

$$\psi(x,t) = A \cdot \cos 2\pi (v \cdot t - k \cdot x) \tag{10.12}$$

is a solution of the partial differential Eq. (10.11):

$$\begin{aligned}\frac{\partial^2 \psi(x,t)}{\partial t^2} &= -A \cdot (2\pi \cdot v)^2 \cdot \cos 2\pi (v \cdot t - k \cdot x) \\ \frac{\partial^2 \psi(x,t)}{\partial x^2} &= -A \cdot (2\pi \cdot k)^2 \cdot \cos 2\pi (v \cdot t - k \cdot x).\end{aligned} \tag{10.13}$$

10.2 Wave Function for Electrons

The more general solution of the partial differential equation for a wave function is

$$\psi(x,t) = A \cdot e^{-2\pi i(v \cdot t - k \cdot x)} \tag{10.14}$$

what can be confirmed by two times differentiation and subsequent insertion into the differential equation.

Schrödinger's equation is the fundamental equation for matter waves. In its general form it is:

$$i \cdot \hbar \cdot \frac{\partial}{\partial t} \psi(\mathbf{r},t) = \left(\frac{-\hbar^2}{2 \cdot m} \Delta + V(\mathbf{r}) \right) \psi(\mathbf{r},t) \tag{10.15}$$

($i^2 = -1$, the imaginary unit, $\hbar = h/2\pi$, h: Planck's constant, Δ: Laplace[2] operator, V: potential energy). At first, for simplification we set V equal to zero, i.e. we discuss the wave function of free electrons moving in x direction. Therewith for Schrödinger's equation we get:

$$i \cdot \frac{\partial \psi(x,t)}{\partial t} = \frac{-\hbar}{2 \cdot m} \cdot \frac{\partial^2 \psi(x,t)}{\partial x^2}. \tag{10.16}$$

We want to check whether the general solution of the wave function (10.14) is a solution of the simplified Schrödinger's equation, too. We differentiate partially:

$$\frac{\partial \psi}{\partial t} = -2\pi i \cdot v \cdot A \cdot e^{-2\pi i(v \cdot t - k \cdot x)}$$

$$\frac{\partial^2 \psi}{\partial x^2} = -(2\pi \cdot k)^2 \cdot A \cdot e^{-2\pi i(v \cdot t - k \cdot x)} \tag{10.17}$$

and insert into Schrödinger's equation:

$$-i^2 \cdot 2\pi \cdot v \cdot A \cdot e^{-2\pi i(v \cdot t - k \cdot x)} = \frac{-\hbar}{2 \cdot m} \cdot (-4\pi^2 \cdot k^2) \cdot A \cdot e^{-2\pi i(v \cdot t - k \cdot x)}$$

$$v = \frac{\hbar}{2 \cdot m} \cdot 2\pi \cdot k^2 = \frac{h}{2 \cdot m} \cdot k^2. \tag{10.18}$$

The energy E is given by (p: momentum)

$$E = h \cdot v = \frac{p^2}{2 \cdot m} \quad \text{i.e.} \quad 2 \cdot m = \frac{p^2}{h \cdot v} \tag{10.19}$$

and it follows

$$v = \frac{h \cdot k^2}{2 \cdot m} = \frac{h \cdot v \cdot h \cdot k^2}{p^2}, \quad \text{i.e.} \quad p = h \cdot k. \tag{10.20}$$

---

[2] Pierre-Simon Laplace, French mathematician, 1749–1827.

With

$$k = \frac{1}{\lambda} \quad \text{it follows} \quad p = \frac{h}{\lambda} \tag{10.21}$$

and with it the well-known formula of de Broglie

$$\lambda = \frac{h}{p}. \tag{10.22}$$

With this we have found the wave function for electrons in a potential-free space.

Now, we would like to think about the changes if the electrons move within a constant potential Φ_0 unequal to zero, e.g. within a crystal with a mean constant potential Φ_0. It is

$$V = -e \cdot \Phi_0 \tag{10.23}$$

with the elementary electric charge e.

The electrons are to be moved in x direction. In this case Schrödinger's equation is

$$i \cdot \hbar \cdot \frac{\partial \psi(x,t)}{\partial t} = \frac{-\hbar^2}{2m} \cdot \frac{\partial^2 \psi(x,t)}{\partial x^2} - e \cdot \Phi_0 \cdot \psi(x,t). \tag{10.24}$$

Under the condition that the same function as stated above is the solution of the partial differential equation we get

$$\hbar \cdot 2\pi \cdot v \cdot A \cdot e^{-2\pi i(v \cdot t - k \cdot x)}$$

$$= \frac{\hbar^2}{2 \cdot m} \cdot (2\pi \cdot k)^2 \cdot A \cdot e^{-2\pi i(v \cdot t - k \cdot x)} - e \cdot \Phi_0 \cdot A \cdot e^{-2\pi i(v \cdot t - k \cdot x)}$$

$$h \cdot v = \frac{h^2}{2 \cdot m} \cdot k^2 - e \cdot \Phi_0 \tag{10.25}$$

$$\frac{p^2}{2 \cdot m} = \frac{h^2}{2 \cdot m} \cdot k^2 - e \cdot \Phi_0$$

$$p^2 = h^2 \cdot k^2 - 2 \cdot m \cdot e \cdot \Phi_0$$

and with it for the wavelength of the electrons within the potential Φ_0:

$$\lambda = \frac{h}{\sqrt{p^2 + 2 \cdot m \cdot e \cdot \Phi_0}}. \tag{10.26}$$

Consequently, a positive potential shortens the electron wavelength, in the potential-free space we get de Broglie's formula.

Finally, an illustrative interpretation of the electron wave function is needed. In the quantum mechanics the product of the complex wave function ψ and its complex conjugate ψ^* is interpreted as the probability of presence. This product is given by

$$\psi(x,t) \cdot \psi^*(x,t) = A \cdot e^{-2\pi i(vt-k\cdot x)} \cdot A \cdot e^{2\pi i(vt-k\cdot x)} = A^2 = |\psi(x,t)|^2. \quad (10.27)$$

The electrons transport a charge, since the square of the absolute value of the electron wave function can be interpreted as charge density.

Commonly, stationary states are contemplated in electron microscopy, i.e. the time dependence is ignored. With it and with transfer of the coordinate x to the space vector $\mathbf{r}$ we get the general electron wave function

$$\psi(\mathbf{r}) = A \cdot e^{-2\pi i \mathbf{k} \cdot \mathbf{r}} \quad (10.28)$$

which is used for calculations in electron microscopy.

10.3 Electron Wavelength Relativistically Calculated

We start with de Broglie's relation (h: Planck's constant):

$$\lambda = \frac{h}{m \cdot v}. \quad (10.29)$$

To calculate the two unknowns mass m and velocity v we use the relativistic law of energy conservation (c: light speed in vacuum, e: elementary electric charge, U_0: acceleration voltage, m_0: rest mass of the electron)

$$(m - m_0) \cdot c^2 = e \cdot U_0 \quad (10.30)$$

and the relativistic mass-velocity relation:

$$m = \frac{m_0}{\sqrt{1 - \left(\frac{v}{c}\right)^2}}. \quad (10.31)$$

Using this it follows

$$m = m_0 + \frac{e \cdot U_0}{c^2} \quad (10.32)$$

and

$$v = c\sqrt{1 - \frac{m_0^2}{\left(m_0 + e \cdot U_0 / c^2\right)^2}}. \quad (10.33)$$

Then, the wavelength is given by

$$\lambda = \frac{h}{\sqrt{e \cdot U_B \left(2 \cdot m_0 + \frac{e \cdot U_0}{c^2}\right)}}. \tag{10.34}$$

For the handling of numerical values, e.g. in computer programs, another equation is often used. From (10.30) and (10.31) it follows

$$\left(\frac{1}{\sqrt{1-\left(\frac{v}{c}\right)^2}} - 1\right) = \frac{e \cdot U_0}{m_0 \cdot c^2} = \frac{U_0}{511.06\,\text{kV}} \tag{10.35}$$

and

$$K_{rel} = \sqrt{1-\left(\tfrac{v}{c}\right)^2} = \frac{511.06\,\text{kV}}{U_0 + 511.06\,\text{kV}} \tag{10.36}$$

as relativistic correction. Then, we can write

$$\lambda = \frac{h}{m_0 \cdot c} \cdot \frac{K_{rel}}{\sqrt{1-K_{rel}^2}} = \frac{K_{rel}}{\sqrt{1-K_{rel}^2}} \cdot 2.4263 \text{ pm} \tag{10.37}$$

instead of (10.34).

10.4 Electron Beam Paths in Rotational-Symmetric Magnetic Fields

The electron beam paths within a magnetic field of the induction **B** are determined by the Lorentz force

$$\mathbf{F} = -e \cdot \mathbf{v} \times \mathbf{B} \tag{10.38}$$

(e: elementary electric charge). For calculation of the paths an initial value of velocity **v** of the electron (magnitude and direction) as well as field form **B** must be known. The Lorentz force has to be inserted into the equation of motion and the acceleration as well as the position-time function of the electrons with the aim of the initial conditions can be calculated as usual in the dynamics.

In the following we are going to confine ourselves to a usual case: The rotational-symmetric pole piece lens. It is mostly used lens in electron microscopes. Because of the rotational symmetry it is advantageous to use cylindrical coordinates r, φ, z (Fig. 10.4).

10.4 Electron Beam Paths in Rotational-Symmetric Magnetic Fields

Fig. 10.4 Cylindrical coordinates in the pole piece area within a rotational-symmetric magnetic lens

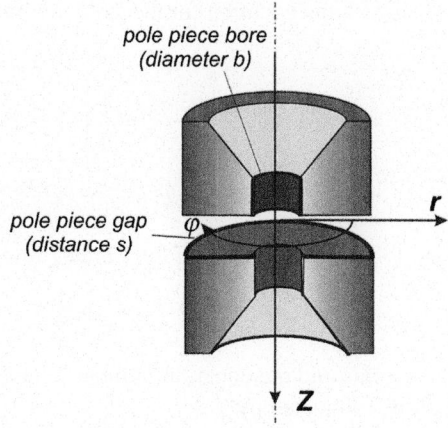

We write the equation of motion in the form

$$\mathbf{F} = \frac{d}{dt}(m \cdot \mathbf{v}). \qquad (10.39)$$

With constant mass m (i.e. in non-relativistic approximation) it is

$$\mathbf{F} = m \cdot \frac{d\mathbf{v}}{dt}. \qquad (10.40)$$

The next step is to calculate the velocity vector in cylindrical coordinates. For the position **s** (do not confuse with the pole piece gap, please) we define:

$$\mathbf{s} = r \cdot \mathbf{e}_r + z \cdot \mathbf{e}_z \qquad (10.41)$$

and for the velocity we then get:

$$\frac{d\mathbf{s}}{dt} = \dot{\mathbf{s}} = \frac{dr}{dt} \cdot \mathbf{e}_r + r \cdot \frac{d\mathbf{e}_r}{dt} + \frac{dz}{dt} \cdot \mathbf{e}_z. \qquad (10.42)$$

To deploy the temporal derivative of the unit vector in **r** direction we utilise Cartesian coordinates:

$$\mathbf{e}_r = \mathbf{e}_x \cdot \cos\varphi + \mathbf{e}_y \cdot \sin\varphi$$
$$\frac{d\mathbf{e}_r}{dt} = \frac{d\varphi}{dt} \cdot \frac{d\mathbf{e}_r}{d\varphi} = \frac{d\varphi}{dt} \cdot [\mathbf{e}_x \cdot \sin\varphi + \mathbf{e}_y \cdot \cos\varphi] = \dot{\varphi} \cdot \mathbf{e}_\varphi, \qquad (10.43)$$

i.e.

$$\dot{\mathbf{s}} = \dot{r} \cdot \mathbf{e}_r + r \cdot \dot{\varphi} \cdot \mathbf{e}_\varphi + \dot{z} \cdot \mathbf{e}_z \qquad (10.44)$$

Therewith the equation for the Lorentz force (10.38) in cylindrical coordinates is:

$$\mathbf{F} = -e \cdot \begin{vmatrix} \mathbf{e}_\ddot{\text{o}} & \mathbf{e}_z & \mathbf{e} \\ \dot{r} & r \cdot \dot{\varphi} & \dot{z} \\ B_r & 0 & B_z \end{vmatrix} \quad (10.45)$$

$$= -e \cdot \mathbf{e}_r \cdot r \cdot \dot{\varphi} \cdot B_z - e \cdot \mathbf{e}_\varphi (\dot{z} \cdot B_r - \dot{r} \cdot B_z) - e \cdot \mathbf{e}_z \cdot (-r \cdot \dot{\varphi} \cdot B_r)$$

and for the components of the Lorentz force, respectively:

$$F_r = -e \cdot r \cdot \dot{\varphi} \cdot B_z, \ F_\varphi = e \cdot \dot{r} \cdot B_z - e \cdot \dot{z} \cdot B_r, \ F_z = e \cdot r \cdot \dot{\varphi} \cdot B_r. \quad (10.46)$$

Using this and the equation of motion the temporal function of the electron motion can be calculated by

$$\dot{r}(t + \Delta t) = \dot{r}(t) + \frac{F_r}{m} \cdot \Delta t = \dot{r}(t) - \frac{e}{m} \cdot r(t) \cdot \dot{\varphi}(t) \cdot B_z(r,z) \cdot \Delta t$$

$$r(t + \Delta t) = r(t) + \dot{r}(t) \cdot \Delta t$$

$$\dot{\varphi}(t + \Delta t) = \dot{\varphi}(t) + \frac{F_\varphi}{m} \cdot \Delta t = \dot{\varphi}(t) + \frac{e}{m} \cdot (\dot{r} \cdot B_z(r,z) - \dot{z} \cdot B_r(r,z)) \cdot \Delta t \quad (10.47)$$

$$\varphi(t + \Delta t) = \varphi(t) + \dot{\varphi}(t) \cdot \Delta t$$

$$\dot{z}(t + \Delta t) = \dot{z}(t) + \frac{F_z}{m} \cdot \Delta t = \dot{z}(t) + \frac{e}{m} \cdot r \cdot \dot{\varphi} \cdot B_r(r,z) \cdot \Delta t$$

$$z(t + \Delta t) = z(t) + \dot{z}(t) \cdot \Delta t$$

with the specific charge of the electron e/m. For the further calculation we need the components B_r and B_z of the magnetic induction. According to Maxwell's equations the magnetic potential Ψ in the current-free space is given by the Laplace equation

$$\Delta \Psi = 0. \quad (10.48)$$

It means in Cartesian coordinates:

$$\frac{\partial^2 \Psi(x,y,z)}{\partial x^2} + \frac{\partial^2 \Psi(x,y,z)}{\partial y^2} + \frac{\partial^2 \Psi(x,y,z)}{\partial z^2} = 0. \quad (10.49)$$

The components B_x, B_y, and B_z of the magnetic induction **B** and the magnetic potential correlate in this way:

$$B_x = -\frac{\partial \Psi}{\partial x}, \ B_y = -\frac{\partial \Psi}{\partial y}, \ B_z = -\frac{\partial \Psi}{\partial z}. \quad (10.50)$$

The Laplace equation in cylindrical coordinates is given by

$$\frac{\partial^2 \Psi(r,\varphi,z)}{\partial r^2} + \frac{\partial^2 \Psi(r,\varphi,z)}{\partial z^2} + \frac{1}{r} \cdot \frac{\partial \Psi(r,\varphi,z)}{\partial r} + \frac{1}{r^2} \cdot \frac{\partial^2 \Psi(r,\varphi,z)}{\partial \varphi^2} = 0. \quad (10.51)$$

10.4 Electron Beam Paths in Rotational-Symmetric Magnetic Fields

In the rotational-symmetric field the potential does not depend on the azimuth φ, hence

$$\frac{\partial^2 \Psi(r,\varphi,z)}{\partial \varphi^2} = 0 \tag{10.52}$$

is valid in this case and the Laplace equation is simplified to

$$\frac{\partial^2 \Psi(r,z)}{\partial r^2} + \frac{\partial^2 \Psi(r,z)}{\partial z^2} + \frac{1}{r} \cdot \frac{\partial \Psi(r,z)}{\partial r} = 0. \tag{10.53}$$

The components of the magnetic induction are

$$B_r = -\frac{\partial \Psi}{\partial r}, \quad B_z = -\frac{\partial \Psi}{\partial z}. \tag{10.54}$$

A progression ansatz is one of the possibilities to solve the *differential equation*

$$\frac{\partial^2 \Psi(r,z)}{\partial r^2} + \frac{1}{r} \cdot \frac{\partial \Psi(r,z)}{\partial r} + \frac{\partial^2 \Psi(r,z)}{\partial z^2} = 0. \tag{10.55}$$

We consider the rotational symmetry, i.e.

$$\Psi(r,z) = \Psi(-r,z). \tag{10.56}$$

Therewith only even-numbered powers of r are possible:

$$\Psi(r,z) = K_0(z) + K_2(z) \cdot r^2 + K_4(z) \cdot r^4 + K_6(z) \cdot r^6 + K_8(z) \cdot r^8 + \ldots \tag{10.57}$$

For $r=0$ we have to set

$$\Psi(0,z) = K_0(z) \tag{10.58}$$

and the terms of the Laplace equation are:

$$\frac{\partial^2 \Psi(r,z)}{\partial r^2} = 2 \cdot K_2(z) + 12 \cdot K_4(z) \cdot r^2 + 30 \cdot K_6(z) \cdot r^4 + 56 \cdot K_8(z) \cdot r^6 + \ldots$$

$$\frac{1}{r} \cdot \frac{\partial \Psi(r,z)}{\partial r} = 2 \cdot K_2(z) + 4 \cdot K_4(z) \cdot r^2 + 6 \cdot K_6(z) \cdot r^4 + 8 \cdot K_8(z) \cdot r^6 + \ldots$$

$$\frac{\partial^2 \Psi(r,z)}{\partial z^2} = \frac{\partial^2 \Psi(0,z)}{\partial z^2} + \frac{\partial^2 K_2(z)}{\partial z^2} \cdot r^2 + \frac{\partial^2 K_4(z)}{\partial z^2} \cdot r^4 + \frac{\partial^2 K_6(z)}{\partial z^2} \cdot r^6$$
$$+ \frac{\partial^2 K_8(z)}{\partial z^2} \cdot r^8 + \ldots$$

$$\tag{10.59}$$

Insertion into the Laplace equation and comparison of the coefficients before r result in the following expressions:

$$r^0: 2\cdot K_2(z) + 2\cdot K_2(z) + \frac{\partial^2 \Psi(0,z)}{\partial z^2} = 0, \text{ i.e. } K_2(z) = -\frac{1}{4}\cdot\frac{\partial^2 \Psi(0,z)}{\partial z^2}$$

$$r^2: 12\cdot K_4(z) + 4\cdot K_4(z) + \frac{\partial^2 K_2(z)}{\partial z^2} = 0,$$

$$\text{i.e. } K_4(z) = -\frac{1}{16}\cdot\frac{\partial^2 K_2(z)}{\partial z^2} = \frac{1}{64}\cdot\frac{\partial^4 \Psi(0,z)}{\partial z^4}$$

$$r^4: 30\cdot K_6(z) + 6\cdot K_6(z) + \frac{\partial^2 K_4(z)}{\partial z^2} = 0,$$

$$\text{i.e. } K_6(z) = -\frac{1}{36}\cdot\frac{\partial^2 K_4(z)}{\partial z^2} = -\frac{1}{2304}\cdot\frac{\partial^6 \Psi(0,z)}{\partial z^6}$$

$$r^6: 56\cdot K_8(z) + 8\cdot K_8(z) + \frac{\partial^2 K_6(z)}{\partial z^2} = 0,$$

$$\text{i.e. } K_8(z) = -\frac{1}{64}\cdot\frac{\partial^2 K_6(z)}{\partial z^2} = \frac{1}{147456}\cdot\frac{\partial^8 \Psi(0,z)}{\partial z^8}.$$

Consequently, the series for the magnetic potential is:

$$\Psi(r,z) = \Psi(0,z) - \frac{r^2}{4}\cdot\frac{\partial^2 \Psi(0,z)}{\partial z^2} + \frac{r^4}{64}\cdot\frac{\partial^4 \Psi(0,z)}{\partial z^4} - \frac{r^6}{2304}\cdot\frac{\partial^6 \Psi(0,z)}{\partial z^6} + \frac{r^8}{147456}\cdot\frac{\partial^8 \Psi(0,z)}{\partial z^8} - +\dots \quad (10.60)$$

Obviously, this series can be generally written as a sum (n: number of term, starting at $n=0$)

$$\Psi(r,z) = \sum_n \frac{(-1)^n \cdot r^{2n}}{4^n \cdot (n!)^2}\cdot\frac{\partial^{2n}\Psi(0,z)}{\partial z^{2n}}. \quad (10.61)$$

We realise that the magnetic field can be completely described with the knowledge of only the function of the magnetic potential along the z axis.

To determine the magnetic field of a rotational-symmetric lens commonly the magnetic induction $B_z(0,z)$ on the optical axis (centreline) of the lens has to be measured using a small induction coil (Fig. 10.5).

10.4 Electron Beam Paths in Rotational-Symmetric Magnetic Fields

Fig. 10.5 Induction coil for the measurement of magnetic lens fields

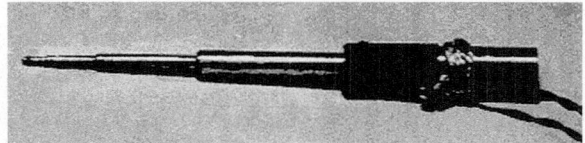

Fig. 10.6 Glaser's bell-shaped curve of the magnetic induction's z component Bz on the centreline of the lens for different widths d

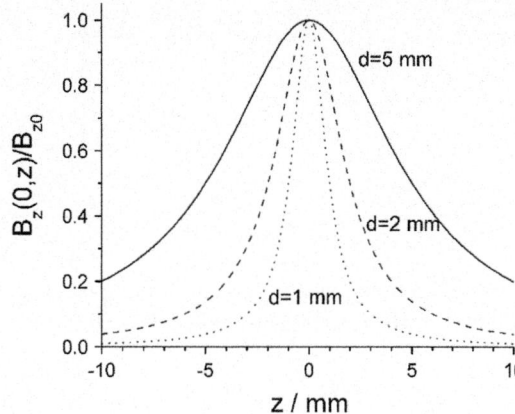

For a saturated magnetic pole piece lens the magnetic induction can be typically described by Glaser[3]'s bell-shaped curve ([1], Fig. 10.6)

$$B_z(0,z) = \frac{B_{z0}}{1+\left(\frac{z}{d}\right)^2} \qquad (10.62)$$

where d characterises the width of the bell-shaped curve. With the knowledge of this shape we are able to calculate the magnetic potential along the centreline of the lens ($r=0$):

$$B_z = -\frac{\partial \Psi}{\partial z}, \quad \text{i.e.} \quad \Psi(0,z) = -\int B_z(0,z) \cdot dz \qquad (10.63)$$

We insert the function (10.62) of Glaser's bell-shaped curve and get

$$\begin{aligned}\Psi(0,z) &= -\int \frac{B_{z0} \cdot dz}{1-\left(\frac{z}{d}\right)^2} = -B_{z0} \cdot d^2 \int \frac{dz}{d^2+z^2} \\ &= -\frac{B_{z0} \cdot d^2}{d} \cdot \arctan\left(\frac{z}{d}\right) + C.\end{aligned} \qquad (10.64)$$

---

[3] Walter Glaser, Austrian physicist and electron optician, 1906–1960.

Fig. 10.7 Curve progression of the magnetic potential on the lens centreline for different Glaser's bell-shaped curves of the z component of the magnetic induction B

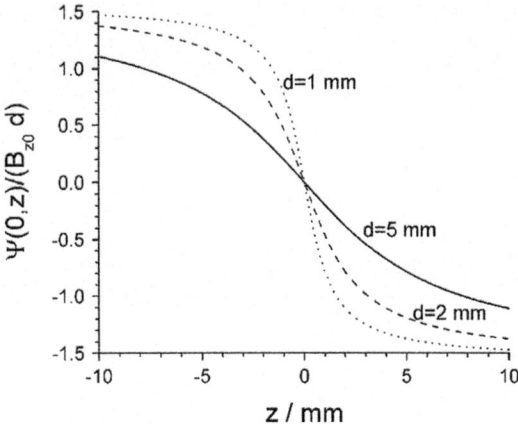

Analogous to the electric potential in a plate capacitor we demand in the centre of the lens ($z=0$) the magnetic potential $\Psi(0,0)=0$ and get (Fig. 10.7):

$$\Psi(0,z) = -B_{z0} \cdot d \cdot \arctan\left(\frac{z}{d}\right). \tag{10.65}$$

For areas close to the axis (*paraxial rays*) it is possible to truncate the series (10.60) after its third term:

$$\Psi(r,z) = \Psi(0,z) - \frac{r^2}{4} \cdot \frac{\partial^2 \Psi(0,z)}{\partial z^2} + \frac{r^4}{64} \cdot \frac{\partial^4 \Psi(0,z)}{\partial z^4}. \tag{10.66}$$

Using Glaser's bell-shaped curve for the derivations one gets

$$B_z(0,z) = \frac{B_{z0} \cdot d^2}{d^2 + z^2} = -\frac{\partial \Psi(0,z)}{\partial z} \tag{10.67}$$

$$\begin{aligned}\frac{\partial^2 \Psi(0,z)}{\partial z^2} &= -\frac{\partial}{\partial z}\left(\frac{d^2 \cdot B_{z0}}{d^2+z^2}\right) \\ &= -d^2 \cdot B_{z0} \cdot \frac{\partial}{\partial z}\left(\frac{1}{d^2+z^2}\right) = \frac{2 \cdot d^2 \cdot B_{z0} \cdot z}{(d^2+z^2)^2}\end{aligned} \tag{10.68}$$

$$\begin{aligned}\frac{\partial^3 \Psi(0,z)}{\partial z^3} &= -\frac{\partial}{\partial z}\left(\frac{2 \cdot d^2 \cdot B_{z0} \cdot z}{(d^2+z^2)^2}\right) \\ &= -2 \cdot d^2 \cdot B_{z0} \cdot \frac{\partial}{\partial z}\left(\frac{z}{(d^2+z^2)^2}\right) = 2 \cdot d^2 \cdot B_{z0} \cdot \left(\frac{z^2-d^2}{(d^2+z^2)^3}\right)\end{aligned} \tag{10.69}$$

10.4 Electron Beam Paths in Rotational-Symmetric Magnetic Fields

$$\frac{\partial^4 \Psi(0,z)}{\partial z^4} = -2 \cdot d^2 \cdot B_{z0} \cdot \frac{\partial}{\partial z}\left(\frac{z^2 - d^2}{(d^2 + z^2)^3}\right)$$

$$= -2 \cdot d^2 \cdot B_{z0} \cdot \left(\frac{4 \cdot z(2 \cdot d^2 - z^2)}{(d^2 + z^2)^4}\right) \quad (10.70)$$

$$= -8 \cdot z \cdot d^2 \cdot B_{z0} \cdot \left(\frac{(2 \cdot d^2 - z^2)}{(d^2 + z^2)^4}\right)$$

$$\frac{\partial^5 \Psi(0,z)}{\partial z^5} = -8 \cdot d^2 \cdot B_{z0} \cdot \frac{\partial}{\partial z}\left(\frac{2 \cdot d^2 \cdot z - z^3}{(d^2 + z^2)^4}\right)$$

$$= -8 \cdot d^2 \cdot B_{z0} \cdot \left(\frac{2 \cdot d^4 - 5 \cdot d^2 \cdot z^2 - z^4}{(d^2 + z^2)^5}\right) \quad (10.71)$$

and

$$\Psi(r,z) = \Psi(0,z) - \frac{r^2}{2} \cdot \frac{d^2 \cdot B_{z0} \cdot z}{(d^2 + z^2)^2} - \frac{r^4}{8} \cdot z \cdot d^2 \cdot B_{z0} \cdot \left(\frac{(2 \cdot d^2 - z^2)}{(d^2 + z^2)^4}\right)$$

$$\Psi(r,z) = \Psi(0,z) - d^2 \cdot B_{z0} \cdot r^2 \cdot z \cdot \left[\frac{1}{2 \cdot (d^2 + z^2)^2} + \left(\frac{r^2 \cdot (2 \cdot d^2 - z^2)}{8 \cdot (d^2 + z^2)^4}\right)\right]. \quad (10.72)$$

Additionally, we can calculate analytically the components $B_z(r, z)$ as well as $B_r(r, z)$ of the magnetic induction important for the Lorentz force:

$$B_r = -\frac{\partial \Psi(r,z)}{\partial r} = -\frac{\partial \Psi(0,z)}{\partial r} + \frac{\partial}{\partial r}\left(\frac{r^2}{4} \cdot \frac{\partial \Psi^2(0,z)}{\partial z^2}\right) - \frac{\partial}{\partial r}\left(\frac{r^4}{64} \cdot \frac{\partial \Psi^4(0,z)}{\partial z^4}\right)$$

$$B_r = \frac{r}{2} \cdot \frac{\partial \Psi^2(0,z)}{\partial z^2} - \frac{r^3}{16} \cdot \frac{\partial \Psi^4(0,z)}{\partial z^4}$$

$$B_r = r \cdot z \cdot d^2 \cdot B_{z0} \cdot \left(\frac{1}{(d^2 + z^2)^2} + \frac{r^2 \cdot (2 \cdot d^2 - z^2)}{2 \cdot (d^2 + z^2)^4}\right) \quad (10.73)$$

and

$$B_z = -\frac{\partial \Psi(r,z)}{\partial z} = -\frac{\partial \Psi(0,z)}{\partial z} + \frac{r^2}{4} \cdot \frac{\partial^3 \Psi(0,z)}{\partial z^3} - \frac{r^4}{64} \cdot \frac{\partial^5 \Psi(0,z)}{\partial z^5}$$

$$B_z = B_z(0,z) + \frac{r^2}{2} \cdot d^2 \cdot B_{z0} \cdot \left(\frac{z^2 - d^2}{\left(d^2 + z^2\right)^3} \right)$$

$$- \frac{r^4}{8} \cdot d^2 \cdot B_{z0} \cdot \left(\frac{2 \cdot d^4 - 5 \cdot d^2 \cdot z^2 - z^4}{\left(d^2 + z^2\right)^5} \right)$$

$$B_z = \frac{B_{z0} \cdot d^2}{d^2 + z^2}$$

$$+ \frac{r^2}{2} \cdot d^2 \cdot B_{z0} \cdot \left[\left(\frac{z^2 - d^2}{\left(d^2 + z^2\right)^3} \right) - \frac{r^2}{4} \cdot \left(\frac{2 \cdot d^4 - 5 \cdot d^2 \cdot z^2 - z^4}{\left(d^2 + z^2\right)^5} \right) \right], \quad (10.74)$$

respectively.

A lens designer needs information about the width d of Glaser's bell-shaped curve. One can derive approximately the relation

$$d \approx 0.35 \cdot b + 0.25 \cdot s \qquad (10.75)$$

from a graphics by Glaser [2] in which the quotient d/b against s/b (b: diameter of the pole piece bore, s: pole piece gap) is drawn. This approximation is valid for a saturated pole piece lens (magnetic field strength so high that the pole piece material is saturated) in the range $s/b = 0.2\ldots 1$.

Realistic values for a pole piece gap and diameter of the pole piece bore are $s = 5.4$ mm and $b = 2.2$ mm. From this it follows with Eq. (10.75) a bell-shaped curve width of $d = 2.1$ mm. The field forms resulting from this are illustrated in Fig. 10.8.

In a transmission electron microscope the specimen is located close to the centre ($z = 0$) of the pole piece. In Fig. 10.8 we can see that the radial component of the magnetic field increases from the centre to the edge. This can lead to a rupture of a magnetic specimen while it is inserted into the pole piece gap. When using a typical clamp-type single tilt holder the specimen can be withdrawn.

Figure 10.9a illustrates the r-z plane of the magnetic potential and shows a two-dimensional view of the field form. Using this form and following the Eqs. (10.54) and (10.47) the electron beam paths drawn in Figs. 10.9b and c have been calculated.

They demonstrate the typical behaviour of an optical lens: Incident parallel paths into the lens meet in the focal point (Fig. 10.9b) and paths originating from one object point meet in the image point (Fig. 10.9c). A closer look shows the problem of the rotational-symmetric magnetic lens: The external zones of the lens refract the

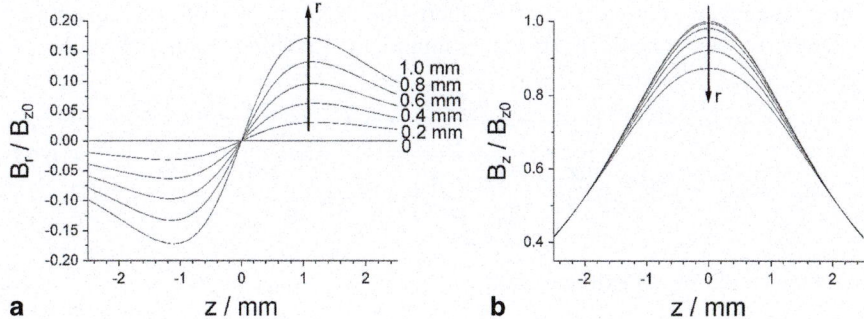

Fig. 10.8 Radial (B_r—**a**) and axial (B_z—**b**) components of the magnetic induction within a saturated pole piece lens

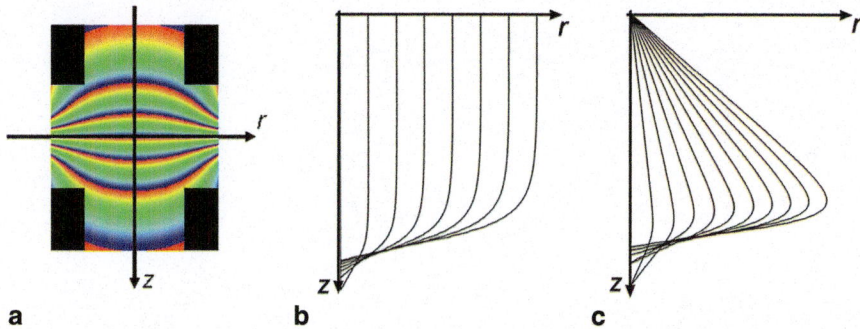

Fig. 10.9 Magnetic potential in a pole piece gap (**a**) and according electron beam path and drawn in the (twisted) *r-z*-plane (**b**) and (**c**). The electrons move on spiral paths across the magnetic lens field

paths more strongly than necessary. This fact was already described in Sect. 2.3 and is called *spherical aberration*.

10.5 Resolution Limit Considering Spherical Aberration

Following the model used here the resolution δ of a transmission electron microscope results from the overlay of diffraction and spherical aberration disk. Simplifying, we use the square root of the sum of the squares of both disks for the calculation, we set the refractive index in the surrounding of the object to $n=1$ and consider the limitation of the aperture to values $\alpha \ll 1$ as necessary because of the spherical aberration (λ: electron wave length, C_S: coefficient of spherical aberration):

$$\delta = \sqrt{\delta_B^2 + \delta_S^2} = \sqrt{\left(\frac{0.6 \cdot \lambda}{\alpha}\right)^2 + \left(C_S \cdot \alpha^3\right)^2}. \tag{10.76}$$

Due to the reverse dependence of the aberration disks on the aperture α an optimal aperture α_{opt} exists at which δ becomes minimum. The differentiation delivers

$$\frac{d\delta}{d\alpha} = \frac{-2 \cdot \frac{0.36 \cdot \lambda^2}{\alpha^3} + 6 \cdot C_S^2 \cdot \alpha^5}{2\sqrt{\left(\frac{0.6 \cdot \lambda}{\alpha}\right)^2 + (C_S \cdot \alpha^3)^2}}. \tag{10.77}$$

With the condition for extremal value

$$\frac{d\delta}{d\alpha}\bigg|_{\alpha_{opt}} = 0, \tag{10.78}$$

i.e.

$$-2 \cdot \frac{0.36 \cdot \lambda^2}{\alpha_{opt}^3} + 6 \cdot C_S^2 \cdot \alpha_{opt}^5 = 0 \tag{10.79}$$

one gets

$$\alpha_{opt} = 0.77 \sqrt[4]{\frac{\lambda}{C_S}} \tag{10.80}$$

and

$$\delta_{min} = \sqrt{\left(\frac{0.6 \cdot \lambda}{\alpha_{opt}}\right)^2 + (C_S \cdot \alpha_{opt}^3)^2} = 0.9 \sqrt[4]{C_S \cdot \lambda^3}. \tag{10.81}$$

At an acceleration voltage of 300 kV (λ=1.97 pm) with C_S=1.2 mm we get: $\alpha_{opt} \approx 4.9$ mrad and $\delta_{min} \approx 2.8$ Å. This is in good agreement with the values shown already in Fig. 2.8 (Sect. 2.4).

10.6 Schottky Effect

We imagine that the thermally emitted electrons from the cathode remain close to the cathode surface for a short time and generate a space charge cloud at this position. The electric field between electron and wire surface is equal to the field between the mentioned electron and a positive point charge in the same distance behind the wire surface (mirror charge—cf. Fig. 10.10).

In this force field an electron with charge $-e$ at distance x to the wire surface experiences the Coulomb force between the charges e and $-e$ in the distance $2 \cdot x$:

10.6 Schottky Effect

Fig. 10.10 Electric field at a metallic surface (mirror charge)

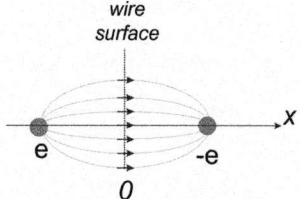

$$F_S(x) = \frac{1}{4 \cdot \pi \cdot \varepsilon_0} \cdot \frac{-e^2}{4 \cdot x^2} = \frac{-e^2}{16 \cdot \pi \cdot \varepsilon_0} \cdot \frac{1}{x^2} \qquad (10.82)$$

(with elementary electric charge e and permittivity of vacuum ε_0).

If an additional electric field E exists at the cathode the electron experiences the force

$$F_E = e \cdot E \qquad (10.83)$$

(the cathode is negative). The potential energy W under the influence of these both forces is

$$W(x) = \int (F_E + F_S(x)) \cdot dx = W_0 + e \cdot E \cdot x + \frac{e^2}{16 \cdot \pi \cdot \varepsilon_0 \cdot x} \qquad (10.84)$$

Let us look at the consequences for the potential wall model. We have to consider that in the graphics of this model the potential becomes more negative in the direction to the top. Additionally, we need a reference point ("zero point") and use the vacuum energy level for it (Fig. 10.11). In this case we define $W_0 = 0$.

It is: $W(x) = W_E + W_S$ for $x \geq 0$
and $W(x) = W_{Fermi}$ for $x < 0$.
The work function is:

$$W_A = W_{Fermi} - (W_E(x=0) + W_S(x=0)) \qquad (10.85)$$

In Fig. 10.11 a reduction of the work function can be seen. For the evaluation of this reduction we firstly determine the position x_{max} of the maximum of the function $W(x) = W_E(x) + W_S(x)$:

$$W(x) = e \cdot E \cdot x + \frac{e^2}{16 \cdot \pi \cdot \varepsilon_0 \cdot x} \qquad (10.86)$$

$$\frac{dW}{dx}\bigg|_{x=x_{max}} = e \cdot E - \frac{e^2}{16 \cdot \pi \cdot \varepsilon_0 \cdot x_{max}^2} = 0 \qquad (10.87)$$

Fig. 10.11 Reduction of the potential wall by the Schottky effect (parameter: $E = 20$ kV/cm)

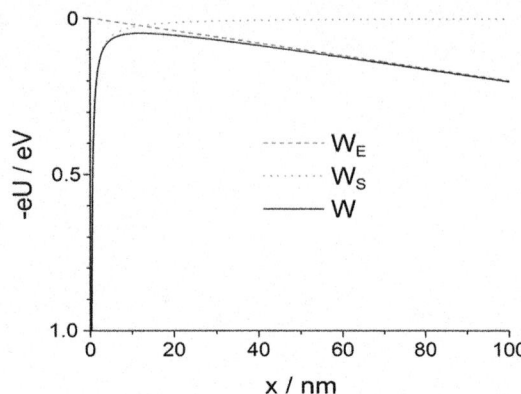

$$x_{max} = \sqrt{\frac{e}{16 \cdot \pi \cdot \varepsilon_0 \cdot E}}. \qquad (10.88)$$

We insert this value into $W(x)$ (10.86) and get for the maximum of the potential energy:

$$W_{max} = W(x_{max}) = e \cdot E \cdot x_{max} + \frac{e^2}{16 \cdot \pi \cdot \varepsilon_0 \cdot x_{max}}$$

$$= e \cdot E \cdot \sqrt{\frac{e}{16 \cdot \pi \cdot \varepsilon_0 \cdot E}} + \frac{e^2}{16 \cdot \pi \cdot \varepsilon_0 \cdot \sqrt{\frac{e}{16 \cdot \pi \cdot \varepsilon_0 \cdot E}}}$$

$$W_{max} = \sqrt{\frac{e \cdot e^2 \cdot E^2}{16 \cdot \pi \cdot \varepsilon_0 \cdot E}} + \sqrt{\frac{e^4}{(16 \cdot \pi \cdot \varepsilon_0)^2 \cdot \frac{e}{16 \cdot \pi \cdot \varepsilon_0 \cdot E}}} = 2 \cdot \sqrt{\frac{e^3 \cdot E}{16 \cdot \pi \cdot \varepsilon_0}} \qquad (10.89)$$

$$W_{max} = \sqrt{\frac{e^3 \cdot E}{4 \cdot \pi \cdot \varepsilon_0}} = 3.8 \cdot 10^{-5} \text{ eV} \cdot \sqrt{\frac{E}{V \cdot m^{-1}}}. \qquad (10.90)$$

It can be seen that the reduction of the work function by the Schottky effect is proportional to the square root of the electric field strength at the cathode tip. Using common units it follows for this reduction:

$$\Delta W_{A, Schottky} = 0.012 \text{ eV} \cdot \sqrt{\frac{E}{kV \cdot cm^{-1}}}. \qquad (10.91)$$

10.7 Electric Potential in Rotational-Symmetric Arrangements of Electrodes

For a better understanding of the functioning of an electron gun it is helpful to know the potential distribution within the gun. We want to calculate it in this paragraph. Starting point is a given arrangement of the electrodes with known constant potentials. In the case of a triode system this is the geometric arrangement of cathode, Wehnelt electrode, and anode as well as the potentials of cathode and Wehnelt electrode. Commonly, the anode potential is equal to zero, i.e. the anode is grounded. In case of the Schottky field emission gun we have the suppressor, the extractor, and the gun lens between cathode and anode instead of the Wehnelt electrode.

We are interested in the potential distribution between the electrodes. There are no charges in this area, the Laplace equation is valid for the potential U:

$$\Delta U = 0. \tag{10.92}$$

In Cartesian coordinates x, y, z that means:

$$\Delta U = \frac{\partial^2 U}{\partial x^2} + \frac{\partial^2 U}{\partial y^2} + \frac{\partial^2 U}{\partial z^2}. \tag{10.93}$$

For rotational symmetry it is advantageous to use cylindrical coordinates r, φ, z:

$$\frac{\partial^2 U(r,\varphi,z)}{\partial r^2} + \frac{\partial^2 U(r,\varphi,z)}{\partial z^2} + \frac{1}{r} \cdot \frac{\partial U(r,\varphi,z)}{\partial r} + \frac{1}{r^2} \cdot \frac{\partial^2 U(r,\varphi,z)}{\partial \varphi^2} = 0. \tag{10.94}$$

Because of the rotational symmetry it is

$$\frac{\partial^2 U(r,\varphi,z)}{\partial \varphi^2} = 0 \tag{10.95}$$

and Laplace's equation simplifies to

$$\frac{\partial^2 U(r,z)}{\partial r^2} + \frac{\partial^2 U(r,z)}{\partial z^2} + \frac{1}{r} \cdot \frac{\partial U(r,z)}{\partial r} = 0. \tag{10.96}$$

We want to solve this partial differential equation numerically. Doing this we convert the differential quotients into difference quotients:

$$\begin{aligned}
\frac{\partial^2 U(r,z)}{\partial r^2} &\rightarrow \frac{\Delta(\Delta U(r,z))}{(\Delta r)^2} = \frac{U(r+\Delta r,z) - 2 \cdot U(r,z) + U(r-\Delta r,z)}{(\Delta r)^2} \\
\frac{\partial^2 U(r,z)}{\partial z^2} &\rightarrow \frac{\Delta(\Delta U(r,z))}{(\Delta z)^2} = \frac{U(r,z+\Delta z) - 2 \cdot U(r,z) + U(r,z-\Delta z)}{(\Delta z)^2} \\
\frac{\partial U(r,z)}{\partial r} &\rightarrow \frac{\Delta U(r,z)}{\Delta r} = \frac{U(r+\Delta r,z) - U(r-\Delta r,z)}{2 \cdot \Delta r}.
\end{aligned} \tag{10.97}$$

For further simplifying we use a net with equal meshes $\Delta r = \Delta z = \Delta$ and label the mesh $U(r, z)$ with $U(i, j)$ using the relations $r = i \cdot \Delta$ as well as $z = j \cdot \Delta$. Now, the Laplace Eq. (10.96) is:

$$\frac{U(r+\Delta,z) - 2\cdot U(r,z) + U(r-\Delta,z)}{\Delta} \\ + \frac{U(r,z+\Delta) - 2\cdot U(r,z) + U(r,z-\Delta)}{\Delta} + \frac{U(r+\Delta,z) - U(r-\Delta,z)}{2\cdot r} = 0, \tag{10.98}$$

and, respectively

$$4 \cdot U(r,z) = \left(1 + \frac{\Delta}{2 \cdot r}\right) \cdot U(r+\Delta,z) + \left(1 - \frac{\Delta}{2 \cdot r}\right) \cdot U(r-\Delta,z) \\ + U(r,z+\Delta) + U(r,z-\Delta) \tag{10.99}$$

$$U(i,j) = \frac{1}{4}\left\{\left(1 + \frac{1}{2\cdot i}\right) \cdot U(i+1,j) + \left(1 - \frac{1}{2\cdot i}\right) \cdot U(i-1,j) \\ + U(i,j+1) + U(i,j-1)\right\}. \tag{10.100}$$

This is a recurrence formula, the result of the kth step can be calculated from the result of the (k−1)th step. The potential distribution at the beginning (0th step) is given by the initial conditions (potentials of the electrodes). The complete recurrence formula is

$$U_k(i,j) = \frac{1}{4}\left\{\left(1 + \frac{1}{2\cdot i}\right) \cdot U_{k-1}(i+1,j) + \left(1 - \frac{1}{2\cdot i}\right) \cdot U_{k-1}(i-1,j) \\ + U_{k-1}(i,j+1) + U_{k-1}(i,j-1)\right\}. \tag{10.101}$$

Since i appears in the denominator there is a problem for $i=0$ that must be separately solved. Doing this we use the difference equation in Cartesian coordinates:

$$\frac{U(x+\Delta x, y, z) - 2 \cdot U(x,y,z) + U(x-\Delta x, y, z)}{(\Delta x)^2} \\ + \frac{U(x, y+\Delta y, z) - 2 \cdot U(x,y,z) + U(x, y-\Delta y, z)}{(\Delta y)^2} \\ + \frac{U(x, y, z+\Delta z) - 2 \cdot U(x,y,z) + U(x, y, z-\Delta z)}{(\Delta z)^2} = 0. \tag{10.102}$$

For $r=0$ is $x=0$ and $y=0$. Additionally, we set $\Delta x = \Delta y = \Delta z = \Delta$:

$$U(\Delta,0,z) - 2\cdot U(0,0,z) + U(-\Delta,0,z) + U(0,\Delta,z) - 2\cdot U(0,0,z) \\ + U(0,-\Delta,z) + U(0,0,z+\Delta) - 2\cdot U(0,0,z) + U(0,0,z-\Delta) = 0. \tag{10.103}$$

Now, we remember the rotational symmetry again and realise:

$$U(x=0, y=0, z) = U(r=0, z)$$
$$U(x=\Delta, y=0, z) = U(x=0, y=\Delta, z) = U(r=\Delta, z) \quad (10.104)$$
$$U(x=-\Delta, y=0, z) = U(x=0, y=-\Delta, z) = U(r=\Delta, z).$$

Therewith it follows from Eq. (10.103):

$$U(\Delta, z) - 2 \cdot U(0, z) + U(\Delta, z) + U(\Delta, z) - 2 \cdot U(0, z) + U(\Delta, z)$$
$$+ U(0, z+\Delta) - 2 \cdot U(0, z) + U(0, z-\Delta) = 0 \quad (10.105)$$

and

$$6 \cdot U(0, z) = 4 \cdot U(\Delta, z) + U(0, z+\Delta) + U(0, z-\Delta) \quad (10.106)$$

as well as the recurrence formula for the calculation of the electric potential on the z axis ($r=0$):

$$U_k(0, j) = \frac{2}{3} \cdot U_{k-1}(1, j) + \frac{1}{6} U_{k-1}(0, j+1) + U_{k-1}(0, j-1). \quad (10.107)$$

Therewith it is possible to calculate the potential distribution: The start distribution is given by the initial conditions, i.e. by the electrode potentials. From this, firstly, the potential distribution on the z axis is calculated step by step following Eq. (10.107) and subsequently the complete potential distribution following Eq. (10.101). This is iteratively repeated until the change between the steps is negligibly small.

Following this method the potential distributions drawn in Fig. 10.12 have been calculated. The focussing effect of both systems can be seen at the shape of the potential lines (the force vectors are perpendicular to the potential lines).

10.8 Laue Equations and Reciprocal Lattice, Ewald Construction

Bragg's law links lattice spacings with diffraction angles. A special lattice spacing generates a diffraction reflection, i.e. a bright spot in a transmission electron microscopic diffraction pattern. What happens if two lattice planes are rotated against each other, e.g. (100) and (010) planes. In cubic crystal systems (all axes are equal and all angles are equal to 90°) these planes are perpendicular and the connecting lines between their diffraction reflections and the centre of the diffraction pattern are expected to be perpendicular, too. In this paragraph we would like the demonstrate how the reflection positions can be calculated.

Fig. 10.12 Distribution of the electric potential in rotational-symmetric electron guns. **a** Triode system. **b** Schottky field emission gun

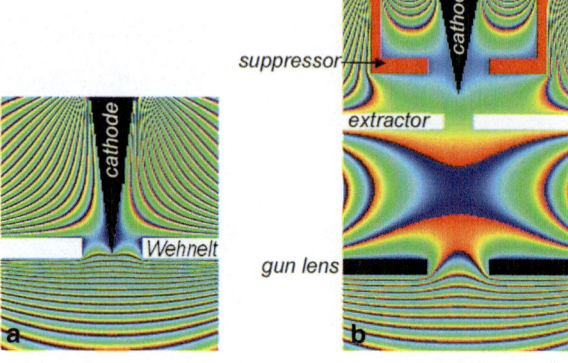

Fig. 10.13 Wave number vectors for the reflection of an electron wave at a lattice plane

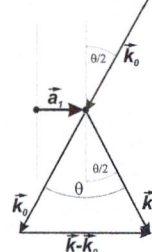

Our starting point is Bragg's law in the transmission electron microscopic form

$$n \cdot \lambda = d \cdot \theta \qquad (10.108)$$

(n: integer number, λ: wavelength, d: lattice spacing, θ: diffraction angle). Let us have a look at Fig. 10.13.

The wave number vector $\mathbf{k_0}$ of the incident wave includes the angle θ with the wave number vector $\mathbf{k}$ of the reflected wave. At small diffraction angles typical for the electron diffraction in transmission electron microscopy we can write for the difference of both wave number vectors ("*scattering vector*") with consideration of Bragg's law (10.108):

$$|\mathbf{k} - \mathbf{k_0}| = \theta \cdot |\mathbf{k}| = \frac{n \cdot \lambda}{d} \cdot |\mathbf{k}|. \qquad (10.109)$$

We postulate elastic scattering, i.e. the moduli of both wave number vectors are equal to:

$$|\mathbf{k_0}| = |\mathbf{k}| = \frac{1}{\lambda}, \qquad (10.110)$$

10.8 Laue Equations and Reciprocal Lattice, Ewald Construction

and

$$d \cdot |\mathbf{k} - \mathbf{k}_0| = n. \qquad (10.111)$$

We substitute d by $\mathbf{a}_1$ and n by the integer number h, consider that $\mathbf{a}_1$ and $\mathbf{k} - \mathbf{k}_0$ are parallel and get

$$\mathbf{a}_1 \cdot (\mathbf{k} - \mathbf{k}_0) = h. \qquad (10.112)$$

Analogous, for both other space directions we get:

$$\begin{aligned}\mathbf{a}_2 \cdot (\mathbf{k} - \mathbf{k}_0) &= k \\ \mathbf{a}_3 \cdot (\mathbf{k} - \mathbf{k}_0) &= l. \end{aligned} \qquad (10.113)$$

The Eqs. (10.112 and 10.113) are known as *Laue equations*.

In Eq. (10.109) it can be seen that the length of the scattering vector and therewith the diffraction angle θ is proportional to the reciprocal value of the lattice spacing d. Therefore it is convenient to use a *"reciprocal lattice"* to describe the atomic arrangement for diffraction calculations. The reciprocal basis vectors $\mathbf{b}_j$ are defined by

$$\mathbf{a}_i \cdot \mathbf{b}_j = \begin{cases} 1 & i = j \\ 0 & i \neq j \end{cases} \text{if} \quad (i, j = 1,2,3). \qquad (10.114)$$

Therewith we can write:

$$\begin{aligned}\mathbf{a}_1 \cdot (h \cdot \mathbf{b}_1 + k \cdot \mathbf{b}_2 + l \cdot \mathbf{b}_3) &= h \\ \mathbf{a}_2 \cdot (h \cdot \mathbf{b}_1 + k \cdot \mathbf{b}_2 + l \cdot \mathbf{b}_3) &= k \\ \mathbf{a}_3 \cdot (h \cdot \mathbf{b}_1 + k \cdot \mathbf{b}_2 + l \cdot \mathbf{b}_3) &= l. \end{aligned} \qquad (10.115)$$

After comparison with the Laue Eqs. (10.112 and 10.113) it follows

$$(h \cdot \mathbf{b}_1 + k \cdot \mathbf{b}_2 + l \cdot \mathbf{b}_3) = \mathbf{k} - \mathbf{k}_0, \qquad (10.116)$$

i.e. the scattering vector has to be a vector in the reciprocal lattice if a diffraction maximum is to be generated. The reciprocal lattice vector characterised by (hkl) is always perpendicular to the lattice planes (hkl).

Let us do a gedankenexperiment: We shift the vector $\mathbf{k}_0$ parallel until its tip hits a reciprocal lattice point and think about the direction of the vector $\mathbf{k}$ that has the same length and meets another reciprocal lattice points. The geometric solution of this problem is a spherical shell; its centre is the origin of the vector $\mathbf{k}_0$ (Fig. 10.14). This method is called *"Ewald[4] construction"*.

---

[4] Paul Peter Ewald, German physicist, 1888–1985.

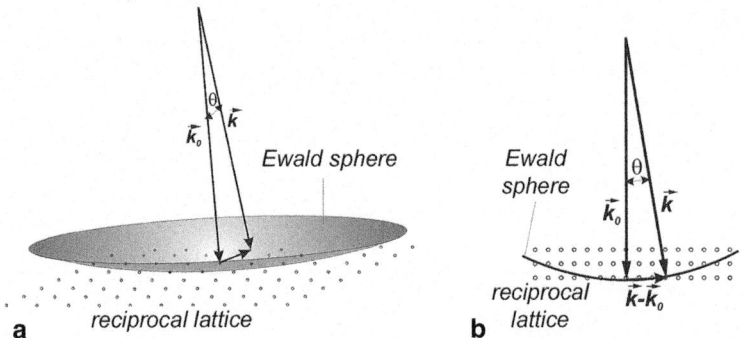

Fig. 10.14 Ewald construction. **a** Perspective illustration of the spherical shell. **b** View on the plane including the wave number vectors **k** and $\mathbf{k}_0$

Therewith we are able to phrase a rule for the construction of the diffraction pattern:

▶ We draw the wave number vector $\mathbf{k}_0$ of the incident wave (length: $1/\lambda$) in a way that its tip (arrow) hits a reciprocal lattice point. All reciprocal lattice points that are hit by the spherical shell around the origin of this vector with the radius $1/\lambda$ generate a diffraction reflection.

For the electron diffraction in the transmission electron microscope the wavelengths are in the order of 3 pm = 0.003 nm, then $1/\lambda$ amounts to about 0.3 pm^{-1} = 0.03 nm^{-1}. The atomic distances in the crystal lattice are typically in the order of 0.3 nm, that means its reciprocal value is about 3 nm^{-1}. The ratio of the radius of the Ewald sphere to the distances within the reciprocal lattice comes to about 100, i.e. it is possible to approximate the part of the spherical shell around the arrow of the vector $\mathbf{k}_0$ by a plane.

We realise that the reciprocal lattice is of fundamental importance for the interpretation of diffraction patterns. But, how can we calculate it from the "real" lattice?

The definition (10.114) of the reciprocal lattice is the starting point of the calculation. The scalar product of two vectors is equal to zero if both vectors are perpendicular to each other and it becomes a maximum for equal directions of both vectors. We realise that

$$\mathbf{b}_1 \parallel \mathbf{a}_1, \mathbf{b}_2 \parallel \mathbf{a}_2, \text{and} \mathbf{b}_3 \parallel \mathbf{a}_3 \qquad (10.117)$$

as well as that

$$\mathbf{b}_1 \perp \mathbf{a}_2, \mathbf{b}_1 \perp \mathbf{a}_3, \mathbf{b}_2 \perp \mathbf{a}_1, \mathbf{b}_2 \perp \mathbf{a}_3, \mathbf{b}_3 \perp \mathbf{a}_1, \mathbf{b}_3 \perp \mathbf{a}_2 \qquad (10.118)$$

must be valid. The cross product (also called: vector product) of two vectors is perpendicular to the plane spanned by both vectors. Using these definitions we can fulfil the requirements (10.118). Additionally, we have to consider normalisation

10.8 Laue Equations and Reciprocal Lattice, Ewald Construction

factors V_1, V_2, and V_3 to get a maximum of 1 for the scalar product as demanded by the definition of the reciprocal lattice:

$$\mathbf{b}_1 = \frac{1}{V_1} \cdot (\mathbf{a}_2 \times \mathbf{a}_3)$$

$$\mathbf{b}_2 = \frac{1}{V_2} \cdot (\mathbf{a}_3 \times \mathbf{a}_1)$$

$$\mathbf{b}_3 = \frac{1}{V_3} \cdot (\mathbf{a}_1 \times \mathbf{a}_2). \tag{10.119}$$

For the calculation of V_1, V_2, and V_3 we insert the Eq. (10.119) into the definition Eq. (10.114) and obtain:

$$\mathbf{a}_1 \cdot \mathbf{b}_1 = \frac{\mathbf{a}_1 \cdot (\mathbf{a}_2 \times \mathbf{a}_3)}{V_1} = 1 \Rightarrow V_1 = \mathbf{a}_1 \cdot (\mathbf{a}_2 \times \mathbf{a}_3)$$

$$\mathbf{a}_2 \cdot \mathbf{b}_2 = \frac{\mathbf{a}_2 \cdot (\mathbf{a}_3 \times \mathbf{a}_1)}{V_2} = 1 \Rightarrow V_2 = \mathbf{a}_2 \cdot (\mathbf{a}_3 \times \mathbf{a}_1) \tag{10.120}$$

$$\mathbf{a}_3 \cdot \mathbf{b}_3 = \frac{\mathbf{a}_3 \cdot (\mathbf{a}_1 \times \mathbf{a}_2)}{V_3} = 1 \Rightarrow V_3 = \mathbf{a}_3 \cdot (\mathbf{a}_1 \times \mathbf{a}_2).$$

Since $\mathbf{a}_1$, $\mathbf{a}_2$, $\mathbf{a}_3$ create by definition a right-handed trihedron the cyclic permutation of the vectors in (10.120) does not change the result, i.e. it is $V_1 = V_2 = V_3 = V$. V is the volume of the unit cell spanned by the three vectors $\mathbf{a}_1$, $\mathbf{a}_2$, and $\mathbf{a}_3$.

The calculation is trivial within a *rectangular (orthogonal) coordinate system*. We get $V = a_1 \cdot a_2 \cdot a_3$, the product of the moduli (lengths) of the three vectors $\mathbf{a}_1$, $\mathbf{a}_2$, and $\mathbf{a}_3$. In this case the basis vectors of the reciprocal lattice are

$$\mathbf{b}_1 = \frac{\mathbf{a}_1}{a_1}, \ \mathbf{b}_2 = \frac{\mathbf{a}_2}{a_2}, \text{ and } \mathbf{b}_3 = \frac{\mathbf{a}_3}{a_3}. \tag{10.121}$$

In *non-orthogonal cases* it becomes a bit more complicated. We want to transfer an arbitrary coordinate system $\mathbf{a}_1$, $\mathbf{a}_2$, $\mathbf{a}_3$ (triclinic lattice) into an orthogonal system with the basis vectors $\mathbf{e}_x$, $\mathbf{e}_y$, $\mathbf{e}_z$ and in this way reduce the calculation to the simple orthogonal case.

From Fig. 10.15a we read:

$$\mathbf{a}_1 = a_1 \cdot \mathbf{e}_x \tag{10.122}$$

and

$$\mathbf{a}_2 = a_2 \cdot (\mathbf{e}_x \cdot \cos \gamma + \mathbf{e}_y \cdot \sin \gamma). \tag{10.123}$$

Fig. 10.15 Triclinic and Cartesian axes of coordinates. The x axis agrees with the $\mathbf{a}_1$-axis, the x-y plane is located in the $\mathbf{a}_1$-$\mathbf{a}_2$ plane. **a** Cartesian coordinates of the $\mathbf{a}_2$ axis. **b** Cartesian coordinates of the $\mathbf{a}_3$ axis

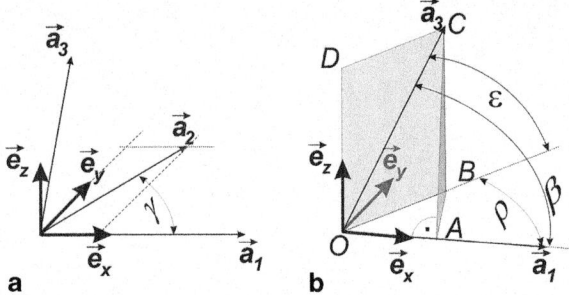

To calculate the Cartesian coordinates of the $\mathbf{a}_3$ axis we introduce the two angles ρ and ε (cf. Fig. 10.15b). The axes $\mathbf{e}_z$ and $\mathbf{a}_3$ lie in the plane OBCD. The angle ε between the x-y plane and the $\mathbf{a}_3$ axis is measured in the plane OBCD, too. The intersecting line of the OBCD plane with the x-y plane is tilted by the angle ρ against the x axis. The length $\overline{OA}$ (x component of $\mathbf{a}_3$) in the rectangular triangle OAC is

$$\overline{OA} = a_3 \cdot \cos \beta. \tag{10.124}$$

In triangle OAB the length $\overline{AB}$ (y component of $\mathbf{a}_3$) is:

$$\overline{AB} = a_3 \cdot \cos \varepsilon \cdot \sin \rho. \tag{10.125}$$

The z component of $\mathbf{a}_3$ is given by the length $\overline{BC}$. In triangle OBC we read

$$\overline{BC} = a_3 \cdot \sin \varepsilon. \tag{10.126}$$

Now, we have to determine both angles ε and ρ. We use that the length $\overline{OA}$ can be calculated in triangle OAC—as done in Eq. (10.124)—as well as in triangle OAB:

$$\overline{OA} = a_3 \cdot \cos \varepsilon \cdot \cos \rho, \tag{10.127}$$

i.e.

$$\cos \beta = \cos \varepsilon \cdot \cos \rho. \tag{10.128}$$

Up to now $\mathbf{a}_1$ was the reference axis. Performing the same calculation with the reference axis $\mathbf{a}_2$ we have to substitute the angle β (between $\mathbf{a}_1$ and $\mathbf{a}_3$ axis) by the angle α (between $\mathbf{a}_2$ and $\mathbf{a}_3$ axis) as well as the angle ρ (between $\mathbf{a}_1$ axis and intersection line of OBCD and x-y plane) by the angle $\gamma - \rho$ (between $\mathbf{a}_2$ axis and intersection line of OBCD and x-y plane), i.e.

$$\cos \alpha = \cos \varepsilon \cdot \cos(\gamma - \rho). \tag{10.129}$$

10.8 Laue Equations and Reciprocal Lattice, Ewald Construction

Therewith we have got two equations for the determination of the angles ε and ρ. With

$$\cos(\gamma - \rho) = \cos\gamma \cdot \cos\rho + \sin\gamma \cdot \sin\rho \qquad (10.130)$$

and Eq. (10.129) it is:

$$\cos\alpha = \cos\beta \cdot \frac{\cos(\gamma - \rho)}{\cos\rho} = \frac{\cos\beta \cdot \cos\gamma \cdot \cos\rho}{\cos\rho} + \frac{\cos\beta \cdot \sin\gamma \cdot \sin\rho}{\cos\rho} \qquad (10.131)$$

$$\cos\alpha = \cos\beta \cdot \cos\gamma + \cos\beta \cdot \sin\gamma \cdot \tan\rho$$

and, respectively,

$$\tan\rho = \frac{\frac{\cos\alpha}{\cos\beta} - \cos\gamma}{\sin\gamma}. \qquad (10.132)$$

With consideration of

$$\cos\rho = \frac{1}{\sqrt{1 + \tan^2\rho}} \qquad (10.133)$$

it follows with relation (10.128):

$$\cos\varepsilon = \frac{\cos\beta}{\cos\rho} = \cos\beta \sqrt{1 + \frac{\left(\frac{\cos\alpha}{\cos\beta} - \cos\gamma\right)^2}{\sin^2\gamma}} \qquad (10.134)$$

$$\cos\varepsilon = \frac{\cos\beta}{\sin\gamma} \cdot \sqrt{\sin^2\gamma + \frac{\cos^2\alpha}{\cos^2\beta} - 2 \cdot \frac{\cos\alpha}{\cos\beta} \cdot \cos\gamma + \cos^2\gamma}$$

as well as

$$(\sin^2\gamma + \cos^2\gamma = 1):$$

$$\cos\varepsilon = \frac{\sqrt{\cos^2\alpha - 2\cos\alpha \cdot \cos\beta \cdot \cos\gamma + \cos^2\beta}}{\sin\gamma}. \qquad (10.135)$$

Additionally it is

$$\cos\varepsilon \cdot \sin\rho = \sqrt{\cos^2\varepsilon \cdot (1 - \cos^2\rho)} = \sqrt{\cos^2\varepsilon - \cos^2\beta}. \qquad (10.136)$$

We summarise the rule for the transformation into the orthogonal system:

$$\mathbf{a}_1 = a_1 \cdot \mathbf{e}_x$$
$$\mathbf{a}_2 = a_2 \cdot (\mathbf{e}_x \cdot \cos\gamma + \mathbf{e}_y \cdot \sin\gamma)$$
$$\mathbf{a}_3 = a_3 \cdot (\mathbf{e}_x \cdot \cos\beta + \mathbf{e}_y \cdot \sqrt{\cos^2\varepsilon - \cos^2\beta} + \mathbf{e}_z \cdot \sin\varepsilon) \quad (10.137)$$

$$\text{with } \cos\varepsilon = \frac{\sqrt{\cos^2\alpha - 2\cos\alpha \cdot \cos\beta \cdot \cos\gamma + \cos^2\beta}}{\sin\gamma}$$

and have got the preconditions to calculate the reciprocal lattice vectors $\mathbf{b}_1$, $\mathbf{b}_2$, $\mathbf{b}_3$ following Eq. (10.119). For the vector products it is:

$$\mathbf{a}_2 \times \mathbf{a}_3 = \begin{vmatrix} \mathbf{e}_x & \mathbf{e}_y & \mathbf{e}_z \\ a_2 \cdot \cos\gamma & a_2 \cdot \sin\gamma & 0 \\ a_3 \cdot \cos\beta & a_3 \cdot \sqrt{\cos^2\varepsilon - \cos^2\beta} & a_3 \cdot \sin\varepsilon \end{vmatrix}, \quad (10.138)$$

$$\mathbf{a}_2 \times \mathbf{a}_3 = \mathbf{e}_x \cdot a_2 \cdot a_3 \cdot \sin\gamma \cdot \sin\varepsilon - \mathbf{e}_y \cdot a_2 \cdot a_3 \cdot \cos\gamma \cdot \sin\varepsilon$$
$$+ \mathbf{e}_z \cdot a_2 \cdot a_3 (\cos\gamma \cdot \sqrt{\cos^2\varepsilon - \cos^2\beta} - \sin\gamma \cdot \cos\beta)$$

$$\mathbf{a}_3 \times \mathbf{a}_1 = \begin{vmatrix} \mathbf{e}_x & \mathbf{e}_y & \mathbf{e}_z \\ a_3 \cdot \cos\beta & a_3 \cdot \sqrt{\cos^2\varepsilon - \cos^2\beta} & a_3 \cdot \sin\varepsilon \\ a_1 & 0 & 0 \end{vmatrix}$$

$$\mathbf{a}_3 \times \mathbf{a}_1 = \mathbf{e}_y \cdot a_1 \cdot a_3 \cdot \sin\varepsilon - \mathbf{e}_z \cdot a_1 \cdot a_3 \cdot \sqrt{\cos^2\varepsilon - \cos^2\beta} \quad (10.139)$$

$$\mathbf{a}_1 \times \mathbf{a}_2 = \begin{vmatrix} \mathbf{e}_x & \mathbf{e}_y & \mathbf{e}_z \\ a_1 & 0 & 0 \\ a_2 \cdot \cos\gamma & a_2 \cdot \sin\gamma & 0 \end{vmatrix}$$

$$\mathbf{a}_1 \times \mathbf{a}_2 = \mathbf{e}_z \cdot a_1 \cdot a_2 \cdot \sin\gamma. \quad (10.140)$$

For the volume of the unit cell it follows

$$V = a_1 \cdot a_2 \cdot a_3 \cdot \sin\varepsilon \cdot \sin\gamma \quad (10.141)$$

and for the reciprocal lattice vectors:

$$\mathbf{b}_1 = \frac{1}{a_1} \cdot \left(\mathbf{e}_x - \mathbf{e}_y \cdot \cot\gamma + \mathbf{e}_z \cdot \frac{\cot\gamma \cdot \sqrt{\cos^2\varepsilon - \cos^2\beta} - \cos\beta}{\sin\varepsilon} \right)$$

$$\mathbf{b}_2 = \frac{1}{a_2 \cdot \sin\gamma} \cdot \left(\mathbf{e}_y - \mathbf{e}_z \cdot \frac{\sqrt{\cos^2\varepsilon - \cos^2\beta}}{\sin\varepsilon} \right) \quad (10.142)$$

10.8 Laue Equations and Reciprocal Lattice, Ewald Construction

$$\mathbf{b}_3 = \frac{\mathbf{e}_z}{a_3 \cdot \sin \varepsilon}. \tag{10.143}$$

We summarise this formalism in the transformation matrix:

$$\begin{pmatrix} \mathbf{b}_1 \\ \mathbf{b}_2 \\ \mathbf{b}_3 \end{pmatrix} = \begin{pmatrix} \dfrac{1}{a_1} & \dfrac{-\cot \gamma}{a_1} & \dfrac{\cot \gamma \cdot \sqrt{\cos^2 \varepsilon - \cos^2 \beta} - \cos \beta}{a_1 \cdot \sin \varepsilon} \\ 0 & \dfrac{1}{a_2 \cdot \sin \gamma} & \dfrac{-\sqrt{\cos^2 \varepsilon - \cos^2 \beta}}{a_2 \cdot \sin \gamma \cdot \sin \varepsilon} \\ 0 & 0 & \dfrac{1}{a_3 \cdot \sin \varepsilon} \end{pmatrix} \cdot \begin{pmatrix} \mathbf{e}_x \\ \mathbf{e}_y \\ \mathbf{e}_z \end{pmatrix} \tag{10.144}$$

and, respectively,

$$\begin{pmatrix} \mathbf{b}_1 \\ \mathbf{b}_2 \\ \mathbf{b}_3 \end{pmatrix} = \begin{pmatrix} b_{1x} & b_{1y} & b_{1z} \\ 0 & b_{2y} & b_{2z} \\ 0 & 0 & b_{3z} \end{pmatrix} \cdot \begin{pmatrix} \mathbf{e}_x \\ \mathbf{e}_y \\ \mathbf{e}_z \end{pmatrix}. \tag{10.145}$$

Therewith we are able to calculate any desired reciprocal lattice vectors. We remember the Eqs. (10.109) and (10.110), set $n=1$ and obtain

$$|\mathbf{k} - \mathbf{k}_0| = \frac{1}{d} \quad \text{and} \quad d = \frac{1}{|\mathbf{k} - \mathbf{k}_0|}. \tag{10.146}$$

With Eq. (10.116) it follows

$$d_{hkl} = \frac{1}{|h \cdot \mathbf{b}_1 + k \cdot \mathbf{b}_2 + l \cdot \mathbf{b}_3|}. \tag{10.147}$$

This is a general equation for the calculation of lattice spacings d_{hkl}. Due to the constraints the axes and angles at the different crystal systems the terms b_{1x}, b_{2x}, b_{3x}, b_{2y}, b_{2z}, b_{3z} of the transformation matrix are simplified in different ways (cf. Table 10.1). The results of the lattice spacings were already listed in Table 5.4 in Chap. 5.

The angle between two point-shaped diffraction reflections (i.e. between two reciprocal lattice vectors $\mathbf{B}_1$, $\mathbf{B}_2$) is equal to the angle between the two lattice planes responsible for the two diffraction reflections. This angle can be calculated by means of the scalar product:

$$\cos \sphericalangle(\mathbf{B}_1, \mathbf{B}_2) = \frac{\mathbf{B}_1 \cdot \mathbf{B}_2}{|\mathbf{B}_1| \cdot |\mathbf{B}_2|}. \tag{10.148}$$

Table 10.1 Transformation matrices for the calculation of orthogonal reciprocal lattice vectors

Cubic	$a_1 = a_2 = a_3$ $\alpha = \beta = \gamma = 90°$ $\varepsilon = 90°$	$\begin{pmatrix} \dfrac{1}{a_1} & 0 & 0 \\ 0 & \dfrac{1}{a_1} & 0 \\ 0 & 0 & \dfrac{1}{a_1} \end{pmatrix}$
Tetragonal	$a_1 = a_2 \neq a_3$ $\alpha = \beta = \gamma = 90°$ $\varepsilon = 90°$	$\begin{pmatrix} \dfrac{1}{a_1} & 0 & 0 \\ 0 & \dfrac{1}{a_1} & 0 \\ 0 & 0 & \dfrac{1}{a_3} \end{pmatrix}$
Orthorhombic	$a_1 \neq a_2 \neq a_3$ $\alpha = \beta = \gamma = 90°$ $\varepsilon = 90°$	$\begin{pmatrix} \dfrac{1}{a_1} & 0 & 0 \\ 0 & \dfrac{1}{a_2} & 0 \\ 0 & 0 & \dfrac{1}{a_3} \end{pmatrix}$
Hexagonal	$a_1 = a_2 \neq a_3$ $\alpha = \beta = 90°$ $\gamma = 120°$ $\varepsilon = 90°$	$\begin{pmatrix} \dfrac{1}{a_1} & \dfrac{1}{\sqrt{3} \cdot a_1} & 0 \\ 0 & \dfrac{2}{\sqrt{3} \cdot a_1} & 0 \\ 0 & 0 & \dfrac{1}{a_3} \end{pmatrix}$
Rhombohedral	$a_1 = a_2 = a_3$ $\alpha = \beta = \gamma \neq 90°$	$\begin{pmatrix} \dfrac{1}{a_1} & \dfrac{-\cot \alpha}{a_1} & \dfrac{-\cos \alpha}{2 \cdot a_1 \cdot \cos\frac{\alpha}{2} \cdot \sin \varepsilon} \\ 0 & \dfrac{1}{a_1 \cdot \sin \alpha} & \dfrac{-\cot \alpha \cdot \tan \frac{\alpha}{2}}{a_1 \cdot \sin \varepsilon} \\ 0 & 0 & \dfrac{1}{a_1 \cdot \sin \varepsilon} \end{pmatrix}$
	with $\cos \varepsilon = \cot \alpha \cdot \sqrt{2 \cdot (1 - \cos \alpha)}$	
Monoclinic	$a_1 \neq a_2 \neq a_3$ $\alpha = \gamma = 90°$ $\beta \neq 90°$ $\varepsilon = \beta$	$\begin{pmatrix} \dfrac{1}{a_1} & 0 & \dfrac{-\cot \beta}{a_1} \\ 0 & \dfrac{1}{a_2} & 0 \\ 0 & 0 & \dfrac{1}{a_3 \cdot \sin \beta} \end{pmatrix}$

10.8 Laue Equations and Reciprocal Lattice, Ewald Construction

Table 10.1 (continued)

Triclinic	$a_1 \neq a_2 \neq a_3$
	$\alpha \neq \beta \neq \gamma$

$$\begin{pmatrix} \dfrac{1}{a_1} & \dfrac{-\cot\gamma}{a_1} & \dfrac{\cot\gamma \cdot \sqrt{\cos^2\varepsilon - \cos^2\beta} - \cos\beta}{a_1 \cdot \sin\varepsilon} \\ 0 & \dfrac{1}{a_2 \cdot \sin\gamma} & \dfrac{-\sqrt{\cos^2\varepsilon - \cos^2\beta}}{a_2 \cdot \sin\gamma \cdot \sin\varepsilon} \\ 0 & 0 & \dfrac{1}{a_3 \cdot \sin\varepsilon} \end{pmatrix}$$

with $\cos\varepsilon = \dfrac{\sqrt{\cos^2\alpha - 2\cdot\cos\alpha\cdot\cos\beta\cdot\cos\gamma + \cos^2}}{\sin\gamma}$

For both reciprocal lattice vectors it is:

$$\begin{aligned}\mathbf{B}_1 &= h_1 \cdot \mathbf{b}_{1,1} + k_1 \cdot \mathbf{b}_{2,1} + l_1 \cdot \mathbf{b}_{3,1} \\ \mathbf{B}_2 &= h_2 \cdot \mathbf{b}_{1,2} + k_2 \cdot \mathbf{b}_{2,2} + l_2 \cdot \mathbf{b}_{3,2}.\end{aligned} \qquad (10.149)$$

For the calculation of the scalar product and the moduli of both reciprocal lattice vectors we use the transformation (10.144), (10.145) into the orthogonal system. Generally, it is

$$\begin{aligned}\mathbf{B}_1 \cdot \mathbf{B}_2 &= h_1 h_2 b_{1x}^2 + (h_1 b_{1y} + k_1 b_{2y}) \cdot (h_2 b_{1y} + k_2 b_{2y}) \\ &+ (h_1 b_{1z} + k_1 b_{2z} + l_1 b_{3z}) \cdot (h_2 b_{1z} + k_2 b_{2z} + l_2 b_{3z})\end{aligned} \qquad (10.150)$$

as well as

$$\begin{aligned}|\mathbf{B}_1||\mathbf{B}_2| &= \sqrt{(h_1 b_{1x})^2 + (h_1 b_{1y} + k_1 b_{2y})^2 + (h_1 b_{1z} + k_1 b_{2z} + l_1 b_{3z})^2} \\ &\cdot \sqrt{(h_2 b_{1x})^2 + (h_2 b_{1y} + k_2 b_{2y})^2 + (h_2 b_{1z} + k_2 b_{2z} + l_2 b_{3z})^2}.\end{aligned} \qquad (10.151)$$

Because of the defaults concerning the axes and angles the transformation matrices are partly simplified for specific crystal systems (cf. Table 10.1). The resulting equations for angles between reflections are listed in Table 5.9.

The point pattern of a single crystal does not only depend on the crystal structure but also on the incident direction of the electron wave into the crystal lattice. Because of the very small diffraction angles in electron diffraction in transmission electron microscopy the incident direction and the zone axis (this is the intersection line of the diffracting lattice planes) agree in practice. Finally, we want to calculate the zone axis with the knowledge of two indexed diffraction reflections.

The reciprocal lattice vectors are perpendicular to the diffracting lattice planes, i.e. the vector product of two reciprocal lattice vectors lies in both responsible lattice planes and characterises the intersection line direction of both planes. This is by definition the zone axis (Fig. 10.16).

Fig. 10.16 Sketch for explanation of the calculation of the zone axis

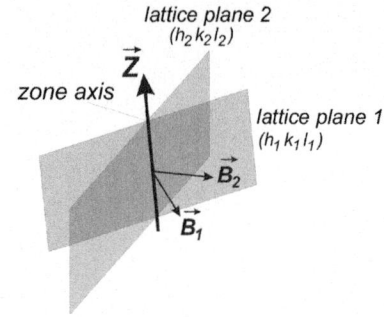

Therefore, the vector Z representing the direction of the zone axis is given by

$$Z = B_1 \times B_2. \tag{10.152}$$

Using

$$\begin{aligned}\mathbf{B_1} &= h_1 \cdot b_{1x} \cdot \mathbf{e_x} + (h_1 \cdot b_{1y} + k_1 \cdot b_{2y})\mathbf{e_y} + (h_1 \cdot b_{1z} + k_1 \cdot b_{2z} + l_1 \cdot b_{3z})\mathbf{e_z} \\ \mathbf{B_2} &= h_2 \cdot b_{1x} \cdot \mathbf{e_x} + (h_2 \cdot b_{1y} + k_2 \cdot b_{2y})\mathbf{e_y} + (h_2 \cdot b_{1z} + k_2 \cdot b_{2z} + l_2 \cdot b_{3z})\mathbf{e_z}.\end{aligned} \tag{10.153}$$

and considering

$$\begin{aligned}\mathbf{e_x} &= b_{1x}\mathbf{a_1} \\ \mathbf{e_y} &= b_{1y}\mathbf{a_1} + b_{2y}\mathbf{a_2} \\ \mathbf{e_z} &= b_{1z}\mathbf{a_1} + b_{2z}\mathbf{a_2} + b_{3z}\mathbf{a_3}\end{aligned} \tag{10.154}$$

we utilise the transformation (10.144), (10.145) into the orthogonal axes system and get for the vector product:

$$\begin{aligned}\mathbf{B_1} \times \mathbf{B_2} &= \begin{vmatrix} b_{1x}\mathbf{a_1} & (b_{1y}\mathbf{a_2} + b_{2y}\mathbf{a_2}) & (b_{1z}\mathbf{a_1} + b_{2z}\mathbf{a_2} + b_{3z}\mathbf{a_3}) \\ h_1 b_{1x} & (h_1 b_{1y} + k_1 b_{2y}) & (h_1 b_{1z} + k_1 b_{2z} + l_1 b_{3z}) \\ h_2 b_{1x} & (h_2 b_{1y} + k_2 b_{2y}) & (h_2 b_{1z} + k_2 b_{2z} + l_2 b_{3z}) \end{vmatrix} \\ &= b_{1x} b_{2y} b_{3z} \begin{bmatrix} (k_1 \cdot l_2 - k_2 \cdot l_1) \cdot \mathbf{a_1} + (h_2 \cdot l_1 - h_1 \cdot l_2) \cdot \mathbf{a_2} + \\ + (h_1 \cdot k_2 - h_2 \cdot k_1) \cdot \mathbf{a_3} \end{bmatrix}.\end{aligned} \tag{10.155}$$

We declare the zone axis as a direction in the crystal axes system $\mathbf{a_1}, \mathbf{a_2}, \mathbf{a_3}$. Common multiples of the components, as the product $b_{1x} \cdot b_{2y} \cdot b_{3z}$, are not considered:

$$\mathbf{Z} = (k_1 \cdot l_2 - k_2 \cdot l_1) \cdot \mathbf{a_1} + (h_2 \cdot l_1 - h_1 \cdot l_2) \cdot \mathbf{a_2} + (h_1 \cdot k_2 - h_2 \cdot k_1) \cdot \mathbf{a_3}. \tag{10.156}$$

10.9 Kinematical Model: Lattice Factor and Structure Factor

According to Bragg's law we are able to determine the positions of the diffraction reflections by means of the reciprocal lattice model ("*geometrical theory*"). To calculate the intensities of the reflections we have to enhance the model: Now, we consider the overlay of all the electron waves scattered at the crystal lattice but neglect furthermore the interactions between waves and crystal lattice exceeding the simple scattering process (e.g. changes of the wavelength within the crystal) and interactions between scattered and initial waves. The intensities of the diffracted waves are much smaller than those of the initial wave which allows us to neglect its intensity change. This idea is called "*kinematical model*". In practice this approximation is only valid for extremely thin samples (thickness of a few 10 nm).

As a result of the overlay we get a sum of electron waves (amplitudes f_j) which are phase-shifted by χ_j against the initial wave (amplitude A_0):

$$\Psi = A_0 \cdot \sum_j f_j \cdot e^{-i(2\pi \mathbf{k}_0 \cdot \mathbf{r} + \chi_j)} = A_0 \cdot e^{-2\pi i \mathbf{k}_0 \cdot \mathbf{r}} \cdot \sum_j f_j \cdot e^{-i \chi_j} \qquad (10.157)$$

(i: imaginary unit). The reason for the phase shift is the path difference Δs experienced by the scattered electron waves. A path difference of the wavelength λ corresponds to a phase shift 2π. Generally it is

$$\chi_j = \frac{2 \cdot \pi}{\lambda} \cdot \Delta s_j. \qquad (10.158)$$

For small scattering vectors $\mathbf{k} - \mathbf{k}_0$ with Eqs. (5.1), (5.2), (5.6), (10.109), and (10.100) we get:

$$\Delta s_j = \mathbf{r}_{gj} \cdot (\mathbf{k} - \mathbf{k}_0) \cdot \lambda \qquad (10.159)$$

and, respectively, with Eq. (10.158):

$$\chi_j = 2\pi \cdot \mathbf{r}_{gj} \cdot (\mathbf{k} - \mathbf{k}_0) \qquad (10.160)$$

as well as with Eq. (10.157):

$$\Psi = A_0 \cdot e^{-2\pi i \mathbf{k}_0 \cdot \mathbf{r}} \cdot \sum_j f_j \cdot e^{-2\pi i \mathbf{r}_{gj} \cdot (\mathbf{k} - \mathbf{k}_0)}. \qquad (10.161)$$

Using the kinematical model we neglect the interaction with the initial wave and, therefore, we can normalise the first term by setting it equal to 1:

$$\Psi = \sum_j f_j \cdot e^{-2\pi i \mathbf{r}_{gj} \cdot (\mathbf{k} - \mathbf{k}_0)}. \qquad (10.162)$$

We separate the distance vector $\mathbf{r}_{gj}$ between the scattering centres into two summands: The translation vector $\mathbf{r}_g$ giving the position of the unit cell within the lattice and a vector pointing from the origin (0,0,0) of the unit cell to the atom j with the coordinates x_j, y_j, z_j within the unit cell:

$$\mathbf{r}_j = x_j \cdot \mathbf{a}_1 + y_j \cdot \mathbf{a}_2 + z_j \cdot \mathbf{a}_3. \tag{10.163}$$

Using these preconditions Eq. (10.162) changes to

$$\Psi = \sum_j f_j \cdot e^{-2\pi \cdot i \cdot (\mathbf{r}_g + \mathbf{r}_j) \cdot (\mathbf{k} - \mathbf{k}_0)} = e^{-2\pi \cdot i \cdot \mathbf{r}_g \cdot (\mathbf{k} - \mathbf{k}_0)} \cdot \sum_j f_j \cdot e^{-2\pi \cdot i \cdot \mathbf{r}_j \cdot (\mathbf{k} - \mathbf{k}_0)}. \tag{10.164}$$

The first term is called *lattice factor G*. We have to consider that each unit cell contributes to this lattice factor. With M as number of the unit cells contributing to the diffraction pattern we define:

$$G(\mathbf{k} - \mathbf{k}_0) = \sum_M e^{-2\pi \cdot i \cdot \mathbf{r}_g \cdot (\mathbf{k} - \mathbf{k}_0)}. \tag{10.165}$$

The second term is called *structure factor F*:

$$F(\mathbf{k} - \mathbf{k}_0) = \sum_j f_j \cdot e^{-2\pi \cdot i \cdot \mathbf{r}_j \cdot (\mathbf{k} - \mathbf{k}_0)}. \tag{10.166}$$

It considers the influence of the atomic arrangement within the unit cell on the intensity of the diffraction reflections.

The product of their squares describes the intensity distribution $I(\mathbf{k} - \mathbf{k}_0)$ in the diffraction pattern:

$$I(\mathbf{k} - \mathbf{k}_0) \propto F^2(\mathbf{k} - \mathbf{k}_0) \cdot G^2(\mathbf{k} - \mathbf{k}_0) = |F(\mathbf{k} - \mathbf{k}_0)|^2 \cdot |G(\mathbf{k} - \mathbf{k}_0)|^2. \tag{10.167}$$

If the structure factor is equal to zero one speaks about "*forbidden reflections*". For a diffraction reflection the scattering vector $\mathbf{k} - \mathbf{k}_0$ is a reciprocal lattice vector. Therefore, it follows from Eqs. (10.116) and (10.163):

$$\begin{aligned}\mathbf{r}_j \cdot (\mathbf{k} - \mathbf{k}_0) &= (x_j \cdot \mathbf{a}_1 + y_j \cdot \mathbf{a}_2 + z_j \cdot \mathbf{a}_3) \cdot (h \cdot \mathbf{b}_1 + k \cdot \mathbf{b}_2 + l \cdot \mathbf{b}_3)\\ &= h \cdot x_j + k \cdot y_j + l \cdot z_j\end{aligned} \tag{10.168}$$

and for the structure factor

$$F(\mathbf{k} - \mathbf{k}_0) = F_{hkl} = \sum_j f_j \cdot e^{-2\pi \cdot i \cdot (h \cdot x_j + k \cdot y_j + l \cdot z_j)}. \tag{10.169}$$

The index hkl is added at the structure factor to indicate its dependency on the diffracting lattice plane. In Sect. 5.2 we had explained the atomic arrangement in salt (NaCl). It is summarised in Table 10.2. In total, the unit cell includes eight atoms.

In the case of NaCl the structure factor F_{hkl} is:

10.9 Kinematical Model: Lattice Factor and Structure Factor

Table 10.2 Atomic arrangement within the NaCl unit cell

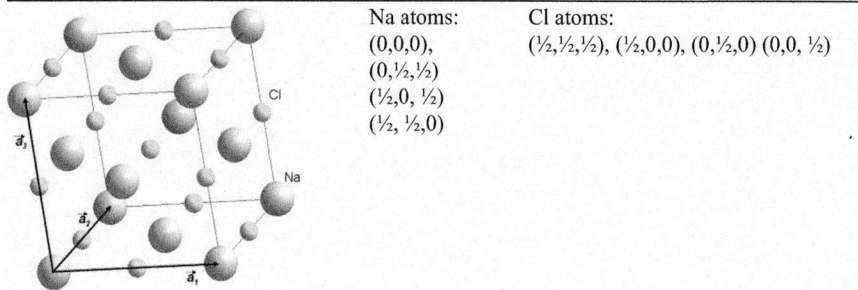

Na atoms:
(0,0,0),
(0,½,½)
(½,0, ½)
(½, ½,0)

Cl atoms:
(½,½,½), (½,0,0), (0,½,0) (0,0, ½)

Table 10.3 Structure factors of NaCl

h	k	l	Na part	Cl part	Sum (10.171)
Even	Even	Even	1+1+1+1	1+1+1+1	$4 \cdot (f_{Na} + f_{Cl})$
Even	Even	Odd	1−1−1+1	−1+1+1−1	0
Even	Odd	Even	1−1+1−1	−1+1−1+1	0
Odd	Even	Even	1+1−1−1	−1−1+1+1	0
Even	Odd	Odd	1+1−1−1	1+1−1−1	0
Odd	Even	Odd	1−1−1+1	1−1+1−1	0
Odd	Odd	Even	1−1−1+1	1−1−1+1	0
Odd	Odd	Odd	1+1+1+1	−1−1−1−1	$4 \cdot (f_{Na} - f_{Cl})$

$$F_{hkl} = f_{Na} \cdot \begin{pmatrix} e^{-2\pi i \cdot (h \cdot 0 + k \cdot 0 + l \cdot 0)} + e^{-2\pi i \cdot (h \cdot 0 + k \cdot \frac{1}{2} + l \cdot \frac{1}{2})} + \\ + e^{-2\pi i \cdot (h \cdot \frac{1}{2} + k \cdot 0 + l \cdot \frac{1}{2})} + e^{-2\pi i \cdot (h \cdot \frac{1}{2} + k \cdot \frac{1}{2} + l \cdot 0)} \end{pmatrix}$$
$$+ f_{Cl} \cdot \begin{pmatrix} e^{-2\pi i \cdot (h \cdot \frac{1}{2} + k \cdot \frac{1}{2} + l \cdot \frac{1}{2})} + e^{-2\pi i \cdot (h \cdot \frac{1}{2} + k \cdot 0 + l \cdot 0)} + \\ + e^{-2\pi i \cdot (h \cdot 0 + k \cdot \frac{1}{2} + l \cdot 0)} + e^{-2\pi i \cdot (h \cdot 0 + k \cdot 0 + l \cdot \frac{1}{2})} \end{pmatrix} \quad (10.170)$$

$$F_{hkl} = f_{Na} \cdot \left(1 + e^{-\pi i \cdot (k+l)} + e^{-\pi i \cdot (h+l)} + e^{-\pi i \cdot (h+k)}\right)$$
$$+ f_{Cl} \cdot \left(e^{-\pi i \cdot (h+k+l)} + e^{-\pi i \cdot h} + e^{-\pi i \cdot k} + e^{-\pi i \cdot l}\right) \quad (10.171)$$

Regarding the Na part we remember the Sect. 5.4.2 where the rules for forbidden reflections of a face centred lattice are derived at the example of gold. Thus, the h, k, l must be all either even- or odd-numbered to result in an intensity greater than zero. We add the Cl influence and get Table 10.3. In comparison with the basic face centred lattice (i.e. lattice with only one type of atoms) we realise an additional difference in the structure factor of reflections with even- and odd-numbered indices and, therefore, in their intensity, too.

The knowledge of the structure factor is of fundamental importance for the calculation of diffraction intensities. They depend on the atomic type, on the packing of the atoms within the unit cell, and on the diffracting lattice planes (hkl).

Fig. 10.17 "Degenerated" reciprocal lattice with labelling of the tolerance **u** of the condition for a diffraction maximum

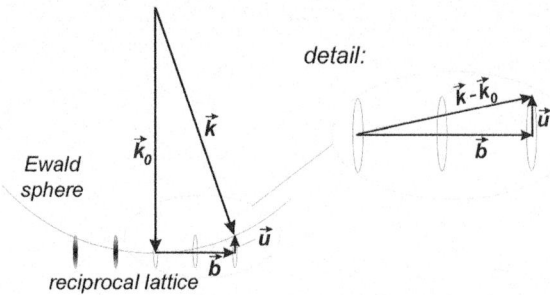

Next, we want to deal with the lattice factor. The Ewald construction means that in case of a diffraction reflection the scattering vector $\mathbf{k}-\mathbf{k}_0$ is a vector $\mathbf{b}$ of the reciprocal lattice. We have to think about the tolerance **u** which is an acceptable deviation from an exact hit. Doing this, we imagine that the reciprocal lattice is (due to the finite size of a crystal) not a "point lattice" but that the lattice points are degenerated into small ellipsoids (Fig. 10.17). Therewith the modulus u of the tolerance would be equal to the "radius" of such an ellipsoid.

From Fig. 10.17 we can derive:

$$\mathbf{k}-\mathbf{k}_0 = \mathbf{b}+\mathbf{u} \tag{10.172}$$

where **b** is a vector of the reciprocal lattice. For the lattice factor (10.165) it is

$$G = \sum_M e^{-2\pi i \mathbf{r}_g \cdot (\mathbf{b}+\mathbf{u})} = \sum_M e^{-2\pi i \mathbf{r}_g \cdot \mathbf{b}} \cdot e^{-2\pi i \mathbf{r}_g \cdot \mathbf{u}}. \tag{10.173}$$

Since $\mathbf{r}_g$ is a vector of the real lattice and **b** is a vector of the reciprocal lattice the scalar product of both results in an integer number. Therewith

$$e^{-2\pi i \mathbf{r}_g \cdot \mathbf{b}} = 1 \tag{10.174}$$

is valid and Eq. (10.173) can be written as

$$G = \sum_M e^{-2\pi i \mathbf{r}_g \cdot \mathbf{u}}. \tag{10.175}$$

We consider the three directions in space and call the number of unit cells in these three directions M_1, M_2, and M_3. The total number of unit cells is $M = M_1 \cdot M_2 \cdot M_3$ and the lattice factor is

$$G = \sum_{M_1} \sum_{M_2} \sum_{M_3} e^{-2\pi i \mathbf{r}_g \cdot \mathbf{u}}. \tag{10.176}$$

Since **u** is a vector in the reciprocal lattice it can be written as a linear combination of the basis vectors of the reciprocal lattice:

10.9 Kinematical Model: Lattice Factor and Structure Factor

$$\mathbf{u} = u_1 \cdot \mathbf{b}_1 + u_2 \cdot \mathbf{b}_2 + u_3 \cdot \mathbf{b}_3. \tag{10.177}$$

Analogous, the vector in the real lattice is:

$$\mathbf{r}_g = m \cdot \mathbf{a}_1 + n \cdot \mathbf{a}_2 + o \cdot \mathbf{a}_3 \tag{10.178}$$

and the lattice factor (10.176) with the reference point in the centre of the crystal:

$$G = \sum_{-M_1/2}^{M_1/2} \sum_{-M_2/2}^{M_2/2} \sum_{-M_3/2}^{M_3/2} e^{-2\pi \cdot i \cdot (m \cdot u_1 + n \cdot u_2 + o \cdot u_3)}. \tag{10.179}$$

For the analytical performance of this summation we assume that a single unit cell is infinitesimally small in comparison with the whole crystal. A single cell with the volume dV delivers the contribution dG to the lattice factor where the position of the unit cell has to be considered which is characterised by m, n, and o. With V_{UC} as volume of the unit cell we get:

$$dG = \frac{dV}{V_{crystal}} \cdot e^{-2\pi \cdot i \cdot (m \cdot u_1 + n \cdot u_2 + o \cdot u_3)}$$
$$= \frac{dm \cdot dn \cdot do}{M_1 \cdot M_2 \cdot M_3 \cdot V_{UC}} \cdot e^{-2\pi \cdot i \cdot (m \cdot u_1 + n \cdot u_2 + o \cdot u_3)}. \tag{10.180}$$

On these bases the sum (10.179) can be written as

$$G = \frac{\int_{-M_1/2}^{M_1/2} \int_{-M_2/2}^{M_2/2} \int_{-M_3/2}^{M_3/2} e^{-2\pi \cdot i \cdot (m \cdot u_1 + n \cdot u_2 + o \cdot u_3)} \cdot dm \cdot dn \cdot do}{M_1 \cdot M_2 \cdot M_3 \cdot V_{UC}}. \tag{10.181}$$

We write the exponential function in a different way and separate the variables:

$$G = \frac{1}{V_{UC}} \cdot \frac{1}{M_1} \int_{-M_1/2}^{M_1/2} e^{-2\pi \cdot i \cdot (m \cdot u_1)} dm \cdot$$
$$\cdot \frac{1}{M_2} \int_{-M_2/2}^{M_2/2} e^{-2\pi \cdot i \cdot (n \cdot u_2)} dn \cdot \tag{10.182}$$
$$\cdot \frac{1}{M_3} \int_{-M_3/2}^{M_3/2} e^{-2\pi \cdot i \cdot (o \cdot u_3)} do.$$

For simplifying (but without loss of generality) we consider an orthogonal unit cell with $V_{UC} = a_1 \cdot a_2 \cdot a_3$ and write for (10.182)

$$G = G_1 \cdot G_2 \cdot G_3 \tag{10.183}$$

with

$$G_1 = \frac{1}{M_1 \cdot a_1} \int_{-M_1/2}^{M_1/2} e^{-2\pi \cdot i \cdot (m \cdot u_1)} \, dm, \tag{10.184}$$

$$G_2 = \frac{1}{M_2 \cdot a_2} \int_{-M_2/2}^{M_2/2} e^{-2\pi \cdot i \cdot (n \cdot u_2)} \, dn, \tag{10.185}$$

$$G_3 = \frac{1}{M_3 \cdot a_3} \int_{-M_3/2}^{M_3/2} e^{-2\pi \cdot i \cdot (o \cdot u_3)} \, do. \tag{10.186}$$

The utilisation of the integral of the term G_1 leads to:

$$\begin{aligned} G_1 &= \frac{1}{M_1 \cdot a_1} \int_{-M_1/2}^{M_1/2} e^{-2\pi \cdot i \cdot u_1 \cdot m} \, dm \\ &= \frac{1}{M_1 \cdot a_1} \left[\frac{e^{-2\pi \cdot i \cdot u_1 \cdot m}}{-2\pi \cdot i \cdot u_1} \right]_{-\frac{M_1}{2}}^{\frac{M_1}{2}} \\ G_1 &= \frac{1}{2\pi \cdot i \cdot M_1 \cdot a_1 \cdot u_1} \left[-e^{-\pi \cdot i \cdot u_1 \cdot M_1} + e^{\pi \cdot i \cdot u_1 \cdot M_1} \right]. \end{aligned} \tag{10.187}$$

We remember Euler's Eq. (5.18) and get:

$$\begin{aligned} G_1 &= \frac{1}{2\pi \cdot i \cdot M_1 \cdot a_1 \cdot u_1} \big[-\cos(\pi \cdot u_1 \cdot M_1) + i \cdot \sin(\pi \cdot u_1 \cdot M_1) \\ &\quad + \cos(\pi \cdot u_1 \cdot M_1) + i \cdot \sin(\pi \cdot u_1 \cdot M_1) \big] \end{aligned} \tag{10.188}$$

$$G_1 = \frac{\sin(\pi \cdot u_1 \cdot M_1)}{\pi \cdot M_1 \cdot a_1 \cdot u_1}.$$

Analogously, it is

$$G_2 = \frac{\sin(\pi \cdot u_2 \cdot M_2)}{\pi \cdot M_2 \cdot a_2 \cdot u_2}$$

$$G_3 = \frac{\sin(\pi \cdot u_3 \cdot M_3)}{\pi \cdot M_3 \cdot a_3 \cdot u_3}.$$
(10.189)

10.9 Kinematical Model: Lattice Factor and Structure Factor

Fig. 10.18 Graph of the function $F(x) = \left(\dfrac{\sin x}{x}\right)^2$ for the discussion of the intensity distribution in a diffraction pattern

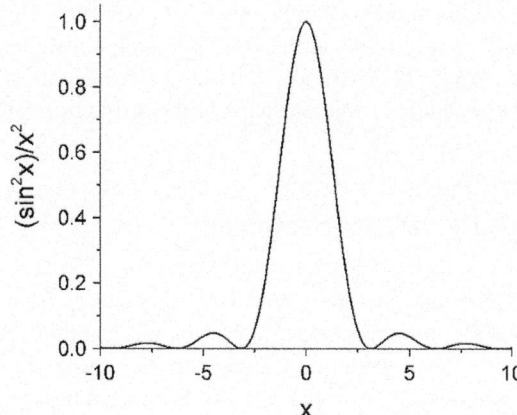

What is the practical consequence of this result for diffraction? We remember: The intensity I of the diffraction reflection is proportional to the square of the lattice factor, i.e.

$$I \propto G_1^2 \cdot G_2^2 \cdot G_3^2$$

$$I \propto \frac{1}{a_1^2 \cdot a_2^2 \cdot a_3^2} \cdot \frac{\sin^2(\pi \cdot u_1 \cdot M_1)}{(\pi \cdot M_1 \cdot u_1)^2} \cdot \frac{\sin^2(\pi \cdot u_2 \cdot M_2)}{(\pi \cdot M_2 \cdot u_2)^2} \cdot \frac{\sin^2(\pi \cdot u_3 \cdot M_3)}{(\pi \cdot M_3 \cdot u_3)^2}$$

$$I \propto \frac{1}{V_{UC}^2} \cdot \frac{\sin^2(\pi \cdot u_1 \cdot M_1)}{(\pi \cdot M_1 \cdot u_1)^2} \cdot \frac{\sin^2(\pi \cdot u_2 \cdot M_2)}{(\pi \cdot M_2 \cdot u_2)^2} \cdot \frac{\sin^2(\pi \cdot u_3 \cdot M_3)}{(\pi \cdot M_3 \cdot u_3)^2}. \quad (10.190)$$

Since u_1, u_2, and u_3 are components in the reciprocal space the Eq. (10.190) describes the intensity distribution in the diffraction pattern. We see that a function of the form

$$F(x) = \left(\frac{\sin x}{x}\right)^2 \quad (10.191)$$

plays an important role and we want to have a look at the graph of this function in Fig. 10.18.

Obviously, the zero points of this function beside the main maximum are at $x = \pm\pi$. We define the enlargement of the ellipsoids in the degenerated reciprocal lattice by these zeros and get the components of our tolerance **u** as:

$$u_1 = \frac{1}{M_1}, \quad u_2 = \frac{1}{M_2}, \quad \text{and} \quad u_3 = \frac{1}{M_3}, \quad (10.192)$$

i.e. the ellipsoids are smaller if the number of unit cells is larger or, respectively, the crystal is larger. The tolerance **u** is also called *excitation error of diffraction*.

If the number of unit cells is especially low in one direction (e.g. plate-like crystals) the ellipsoids are especially long in this direction, i.e. the diffraction reflections are streak-like ("streaks"), too. In certain circumstances the shape of the crystal can be concluded from the shape of the diffraction reflections.

10.10 Debye Scattering

For an analysis of electron scattering curves from amorphous materials generally a formalism is used which bases on an idea by P. Debye[5]. The original goal was the analysis of scattering curves obtained for transmission of light through colloidal solutions [3]. One speaks about Debye scattering if the particles are larger than 1/20 of the wavelength of the scattered waves. This is also true for atoms and electron waves in transmission electron microscopy.

The scattering curve is a result of the overlay of electron waves scattered at an atomic arrangement or, more generally, at an arrangement of N scattering centres. The intensity I in dependence on the scattering direction (represented by the wave number vector $\mathbf{k}$) is given by:

$$I(\mathbf{k}) \propto |\psi(\mathbf{k})|^2 \quad \text{with} \quad \psi = \sum_{m=1}^{N} \sum_{n=1}^{N} f_m \cdot f_n \cdot e^{-2\pi i (\mathbf{k}-\mathbf{k}_0) \cdot \mathbf{r}_{mn}} \qquad (10.193)$$

(f_m, f_n: cross sections for scattering (atomic scattering factors) of the scattering centres m and n, respectively,

$\mathbf{k}, \mathbf{k}_0$: wave number vectors of the scattered and the initial wave,

$\mathbf{r}_{mn}$: distance vector between the scattering centres m and n (cf. Fig. 10.19)).

The modulus s of the scattering vector is (assumption $\theta \ll 1$):

$$|\mathbf{s}| = s = |\mathbf{k}| \cdot \theta = k \cdot \theta = \frac{\theta}{\lambda}. \qquad (10.194)$$

We assume isotropy for the directions of $\mathbf{r}_{mn}$, i.e. all directions are present with equal probability. Thus, the atom n is arranged with the same probability at point (1) or point (2) or at each other position on the spherical surface with radius r_{mn} (Fig. 10.20a).

We read from this figure:

$$\mathbf{s} \cdot \mathbf{r}_{mn} = s \cdot r_{mn} \cdot \cos \gamma. \qquad (10.195)$$

Now, we think about the radius directions having the same angle γ against a (regarding length and direction) fixed scattering vector $\mathbf{s}$. These directions result in the

---
[5] Peter Debye, Dutch physicist, 1884–1966, Nobel prize in chemistry in 1936.

10.10 Debye Scattering

Fig. 10.19 Scattering of a wave at two centres m and n

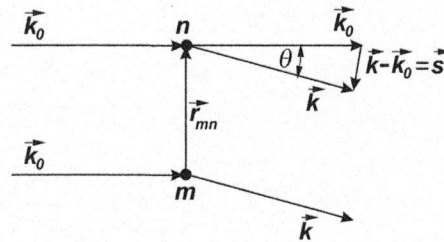

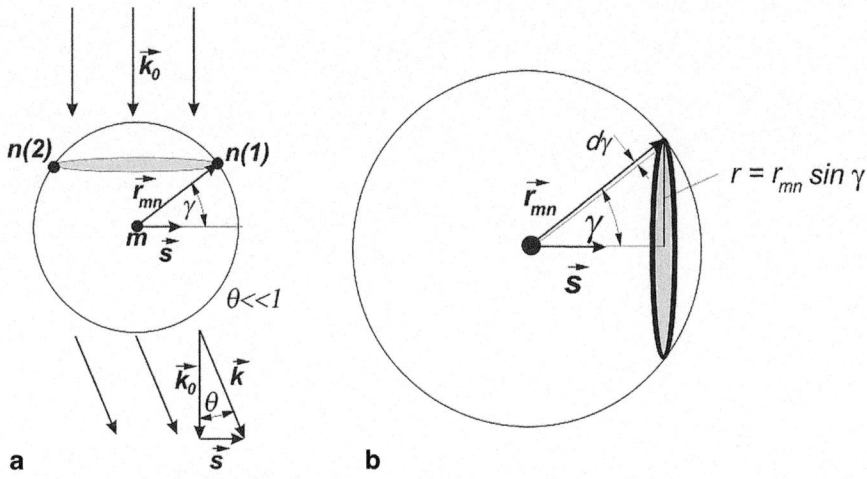

Fig. 10.20 Sketches for the explanation of the calculation. **a** Isotropic distribution of the directions. **b** Equal radius directions

same contribution to the scalar product (10.195) and generate a solid angle segment $d\Omega$ (Fig. 10.20b):

$$d\Omega = \frac{df}{r_{mn}^2} = \frac{2 \cdot \pi \cdot r_{mn} \cdot \sin\gamma \cdot r_{mn} \cdot d\gamma}{r_{mn}^2} = 2 \cdot \pi \cdot \sin\gamma \cdot d\gamma. \qquad (10.196)$$

The atoms within the solid angle $d\Omega$ with distance r_{mn} from the reference atom in the centre contribute the quantity

$$d\psi(r_{mn}, s) = \frac{d\Omega}{\Omega} \cdot f_m \cdot f_n \cdot e^{-2\pi \cdot i \cdot s \cdot r_{mn} \cdot \cos\gamma} \qquad (10.197)$$

to the scattered wave function, i.e. for the scattered wave contribution of all atoms being in distance r_{mn} it is with consideration of Eq. (10.196):

$$\psi(r_{mn}, s) = f_m \cdot f_n \cdot \frac{2 \cdot \pi}{\Omega} \cdot \int_0^\pi e^{-2\pi \cdot i \cdot s \cdot r_{mn} \cdot \cos\gamma} \cdot \sin\gamma \cdot d\gamma. \qquad (10.198)$$

To solve this integral we substitute

$$x = -2\cdot\pi\cdot i\cdot s\cdot r_{mn}\cdot\cos\gamma$$
$$\frac{dx}{d\gamma} = 2\cdot\pi\cdot i\cdot s\cdot r_{mn}\cdot\sin\gamma \qquad (10.199)$$
$$dy = \frac{dx}{2\cdot\pi\cdot i\cdot s\cdot r_{mn}\cdot\sin\gamma}$$

and obtain with consideration of $\Omega = 4\pi$:

$$\psi(r_{mn},s) = \frac{f_m\cdot f_n}{4\cdot\pi\cdot i\cdot s\cdot r_{mn}}\cdot\int_{x_0}^{x_1}e^x\cdot dx. \qquad (10.200)$$

The bounds of integration are:

$$\begin{aligned}x_0 &= x(\gamma = 0) = -2\cdot\pi\cdot i\cdot s\cdot r_{mn},\\ x_1 &= x(\gamma = \pi) = 2\cdot\pi\cdot i\cdot s\cdot r_{mn}\end{aligned} \qquad (10.201)$$

and Eq. (10.200) becomes

$$\begin{aligned}\psi(r_{mn},s) &= \frac{f_m\cdot f_n}{4\cdot\pi\cdot i\cdot s\cdot r_{mn}}\cdot\left[e^{2\cdot\pi\cdot i\cdot s\cdot r_{mn}} - e^{-2\cdot\pi\cdot i\cdot s\cdot r_{mn}}\right]\\ &= \frac{f_m\cdot f_n}{4\cdot\pi\cdot i\cdot s\cdot r_{mn}}\cdot\left[\cos(2\cdot\pi\cdot s\cdot r_{mn}) + i\cdot\sin(2\cdot\pi\cdot s\cdot r_{mn})\right.\\ &\quad\left. -\cos(2\cdot\pi\cdot s\cdot r_{mn}) + i\cdot\sin(2\cdot\pi\cdot s\cdot r_{mn})\right]\end{aligned} \qquad (10.202)$$

$$\psi(r_{mn},s) = f_m\cdot f_n\cdot\frac{\sin(2\cdot\pi\cdot s\cdot r_{mn})}{2\cdot\pi\cdot s\cdot r_{mn}}.$$

To get the total scattering wave needs the summation over all distances r_{mn}:

$$\psi(s) = \sum_{m=1}^{N}\sum_{n=1}^{N}f_m\cdot f_n\cdot\frac{\sin(2\cdot\pi\cdot s\cdot r_{mn})}{2\cdot\pi\cdot s\cdot r_{mn}}. \qquad (10.203)$$

The intensity of the scattering curve is equal to the square of the scattering wave function. It can be calculated by Eq. (10.203) with the knowledge of the distribution of the distances r_{mn} between the scattering centres. In practice, the problem is inverse: We want to calculate the distance distribution from the measured scattering curve.

Following our model of the amorphous state the directions between the atoms are isotropically arranged. Let their (to be determined) radial density distribution be $\rho(r)$. A spherical shell with radius r and thickness dr includes

$$N(r) = 4\cdot\pi\cdot r^2\cdot\rho(r)\cdot dr \qquad (10.204)$$

10.10 Debye Scattering

atoms. We simplify the Eq. (10.203) by shifting the influence of the atomic scattering factors into the background of the scattering curve. The quotient with the sine term encloses the structure information. Additionally, we use the modulus of the sum of all scattered waves for the intensity of the scattering curve. By use of Eq. (10.204) the result is:

$$I(s) = 2 \cdot \int_0^\infty r \cdot \rho(r) \cdot \frac{\sin(2 \cdot \pi \cdot s \cdot r)}{s} \cdot dr. \tag{10.205}$$

The scattering curve includes the overlaid scattering wave contributions of all atomic distances. From the mathematical point of view a deconvolution is necessary. Doing this, we use the Fourier integral for odd functions

$$f(x) = \frac{2}{\pi} \cdot \int_0^\infty \sin(u \cdot x) \cdot du \cdot \int_0^\infty f(t) \cdot \sin(u \cdot t) \cdot dt \tag{10.206}$$

and set the variables to

$$x = t = r \quad \text{and} \quad u = 2 \cdot \pi \cdot s \tag{10.207}$$

as well as the functions to

$$f(x) = f(t) = 2 \cdot r \cdot \rho(r) \tag{10.208}$$

and get therewith from integral (10.206):

$$2 \cdot r \cdot \rho(r) = \frac{2}{\pi} \cdot \int_0^\infty \sin(2 \cdot \pi \cdot s \cdot r) \cdot 2 \cdot \pi \cdot ds \cdot \int_0^\infty 2 \cdot r \cdot \rho(r) \cdot \sin(2 \cdot \pi \cdot s \cdot r) \cdot dr$$

$$r \cdot \rho(r) = 2 \cdot \int_0^\infty \sin(2 \cdot \pi \cdot s \cdot r) \cdot ds \cdot s \cdot \int_0^\infty 2 \cdot r \cdot \rho(r) \cdot \frac{\sin(2 \cdot \pi \cdot s \cdot r)}{s} \cdot dr \tag{10.209}$$

and, after comparison with Eq. (10.205),

$$r \cdot \rho(r) = 2 \cdot \int_0^\infty I(s) \cdot s \cdot \sin(2 \cdot \pi \cdot s \cdot r) \cdot ds \tag{10.210}$$

as well as

$$\rho(r) = \frac{2}{r} \cdot \int_0^\infty I(s) \cdot s \cdot \sin(2 \cdot \pi \cdot s \cdot r) \cdot ds, \tag{10.211}$$

respectively. In this way it is possible to determine the (isotropic) distribution of the atomic distances on the basis of a measured and background-corrected scattering curve as it was shown at an example in Sect. 5.6.

10.11 Electrons Within a Field of a Central Force

We would like to think about the dependencies of the electron deflections in the force field of an atomic nucleus. We neglect the influence of the neighboured atoms. Furthermore, we assume only elastic interaction, i.e. the modulus of the velocity and, with it, the electron energy does not change. Let us illustrate this elastic interaction more accurately with the help of Fig. 10.21a.

The electron moves to the atomic nucleus on a straightforward path in the (perpendicular) distance a from the nucleus centre. Let the current distance between electron and nucleus be r and the angle between the vector $\mathbf{r}$ and the connecting line between the centre of the nucleus and the vertex of the electron trajectory be φ. Electron and atomic nucleus are electrically charged, the electron carriers the (negative) elementary electric charge $-e$, the nucleus carriers the (positive) charge $Z \cdot e$ with Z as atomic number in the periodic table of elements. Both of them attract each other with the Coulomb force

$$|\mathbf{F_c}| = \frac{1}{4\pi \cdot \varepsilon_0} \cdot \frac{Z \cdot e^2}{r^2} \tag{10.212}$$

(ε_0: permittivity of vacuum). We neglect the partial shielding of the nucleus charge by the electron shell. The component $\mathbf{F_S}$ is responsible for the deflection:

$$|\mathbf{F_S}| = |\mathbf{F_c}| \cdot \cos\varphi = \frac{1}{4\pi \cdot \varepsilon_0} \cdot \frac{Z \cdot e^2}{r^2} \cdot \cos\varphi \tag{10.213}$$

This force component leads to a direction inversion of the momentum component $\mathbf{p_S}$ of the electron (Fig. 10.21b). The force component $\mathbf{F_W}$ inverses after passing the trajectory vertex, in the sum it rescinds between the electron positions 1 and 2. The modulus of the momentum change between positions 1 and 2 is

$$|\Delta\mathbf{p}| = 2 \cdot |\mathbf{p_S}| = 2 \cdot |\mathbf{p}| \cdot \sin\frac{\theta}{2} = 2 \cdot m \cdot v \cdot \sin\frac{\theta}{2} \tag{10.214}$$

(m: electron mass, v: electron velocity). For infinitesimally small changes it is

$$d\mathbf{p} = \mathbf{v} \cdot dm + m \cdot d\mathbf{v} \tag{10.215}$$

and in non-relativistic approximation (electron mass constant), respectively:

$$d\mathbf{p} = m \cdot d\mathbf{v}. \tag{10.216}$$

This change of the momentum is caused by the force $\mathbf{F}$:

$$\mathbf{F} = m \cdot \frac{d\mathbf{v}}{dt}, \quad \text{respectively} \quad d\mathbf{v} = \frac{\mathbf{F}}{m} \cdot dt. \tag{10.217}$$

10.11 Electrons Within a Field of a Central Force

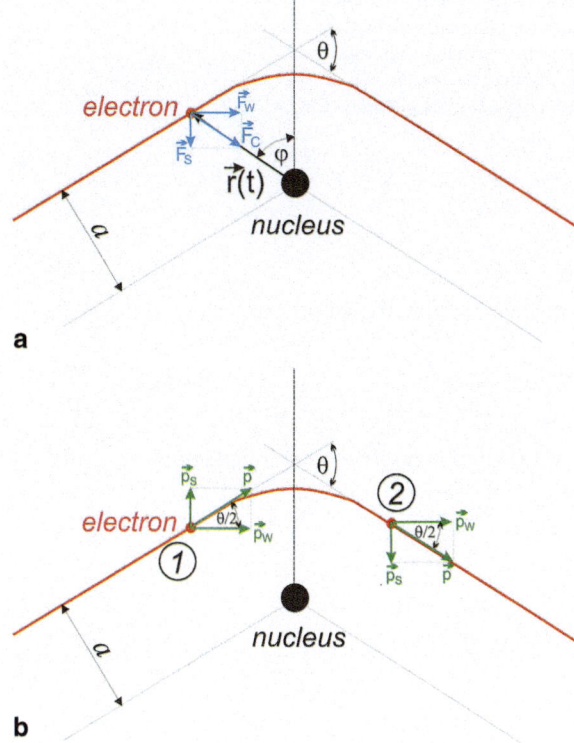

Fig. 10.21 a Elastic deflection of an electron in the force field of an atomic nucleus. **b** Change of the momentum at elastic interaction between electron and atomic nucleus

Therewith the change of the momentum is:

$$d\mathbf{p} = \mathbf{F} \cdot dt, \quad \text{respectively} \quad \Delta \mathbf{p} = \int_{1}^{2} \mathbf{F} \cdot dt \qquad (10.218)$$

and for the electron with $\mathbf{F} = \mathbf{F}_s$:

$$2 \cdot m \cdot v \cdot \sin\frac{\theta}{2} = \frac{Z \cdot e^2}{4\pi \cdot \varepsilon_0} \int_{0}^{\infty} \frac{\cos\varphi}{r^2} \cdot dt. \qquad (10.219)$$

There are no external angular momenta during the motion within the central force field, therefore the conservation law of the angular momentum is valid:

$$\mathbf{L} = \mathbf{r} \times \mathbf{p} = \text{const.} \qquad (10.220)$$

The modulus of the angular momentum is

$$|\mathbf{L}| = m \cdot r^2 \cdot \frac{d\varphi}{dt} \qquad (10.221)$$

Fig. 10.22 Electrons move to the scattering atom through an infinitesimally thin ring in distance a from the axis at which the atom is positioned

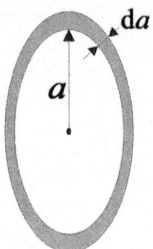

and, compared with that of the electron at the begin of its trajectory,

$$a \cdot m \cdot v = m \cdot r^2 \cdot \frac{d\varphi}{dt}. \tag{10.222}$$

Therewith it is possible to eliminate dt in Eq. (10.219):

$$dt = \frac{r^2}{a \cdot v} \cdot d\varphi \tag{10.223}$$

to get

$$m \cdot v \cdot \sin\frac{\theta}{2} = \frac{Z \cdot e^2}{8\pi \cdot \varepsilon_0} \int_{-(\pi+\theta)/2}^{(\pi+\theta)/2} \frac{\cos\varphi}{a \cdot v} \cdot d\varphi. \tag{10.224}$$

Integration and transformation yield:

$$\sin\frac{\theta}{2} = \frac{Z \cdot e^2}{8\pi \cdot \varepsilon_0 \cdot m \cdot v^2 \cdot a} \sin((\pi+\theta)/2) - \sin(-(\pi+\theta)/2)$$

$$\sin\frac{\theta}{2} = \frac{Z \cdot e^2}{8\pi \cdot \varepsilon_0 \cdot m \cdot v^2 \cdot a} \cdot 2 \cdot \sin((\pi+\theta)/2) \tag{10.225}$$

$$\sin\frac{\theta}{2} = \frac{Z \cdot e^2}{4\pi \cdot \varepsilon_0 \cdot m \cdot v^2 \cdot a} \cdot \cos\frac{\theta}{2}.$$

For the deflection angle θ it follows:

$$\tan\frac{\theta}{2} = \frac{Z \cdot e^2}{4\pi \cdot \varepsilon_0 \cdot m \cdot v^2 \cdot a}. \tag{10.226}$$

On a first view, the handling of the parameter a, this is the distance between trajectory and nucleus, seems to be a problem. These distances are random. Therefore we select a distance a and construct a circular ring with radius a and thickness da (Fig. 10.22).

10.11 Electrons Within a Field of a Central Force

The area of this ring is

$$d\sigma = 2 \cdot \pi \cdot a \cdot da. \quad (10.227)$$

Electrons which move through this area are deflected into a solid angle interval $d\Omega$. This solid angle interval can be calculated from the deflection angle θ (cf. Eq. (10.196):

$$d\Omega = 2 \cdot \pi \cdot \sin\theta \cdot d\theta \quad (10.228)$$

and after division we get:

$$\frac{d\sigma}{d\Omega} = \frac{a}{\sin\theta} \cdot \frac{da}{d\theta}, \quad (10.229)$$

known as *differential cross section of scattering*. It describes the "strength of the scattering" and is a criterion for the probability of the scattering of an electron into the direction marked by the deflection angle θ.

In the next step we consider Eq. (10.226) and get

$$\frac{d\sigma}{d\Omega} = \frac{Z \cdot e^2}{4\pi \cdot \varepsilon_0 \cdot m \cdot v^2 \cdot \tan\frac{\theta}{2} \cdot \sin\theta} \cdot \frac{da}{d\theta} \quad (10.230)$$

as well as

$$\frac{da}{d\theta} = \frac{Z \cdot e^2}{4\pi \cdot \varepsilon_0 \cdot m \cdot v^2} \cdot \frac{d}{d\theta}\left(\frac{1}{\tan\frac{\theta}{2}}\right) = \frac{Z \cdot e^2}{4\pi \cdot \varepsilon_0 \cdot m \cdot v^2} \cdot \frac{-1}{2 \cdot \sin^2\frac{\theta}{2}}. \quad (10.231)$$

For the modulus it is:

$$\frac{d\sigma}{d\Omega} = \left(\frac{Z \cdot e^2}{4\pi \cdot \varepsilon_0 \cdot m \cdot v^2}\right)^2 \cdot \frac{1}{\tan\frac{\theta}{2} \cdot \sin\theta \cdot 2 \cdot \sin^2\frac{\theta}{2}}$$

$$= \left(\frac{Z \cdot e^2}{4\pi \cdot \varepsilon_0 \cdot m \cdot v^2}\right)^2 \cdot \frac{\cos\frac{\theta}{2}}{2 \cdot \sin\theta \cdot \sin^3\frac{\theta}{2}}. \quad (10.232)$$

Using the known trigonometric relation

$$\sin(2 \cdot \alpha) = 2 \cdot \sin\alpha \cdot \cos\alpha \quad (10.233)$$

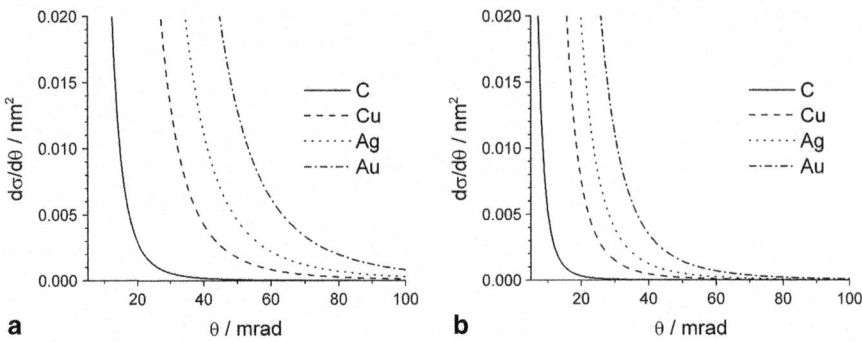

Fig. 10.23 Differential scattering cross section after Eq. (10.236) for the elements C, Cu, Ag, and Au at electron energies of 100 keV (**a**) and 300 keV (**b**)

we simplify to:

$$\frac{d\sigma}{d\Omega} = \left(\frac{Z \cdot e^2}{4\pi \cdot \varepsilon_0 \cdot m \cdot v^2}\right)^2 \cdot \frac{\cos\frac{\theta}{2}}{4 \cdot \sin\frac{\theta}{2} \cdot \cos\frac{\theta}{2} \cdot \sin^3\frac{\theta}{2}}$$

$$= \left(\frac{Z \cdot e^2}{8\pi \cdot \varepsilon_0 \cdot m \cdot v^2}\right)^2 \cdot \frac{1}{\sin^4\frac{\theta}{2}}$$

(10.234)

and with consideration of

$$m \cdot v^2 = 2 \cdot E_0$$

(10.235)

(E_0: electron energy):

$$\frac{d\sigma}{d\Omega} = \left(\frac{1}{4\pi \cdot \varepsilon_0} \cdot \frac{Z \cdot e^2}{4 \cdot E_0}\right)^2 \cdot \frac{1}{\sin^4\frac{\theta}{2}},$$

(10.236)

a relation known as *Rutherford's scattering formula*. Figure 10.23 illustrates this equation for two electron energies and four elements.

Following this equation we realise that for a single elastic scattering process in transmission electron microscopy deflection angles of less than 100 mrad (=5.7°) have to be expected. Since we neglected the partial shielding of the nucleus charge by the electron shell the deflection angles are expected to be lower in reality.

10.12 Mean Free Path for Elastic Scattering

From the previous paragraph we know the probability for elastic scattering of an electron into a certain solid angle interval. Now, we would like to calculate the probability $\sigma(\alpha)$ for the scattering of an electron into a random angle $\theta \geq \alpha$. For this purpose we consider Eq. (10.236) and integrate:

$$\sigma(\alpha) = \left(\frac{Z \cdot e^2}{16 \cdot \pi \cdot \varepsilon_0 \cdot E_0}\right)^2 \cdot \int_\alpha^\pi \frac{2 \cdot \pi \cdot \sin\theta}{\sin^4 \frac{\theta}{2}} \cdot d\theta \qquad (10.237)$$

(Z: atomic number, e: elementary electric charge, ε_0: permittivity of vacuum, E_0: electron energy). With (10.233) it follows:

$$\sigma(\alpha) = \left(\frac{Z \cdot e^2}{16 \cdot \pi \cdot \varepsilon_0 \cdot E_0}\right)^2 \cdot 2 \cdot \pi \cdot \int_\alpha^\pi \frac{2 \cdot \sin\frac{\theta}{2} \cdot \cos\frac{\theta}{2}}{\sin^4 \frac{\theta}{2}} \cdot d\theta$$

$$\sigma(\alpha) = \left(\frac{Z \cdot e^2}{8 \cdot \pi \cdot \varepsilon_0 \cdot E_0}\right)^2 \cdot \pi \cdot \int_\alpha^\pi \frac{\cos\frac{\theta}{2}}{\sin^3 \frac{\theta}{2}} \cdot d\theta$$

$$= \left(\frac{Z \cdot e^2}{8 \cdot \pi \cdot \varepsilon_0 \cdot E_0}\right)^2 \cdot \pi \cdot \left[-\frac{1}{\sin^2 \frac{\theta}{2}}\right]_\alpha^\pi \qquad (10.238)$$

$$= \left(\frac{Z \cdot e^2}{8 \cdot \pi \cdot \varepsilon_0 \cdot E_0}\right)^2 \cdot \pi \cdot \left[-1 + \frac{1}{\sin^2 \frac{\alpha}{2}}\right]$$

$$= \left(\frac{Z \cdot e^2}{8 \cdot \varepsilon_0 \cdot E_0}\right)^2 \cdot \frac{1 - \sin^2 \frac{\alpha}{2}}{\pi \cdot \sin^2 \frac{\alpha}{2}}$$

and finally

$$\sigma(\alpha) = \left(\frac{Z \cdot e^2}{8 \cdot \varepsilon_0 \cdot E_0}\right)^2 \cdot \frac{1}{\pi \cdot \tan^2 \frac{\alpha}{2}}. \qquad (10.239)$$

What happens if the electrons hit a variety of atoms? We imagine a thin layer with the cross sectional area A and the thickness ds. This layer includes N atoms:

$$N = N_A \cdot \rho \cdot A \cdot ds \qquad (10.240)$$

(N_A: Avogadro's constant, ρ: density). N_E electrons incide into this layer per time and area unit. During penetrating the layer a number of dN_E electrons are scattered into an angle $\geq \alpha$ (*acceptance angle*) as sketched in Fig. 10.24.

Fig. 10.24 Balance of the electron number during scattering in a thin layer

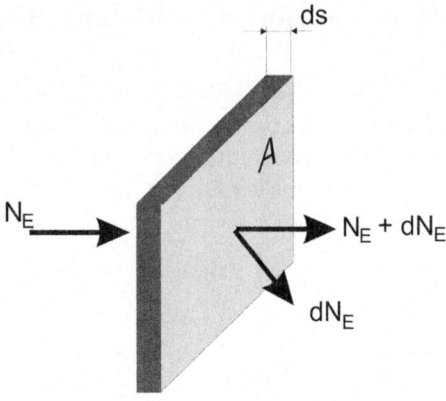

Thus, the number of electrons scattered into an angle $<\alpha$ changes by

$$dN_E = -N_E \cdot \frac{N}{A} \cdot \sigma(\alpha) = -N_E \cdot N_A \cdot \rho \cdot \sigma(\alpha) \cdot ds. \quad (10.241)$$

This differential equation is solved by separation of the variables:

$$\frac{dN_E}{N_E} = -N_A \cdot \rho \cdot \sigma(\alpha) \cdot ds$$
$$N_E = C_1 + C_2 \cdot e^{-N_A \cdot \rho \sigma(\alpha) \cdot s}. \quad (10.242)$$

The integration constants C_1 and C_2 are determined using the boundary conditions:

$$N_E(s=0) = N_{E,0} \quad \text{und} \quad N_E(s \to \infty) = 0 \quad (10.243)$$

resulting in

$$N_E = N_{E,0} \cdot e^{-N_A \cdot \rho \sigma(\alpha) \cdot s}. \quad (10.244)$$

We set

$$\Lambda_{el} = \left(\frac{8 \cdot \varepsilon_0 \cdot E_0}{Z \cdot e^2}\right)^2 \cdot \frac{\pi \cdot \tan^2 \frac{\alpha}{2}}{N_A \cdot \rho} \quad (10.245)$$

(Λ_{el}: mean free path of elastic scattering—Fig. 10.25) and write with it

$$N_E = N_{E,0} \cdot e^{-s/\Lambda_{el}}. \quad (10.246)$$

10.13 Distances in Moiré Patterns

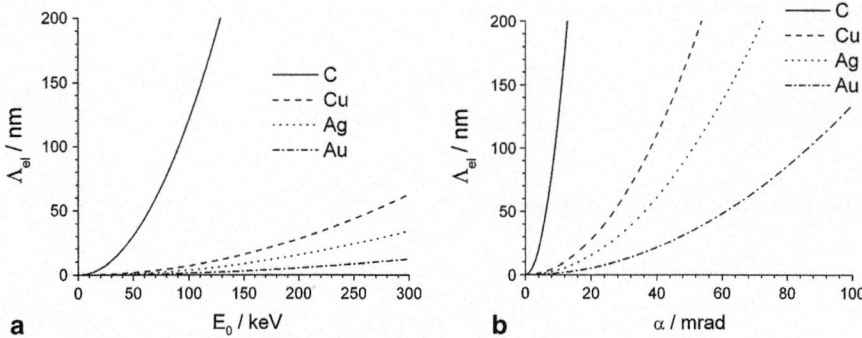

Fig. 10.25 Mean free path of elastic scattering after Eq. (10.245) for the elements C, Cu, Ag, and Au in dependence on the electron energy E_0 at $\alpha = 20$ mrad (**a**) and on the acceptance angle α at an electron energy of $E_0 = 200$ keV, respectively (**b**)

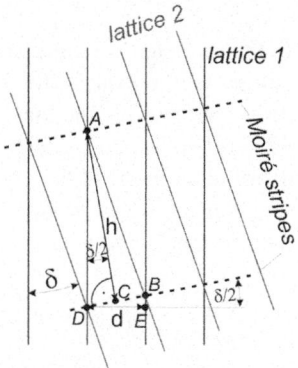

Fig. 10.26 Distances of Moiré stripes in the case of rotation of two equal lattices by angle δ

10.13 Distances in Moiré Patterns

Moiré stripes are generated as a consequence of an overlay of two crystal lattices. The lattices can rotate against each other and can have different lattice spacings. At first, we think about the possible distances of the stripes in Moiré patterns if two lattices with the lattice constant d are rotated by the angle δ against each other (Fig. 10.26).

The Moiré stripes arise by the overlay of both lattices resulting in higher intensities at the cross points. The distance between the stripes is labelled by h in Fig. 10.26. This is the height in the isosceles triangle ABD. This height is equal to the heigth of the right-angled triangle ACD:

$$h = \frac{\overline{CD}}{\tan\left(\dfrac{\delta}{2}\right)}. \tag{10.247}$$

On the other hand, in triangle BDE the length between B and D is:

$$\overline{BD} = 2 \cdot \overline{CD} = \frac{d}{\cos\left(\frac{\delta}{2}\right)} \tag{10.248}$$

and, therewith, the stripe distance h is:

$$h = \frac{d}{2 \cdot \tan\left(\frac{\delta}{2}\right) \cdot \cos\left(\frac{\delta}{2}\right)} = \frac{d}{2 \cdot \sin\left(\frac{\delta}{2}\right)}. \tag{10.249}$$

Therefore, for small rotation angles $\delta \ll 1$ this simplifies to

$$h = \frac{d}{\delta}. \tag{10.250}$$

In the next step we think about the stripe distance for an unrotated overlay of two lattices with unequal lattice constants d_1 and d_2. The line density at the overlay is decisive; therefore the localisation of the stripe distance is not so simple as for the rotated lattices. To calculate the line density we assume a sine-shaped brightness distribution in the lattice pictures. The lattice constant gives the period of the sine function and we write for the brightness distribution of both lattices with an amplitude normalised to one:

$$H_1(x) = \sin\left(2 \cdot \pi \cdot \frac{x}{d_1}\right) \quad \text{and} \quad H_2(x) = \sin\left(2 \cdot \pi \cdot \frac{x}{d_2}\right). \tag{10.251}$$

The reciprocal values of the lattice constants are equivalent to *space frequencies* already known from Chap. 7. For the overlay $H(x)$ we get:

$$H(x) = H_1(x) + H_2(x) = \sin\left(2 \cdot \pi \cdot \frac{x}{d_1}\right) + \sin\left(2 \cdot \pi \cdot \frac{x}{d_2}\right) \tag{10.252}$$

and with consideration of the addition theorem for the sum of two sine functions:

$$H = 2 \cdot \sin\left(\pi \cdot x \cdot \left(\frac{1}{d_1} + \frac{1}{d_2}\right)\right) \cdot \cos\left(\pi \cdot x \cdot \left(\frac{1}{d_1} - \frac{1}{d_2}\right)\right), \tag{10.253}$$

i.e. we get an overlay of a high and a low space frequency (Fig. 10.27). In acoustics this fact is known as beat. The low space frequency describes the distance h of the Moiré stripes:

$$\frac{1}{h} = \frac{1}{d_1} - \frac{1}{d_2} \quad \text{and} \quad h = \frac{d_1 \cdot d_2}{d_2 - d_1}, \text{ respectively.} \tag{10.254}$$

10.13 Distances in Moiré Patterns

Fig. 10.27 Moiré stripes at an unrotated overlay of two lattices with the spacings d1 = 1/1 nm and d2 = 1/10 nm

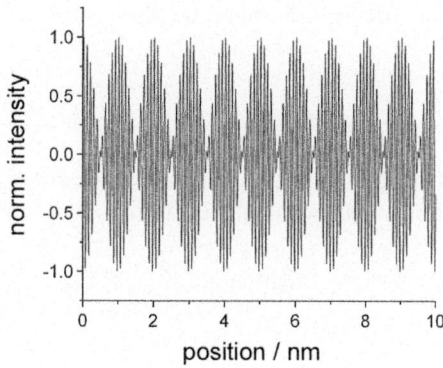

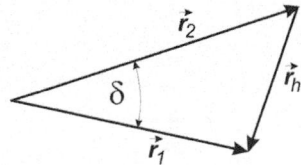

Fig. 10.28 Balance of the space frequencies at rotated lattices with different lattice constants

In the example shown in Fig. 10.27 two lattices with spacings $d_1 = 1/11$ nm ≈ 0.091 nm and $d_2 = 1/10$ nm = 0.1 nm are overlaid. The result is a brightness modulation (Moiré stripes) with a distance of 1 nm.

For the discussion of the general case with rotated lattices and different spacings we add our "space frequency model" to the direction model, i.e. we handle the two space frequencies characterising the periodicities of both lattices as vectors $\mathbf{r}_1$ and $\mathbf{r}_2$. Let the space frequency describing the distance of the Moiré stripes be $\mathbf{r}_h$. Obviously, in general it is:

$$\mathbf{r}_h = \mathbf{r}_1 - \mathbf{r}_2. \tag{10.255}$$

Then, the distance of the Moiré stripes is

$$h = \frac{1}{|\mathbf{r}_h|}. \tag{10.256}$$

From the sketch in Fig. 10.28 it follows for the modulus of the vector $\mathbf{r}_h$ using the cosine rule:

$$|\mathbf{r}_h|^2 = \frac{1}{h^2} = |\mathbf{r}_1|^2 + |\mathbf{r}_2|^2 - 2 \cdot |\mathbf{r}_1| \cdot |\mathbf{r}_2| \cdot \cos \delta$$

$$\frac{1}{h^2} = \frac{1}{d_1^2} + \frac{1}{d_2^2} - \frac{2 \cdot \cos \delta}{d_1 \cdot d_2} \tag{10.257}$$

and for the distance of the Moiré stripes:

$$h = \frac{d_1 \cdot d_2}{\sqrt{d_1^2 + d_2^2 - 2 \cdot d_1 \cdot d_2 \cdot \cos \delta}}. \tag{10.258}$$

Both previously deduced equations are special cases of Eq. (10.258). For the rotation of two equal lattices it is $d_1 = d_2 = d$ and therewith

$$h = \frac{d^2}{\sqrt{2 \cdot d^2 (1 - \cos \delta)}} = \frac{d}{2 \cdot \sqrt{\frac{1}{2}(1 - \cos \delta)}} = \frac{d}{2 \cdot \sin\left(\frac{\delta}{2}\right)}, \tag{10.259}$$

i.e. equal to Eq. (10.249). For the unrotated overlay of two unequal lattices ($\delta = 0$) we get

$$h = \frac{d_1 \cdot d_2}{\sqrt{d_1^2 + d_2^2 - 2 \cdot d_1 \cdot d_2}} = \frac{d_1 \cdot d_2}{\sqrt{(d_1 - d_2)^2}} = \frac{d_1 \cdot d_2}{|d_1 - d_2|}, \tag{10.260}$$

i.e. identical to Eq. (10.254).

10.14 Contrast Transfer Function

In Sect. 7.2 it was shown by qualitative deductions that it is possible that the objective lens causes the phase shift of a diffracted wave which transform a phase contrast into an amplitude contrast. We remember: The phase shift ϕ depends on the structure size d, on the coefficient of the spherical aberration C_S, and on the change Δf of the focal length in comparison with the exact focussing in the Gaussian image plane:

$$\phi = \phi\left(\frac{1}{d}, C_S, \Delta f\right). \tag{10.261}$$

Now, we try to understand this relation quantitatively. Figure 10.29 illustrates the role of spherical aberration (Fig. 10.29a) and defocussing (Fig. 10.29b).

The abbreviations in Fig. 10.29 and in the next equations mean the following: P: object point, P': image point for aberration-free imaging, P'': image point for imaging with aberration, θ: aperture angle of the beam incident into the lens, θ': aperture angle of the refracted beam, R: maximum distance of the incident beam from the optical axis, F: focal point on the object side, f: focal length, Δf: change of the focal length at defocussing, g: object distance, b: image distance, Δb: change of the image distance at defocussing, C_S: coefficient of spherical aberration, and M: magnification.

10.14 Contrast Transfer Function

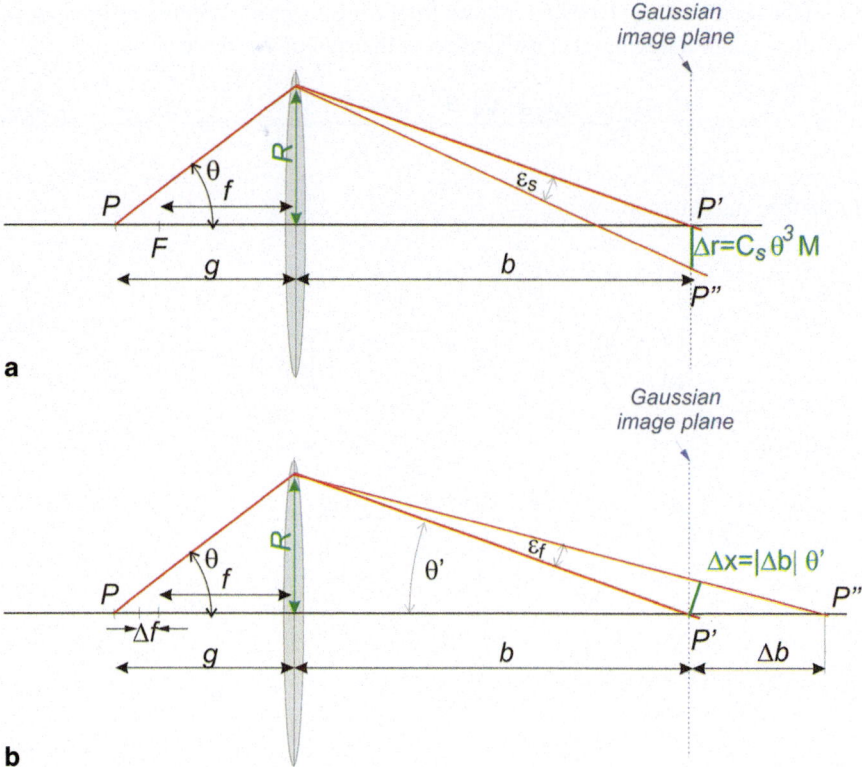

Fig. 10.29 Influence of spherical aberration (**a**) and defocussing (**b**) on the ray path. (see also [4])

The sketch in Fig. 10.29a shows that the spherical aberration leads to an angle deviation ε_s of the refracted beam which is given by

$$\varepsilon_S = \frac{\Delta r}{b} = \frac{C_S \cdot \theta^3 \cdot M}{b} \tag{10.262}$$

as well as with

$$\theta = \frac{R}{g}, \quad g \approx f \text{ (at high magnification) and } M = \frac{b}{g}: \tag{10.263}$$

$$\varepsilon_S = \frac{C_S \cdot R^3 \cdot b}{g^3 \cdot g \cdot b} = C_S \frac{R^3}{f^4}. \tag{10.264}$$

From the sketch in Fig. 10.29b it follows for the influence of defocussing (change of the focal length Δf) without consideration of the sign of the angle change:

$$\varepsilon_f = \frac{\Delta x}{b} = \frac{|\Delta b| \cdot \theta'}{b} = \frac{|\Delta b| \cdot R}{b \cdot b} = \frac{|\Delta b| \cdot R}{b^2}. \tag{10.265}$$

From the lens equation

$$\frac{1}{f+\Delta f} = \frac{1}{g} + \frac{1}{b+\Delta b} \quad \text{and}$$

$$\frac{1}{f}\left(1 - \frac{\Delta f}{f} + \ldots\right) = \frac{1}{g} + \frac{1}{b}\left(1 - \frac{\Delta b}{b} + \ldots\right), \text{ respectively} \tag{10.266}$$

it follows

$$\frac{1}{f} - \frac{\Delta f}{f^2} = \frac{1}{g} + \frac{1}{b} - \frac{\Delta b}{b^2}$$

$$\text{and with} \quad \frac{1}{f} = \frac{1}{g} + \frac{1}{b}: \quad \frac{\Delta f}{f^2} = \frac{\Delta b}{b^2} \tag{10.267}$$

as well as for the angle change

$$\varepsilon_f = \frac{\Delta f \cdot R}{f^2}. \tag{10.268}$$

Now, let us discuss the consequences of the total angle

$$\varepsilon = \varepsilon_s - \varepsilon_f$$

for the phase shift

$$\phi(\theta) = 2\pi \cdot \frac{\Delta s}{\lambda} \tag{10.269}$$

with Δs as path difference between the wave fronts generated by the image aberrations and λ the electron wavelength. For the calculation of Δs we look at Fig. 10.30.

The change of the aperture angle θ by an infinitesimally small value $d\theta$ leads to the path difference

$$ds = \varepsilon \cdot dR. \tag{10.270}$$

The total path difference Δs is calculated by integration over all ds:

$$\Delta s = \int ds = \int_0^R \varepsilon \cdot dR = \int_0^R (\varepsilon_s - \varepsilon_f) \cdot dR. \tag{10.271}$$

10.14 Contrast Transfer Function

Fig. 10.30 Deduction of the path difference between the wave fronts as consequence of image aberration

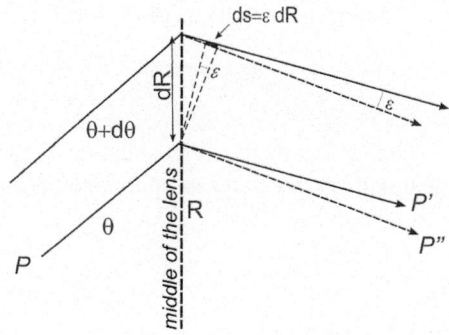

After insertion of the relations (10.264), (10.268), and (10.269) mentioned above it follows for the phase difference between initial and diffracted wave generated by the lens:

$$\phi(R) = 2\pi \cdot \frac{\Delta s}{\lambda} = \frac{2\pi}{\lambda} \int_0^R (C_S \cdot \frac{R^3}{f^4} - \Delta f \cdot \frac{R}{f^2}) \cdot dR$$

$$= \frac{\pi}{\lambda} \left(\frac{C_S}{2} \cdot \frac{R^4}{f^4} - \Delta f \cdot \frac{R^2}{f^2} \right). \tag{10.272}$$

With consideration of

$$\theta = \frac{R}{g} \approx \frac{R}{f} \quad \text{and} \quad \theta = \frac{\lambda}{d} = \lambda \cdot q \quad \text{(Bragg's law)}$$

it follows:

$$\phi(\theta) = \frac{\pi}{\lambda} \left(\frac{C_S}{2} \cdot \theta^4 - \Delta f \cdot \theta^2 \right) \tag{10.273}$$

and

$$\phi(q) = \frac{\pi}{2}\left(C_S \cdot \lambda^3 \cdot q^4 - 2 \cdot \Delta f \cdot \lambda \cdot q^2\right), \text{ respectively.} \tag{10.274}$$

This is the phase shift function (7.14).

Let the incident wave be a plane wave of the form

$$\psi_{in}(\mathbf{r}) = A \cdot e^{-2\pi i \mathbf{k} \cdot \mathbf{r}}. \tag{10.275}$$

During its transmission through the crystal lattice it experiences a position-dependent phase shift (cf. Sect. 7.1)

$$\varphi(x, y) = \frac{\pi \cdot t}{\lambda \cdot U_B} \cdot \Phi(x, y), \tag{10.276}$$

i.e. the object exit wave is of the form

$$\psi_{exit}(\mathbf{r}) = A \cdot e^{-i(2\pi \mathbf{k} \cdot \mathbf{r} + \varphi(x,y))} = A \cdot e^{-2\pi i \mathbf{k} \cdot \mathbf{r}} \cdot e^{-i\varphi(x,y)}. \qquad (10.277)$$

Assuming that the phase modulation by the crystal lattice is small against 1 we can approximate (weak-phase approximation)

$$e^{-i\varphi(x,y)} \approx 1 - i \cdot \varphi(x,y) \qquad (10.278)$$

and therewith

$$\psi_{exit}(\mathbf{r}) = A \cdot e^{-2\pi i \mathbf{k} \cdot \mathbf{r}} \cdot (1 - i \cdot \varphi(x,y)). \qquad (10.279)$$

This wave is overlaid with a plane wave which is phase-shifted by $\phi(q)$ by the lens (Eq. (10.274)). In the image plane we get the wave function

$$\psi_{image}(\mathbf{r}) = \psi_{exit}(\mathbf{r}) + A \cdot e^{-i(2\pi \mathbf{k} \cdot \mathbf{r} + \phi(q))} \qquad (10.280)$$

and with Eq. (10.279):

$$\psi_{image}(\mathbf{r}) = A \cdot e^{-2\pi i \mathbf{k} \cdot \mathbf{r}} \cdot (1 - i \cdot \varphi(x,y)) + A \cdot e^{-2\pi i \mathbf{k} \cdot \mathbf{r}} \cdot e^{-i\phi(q)}$$

$$\psi_{image}(\mathbf{r}) = A \cdot e^{-2i \mathbf{k} \cdot \mathbf{r}} \cdot \left[1 + \cos\phi(q) - i \cdot (\varphi(x,y) + \sin\phi(q))\right]. \qquad (10.281)$$

The intensity $I(x, y)$ in the image is equal to the wave function multiplied by its conjugate:

$$I(x,y) = \psi_{image} \cdot \psi^*_{image}$$
$$I(x,y) = A \cdot e^{-i\mathbf{k}\cdot\mathbf{r}} \cdot 1 + \cos\phi(q) - i \cdot (\varphi(x,y) + \sin\phi(q)) \cdot$$
$$\cdot A \cdot e^{+i\mathbf{k}\cdot\mathbf{r}} \cdot 1 + \cos\phi(q) + i \cdot (\varphi(x,y) + \sin\phi(q)) \qquad (10.282)$$

$$I(x,y) = A^2 \cdot \left[(1+\cos\phi(q))^2 + (1+\cos\phi(q)) \cdot i \cdot (\varphi(x,y) + \sin\phi(q)) - \right.$$
$$\left. - (1+\cos\phi(q)) \cdot i \cdot (\varphi(x,y) + \sin\phi(q)) + (\varphi(x,y) + \sin\phi(q))^2 \right]$$

$$I(x,y) = A^2 \cdot \left[(1+\cos\phi(q))^2 + (\varphi(x,y) + \sin\phi(q))^2 \right]. \qquad (10.283)$$

In the quadratic term containing the cosine function of $\phi(q)$ the constant number 1 is added, i.e. this term does not contribute to the brightness modulation in the image, it is part of the (constant) image background. Contrary, in the second quadratic term

10.14 Contrast Transfer Function

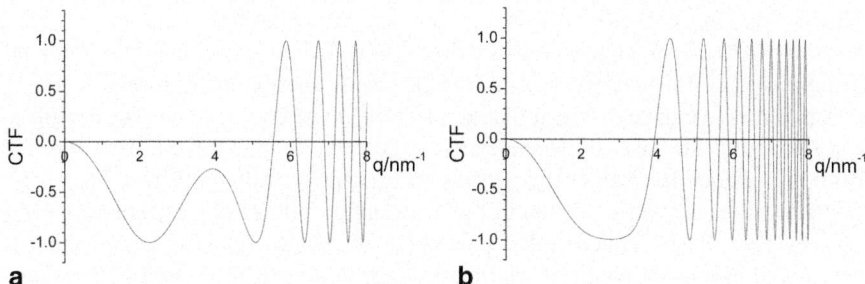

Fig. 10.31 Examples for contrast transfer functions $CTF(q)$ at $U_B=300$ kV, $\Delta f=60$ nm and two different coefficients of spherical aberration: **a** $C_S=1$ mm. **b** $C_S=2$ mm

the position-dependent phase modulation $\varphi(x,y)$ is combined with the sine function of $\phi(q)$, i.e. the function

$$CTF = \sin \phi(q) \qquad (10.284)$$

is essential for the phase contrast transfer by the lens, therefore it is called "*contrast transfer function*" (*CTF*), more precisely: "*phase contrast transfer function*", and depends on the space frequency q, the spherical aberration (coefficient C_S), the wavelength λ, and the defocus Δf (cf. Eq. (10.274)):

$$CTF(q,C_S,\lambda,\Delta f) = \sin\left(\frac{\pi}{2}(C_S \cdot \lambda^3 \cdot q^4 - 2 \cdot \Delta f \cdot \lambda \cdot q^2)\right) \qquad (10.285)$$

(Fig. 10.31). The contrast transfer function starts at (0,0). Up to the next zero point it affiliates a "band" of space frequencies where the contrast can be higher and lower but its sign does not change, i.e. the contrast does not invert. In this band of space frequencies the image interpretation becomes simple. Figure 10.31 shows that this zero point shifts to higher space frequencies (i.e. to smaller structures) at smaller spherical aberrations. The reduction of the spherical aberration does not only improve the resolution limit, it simplifies the image interpretation, too.

We also see that some contrast is also transferred for higher space frequencies but not monotonously. The transfer function oscillates, i.e. some space frequencies are transferred with "positive", others with "negative" contrast and still others not at all. In this region the image interpretation is only possible with an exact knowledge of the contrast transfer function.

It is apparent that the contrast transfer functions in Fig. 10.31 are undamped. Thus, the contrast transfer would be possible up to unlimited high space frequencies, although with some gaps. This disagrees with the experience.

In reality the contrast transfer function is damped, i.e. its amplitude decreases with increasing space frequency. There are different reasons for that: Mechanical vibrations of the specimen "blur" the small structures, i.e. the high space frequencies.

They are no longer visible in the image. But that does not directly correlate with the transfer properties of the lens. Let us closer deal with another reason of the damping being directly influenced by the lens: the chromatic aberration of the lens.

Up to now we have assumed that all electrons have one constant wavelength λ. But this is not the case. In dependence on the cathode type the electrons emitted from the cathode have an energy spread up to some electron volts (see Sect. 2.5). The consequence is a fluctuation of the wavelength followed by chromatic aberration (cf. Sect. 2.3). Similar to the spherical aberration the chromatic aberration is considered by the chromatic aberration disk with the radius (referred to the image plane)

$$\Delta r_C = C_C \cdot \theta \cdot \frac{\Delta E}{E_0} \cdot M \quad (10.286)$$

(ΔE: energy spread of the electrons, E_0: energy of the primary electrons, C_C: coefficient of the chromatic aberration, M: magnification).

Analogously to the discussion about the influence of the spherical aberration a dependence for the phase shift on the chromatic aberration can be deduced:

$$\phi_C(q) = \pi \cdot C_C \cdot \frac{\Delta E}{E_0} \lambda \cdot q^2. \quad (10.287)$$

However, this is the maximum possible phase shift. The actual energy E of an electron is between $E_0 - \frac{1}{2}\Delta E$ and $E_0 + \frac{1}{2}\Delta E$. Different phase shifts result from that and lead to "smeared" interferences and therefore a decrease of the resulting wave amplitude. Commonly, such dampings are described by exponential functions. Hence, we use a damping function of the form

$$D(q) = e^{-(\phi_C(q))^2} = e^{-\pi^2 \cdot C_C^2 \cdot \left(\frac{\Delta E}{E_0}\right)^2 \lambda^2 \cdot q^4} \quad (10.288)$$

to get the damped contrast transfer function

$$CTF\left(q, C_S, \lambda, \Delta f, C_C, \frac{\Delta E}{E_0}\right) =$$

$$\sin\left(\frac{\pi}{2}(C_S \cdot \lambda^3 \cdot q^4 - 2 \cdot \Delta f \cdot \lambda \cdot q^2)\right) \cdot e^{-\pi^2 \cdot C_C^2 \cdot \left(\frac{\Delta E}{E_0}\right)^2 \lambda^2 \cdot q^4}. \quad (10.289)$$

Since the richtstrahlwert and therewith the energy spread ΔE depend on the cathode its type influences the contrast transfer, too (Fig. 10.32).

For the Schottky field emission cathode the first zero point is at a space frequency of 5.5 nm^{-1}, i.e. at structure sizes of 0.18 nm, higher space frequencies are transferred oscillatory. Using the same objective lens but an LaB$_6$ cathode there

10.14 Contrast Transfer Function

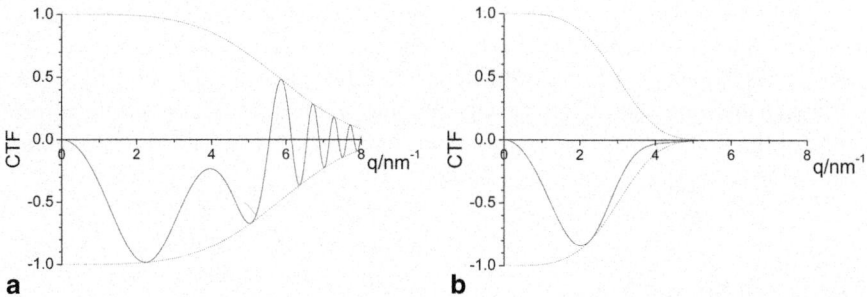

Fig. 10.32 Influence of the cathode type on the contrast transfer (U_0=300 kV, Δf=60 nm, C_S=1 mm, C_C=1.5 mm): **a** ΔE=0.7 eV (Schottky field emission cathode), **b** ΔE=3 eV (LaB$_6$ cathode), *dashed lines*: damping function

is no zero. The damping is so strong that the contrast transfer function asymptotically converges the abscissa and touches it at about 4.5 nm^{-1} corresponding to a structure size of 0.22 nm. There are no oscillations. On the one hand this has the advantage of a simple image interpretation, on the other hand the disadvantage of a worse resolution limit.

Obviously, the damping influences the maximal transferable space frequency. It is useful to indicate this by a value which is termed *information limit*. In our model we take that space frequency as a basis at which the damped amplitude decreases to $1/e^2 = 0.135$ (e=2.7182..., Euler's number):

$$D(q_{\lim}) = e^{-2} = e^{-\pi^2 \cdot C_C^2 \cdot \left(\frac{\Delta E}{E_0}\right)^2 \lambda^2 \cdot q_{\lim}^4}$$

$$2 = \pi^2 \cdot C_C^2 \cdot \left(\frac{\Delta E}{E_0}\right)^2 \lambda^2 \cdot q_{\lim}^4$$

$$q_{\lim} = \sqrt{\frac{\sqrt{2}}{\pi \cdot C_C \cdot \left(\frac{\Delta E}{E_0}\right) \cdot \lambda}} \tag{10.290}$$

The information limit in the real space is:

$$\delta_{\lim} = 1.49 \cdot \sqrt{C_C \cdot \left(\frac{\Delta E}{E_0}\right) \cdot \lambda} \tag{10.291}$$

For a microscope which is described by the parameters mentioned in Fig. 10.32b (LaB$_6$ cathode) the information limit is 0.26 nm, however using a Schottky field emission cathode (Fig. 10.32a) it reaches 0.12 nm. This is an improvement of more than factor 2.

Finally, we want to hint to the importance of the first zero of the contrast transfer function: Space frequencies up to this value are transferred without any contrast oscillations. This opens the possibility to interpret the resolution limit by a third variant on the wave-optical basis using the contrast transfer function. The reciprocal value of the first zero is, contrary to the information limit, termed *point resolution* (see also Sect. 7.3).

10.15   Scherzer Focus

As it follows from the argument

$$\phi(q) = \frac{\pi}{2}(C_S \cdot \lambda^3 \cdot q^4 - 2 \cdot \Delta f \cdot \lambda \cdot q^2) \tag{10.292}$$

of the contrast transfer function the influence of the spherical aberration can be compensated by a suitable defocus Δf. However, the defocus for the accurate compensation depends on the space frequency.

Figure 10.33 illustrates the undamped phase contrast transfer function for three different defoci at equal other parameters. For the function of Fig. 10.33b the first zero point lies at the highest space frequency, there is no contrast inversion in the space frequency range up to this point. We determine the defocus value Δf_r at which the first zero point is shifted farthest to the right side.

For the zero points $q_{0,n}$ of the contrast transfer function we get

$$\sin \phi(q_{0,n}) = 0 \text{ bzw. } \phi(q_{0,n}) = \pm n \cdot \pi \text{ with n = 0, 1, 2, 3, ...} \tag{10.293}$$

and with consideration of (10.292):

$$C_S \cdot \lambda^3 \cdot q^4 - 2 \cdot \Delta f \cdot \lambda \cdot q^2 = \pm m \text{ with m = 0, 2, 4, ... even-numbered} \tag{10.294}$$

The zero point for m=0 lies at q=0. However, this can also be reached for space frequencies $q>0$ at the condition

$$C_S \cdot \lambda^3 \cdot q^4 - 2 \cdot \Delta f \cdot \lambda \cdot q^2 = 0, \text{ i.e. } q = \sqrt{\frac{2 \cdot \Delta f}{C_S \cdot \lambda^2}} \tag{10.295}$$

Thus, the zero point following on $q=0$ would be more correct at larger defocus values. This simple monotonous relation is avoided by the periodicity of the sine function. Therefore, we discuss m=−2 as the next case:

$$C_S \cdot \lambda^3 \cdot q^4 - 2 \cdot \Delta f_r \cdot \lambda \cdot q^2 = -2 \tag{10.296}$$

10.15 Scherzer Focus

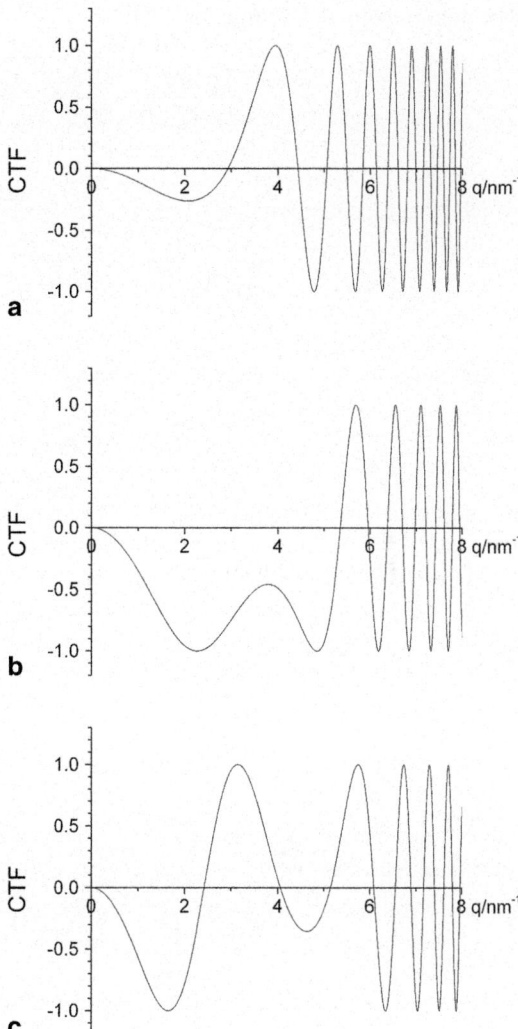

Fig. 10.33 Undamped contrast transfer functions at $U_0=300$ kV and $C_S=1.2$ mm for **a** $\Delta f=20$ nm, **b** $\Delta f=60$ nm, **c** $\Delta f=100$ nm

This is a biquadratic equation for q which can be transformed into the normal form of a quadratic equation using the substitution $u=q^2$:

$$u^2 - 2\cdot\frac{\Delta f_r}{C_S\cdot\lambda^2}\cdot u + \frac{2}{C_S\cdot\lambda^3} = 0 \qquad (10.297)$$

There are two solutions of this equation:

$$u_{1,2} = \frac{\Delta f_r}{C_S\cdot\lambda^2} \pm \sqrt{\frac{(\Delta f_r)^2}{C_S^2\cdot\lambda^4} - \frac{2}{C_S\cdot\lambda^3}} \qquad (10.298)$$

The non-monotonous solution is

$$u = \frac{\Delta f_r}{C_S \cdot \lambda^2} - \sqrt{\frac{(\Delta f_r)^2}{C_S^2 \cdot \lambda^4} - \frac{2}{C_S \cdot \lambda^3}} \qquad (10.299)$$

which reaches a maximum

$$\frac{(\Delta f_r)^2}{C_S^2 \cdot \lambda^4} = \frac{2}{C_S \cdot \lambda^3}. \qquad (10.300)$$

Therewith it follows for the optimal defocus

$$\Delta f_r = \sqrt{2 \cdot C_S \cdot \lambda}, \qquad (10.301)$$

i.e. at an accelerating voltage of $U_0=300$ kV and a coefficient of the spherical aberration of $C_S=1.2$ mm an optimal defocussing $\Delta f_r=68.8$ nm exists.

In Fig. 10.34 the contrast transfer function is drawn for this optimal defocussing. We see the problem: the dramatic contrast reduction to zero at $q \approx 3.8$ nm^{-1} which has to be avoided. Therefore we use another optimisation criterion.

We require that the contrast transfer function only converges to the abscissa up to 80% before the first zero:

$$|\sin \phi(q)| \geq 0.8. \qquad (10.302)$$

This circumstance is illustrated in Fig. 10.35. We select the marked, highest possible abscissa value, i.e.

$$|\phi(q_{max})| = 2.214. \qquad (10.303)$$

The space frequency q_{max} at which this extreme is reached can be calculated by

$$\left. \frac{d\phi}{dq} \right|_{q_{max}} = \frac{\pi}{2}(4 \cdot C_S \cdot \lambda^3 \cdot q^3_{max} - 4 \cdot \Delta f \cdot \lambda \cdot q_{max}) = 0$$

$$q^2_{max} = \frac{\Delta f}{C_S \cdot \lambda^2}. \qquad (10.304)$$

After insertion in Eq. (10.292) we find:

$$|\phi(q_{max})| = \frac{\pi}{2} \left| C_S \cdot \lambda^3 \cdot \left(\frac{\Delta f}{C_S \cdot \lambda^2} \right)^2 - 2 \cdot \Delta f \cdot \lambda \cdot \frac{\Delta f}{C_S \cdot \lambda^2} \right| = 2.214 \qquad (10.305)$$

10.15 Scherzer Focus

Fig. 10.34 Contrast transfer function for $U_0 = 300$ kV, $C_S = 1.2$ mm at optimal defocus of 68.8 nm

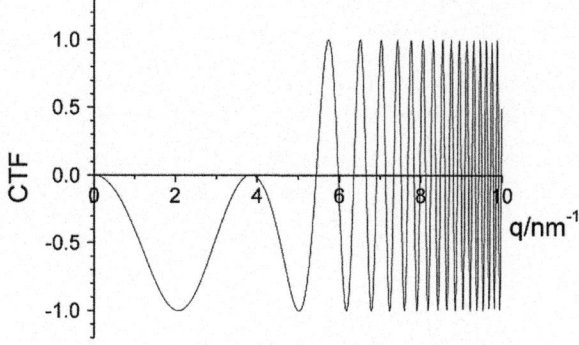

Fig. 10.35 Sine function with illustration of the "80%" value

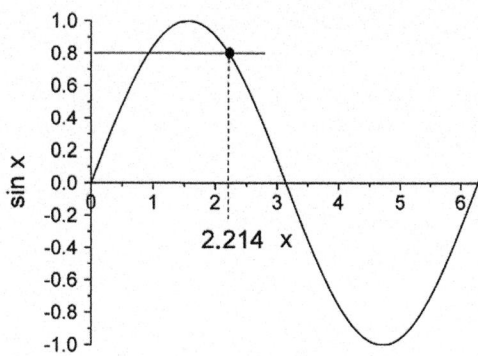

and

$$\left| \frac{(\Delta f)^2}{C_S \cdot \lambda} - 2 \cdot \frac{(\Delta f)^2}{C_S \cdot \lambda} \right| = \frac{(\Delta f)^2}{C_S \cdot \lambda} = 1.41 \quad (10.306)$$

The consideration of the focus change which is calculated from Eq. (10.306) leads to the so-called *Scherzer focus (or defocus)* Δf_{Sch}:

$$\Delta f_{Sch} = \sqrt{1.41 \cdot C_S \cdot \lambda} = 1.2 \cdot \sqrt{C_S \cdot \lambda} \quad (10.307)$$

Figure 10.36 shows the contrast transfer function for this defocus.

As expected the function does go below the distance of 0.8 to the abscissa up to the first zero. This distance of 0.8 was some randomly chosen. Insofar the factor 1.2 in Eq. (10.307) is random, too. However, it agrees with the value commonly used in the literature which is based on a maximum area rectangle that can be put between the curve and the space frequency axis.

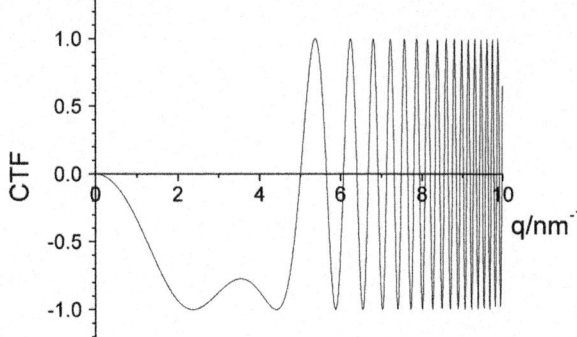

Fig. 10.36 Contrast transfer function for $U_0 = 300$ kV, $C_S = 1.2$ mm at an optimal defocus of 58.3 nm (Scherzer focus)

10.16 Delocalisation

From Chaps. 5 and 7 we know that for the imaging of very small structures by phase contrast the diffraction angles θ of the waves needed for the interference can be larger than 10 mrad. They have to be transferred by the objective lens where the influence of the spherical aberration is larger than for the imaging of coarser structures generating smaller diffraction angles. The sketches in Fig. 10.37 illustrate this.

For the imaging with electron waves or rays, respectively, which run close to the optical axis as typical for mass thickness contrast the spherical aberration does not play any role (Fig. 10.37a). The Gaussian image plane is identical with the focus plane. This changes completely if more strongly inclined waves are used as given in the diffraction maxima of crystal lattices. In this case the spherical aberration causes a stronger ray refraction and these rays cross each other in a point already in front of the Gaussian image plane. Additionally, this point is shifted perpendicular to the optical axis z (Fig. 10.37b). This shift between the pictures for imaging with rays close to the optical axis (mass thickness contrast) and farther from the axis (crystal lattice imaging) is called *delocalisation*.

We use our simple geometric model of Fig. 10.37b to evaluate the influence parameters and the magnitude of the delocalisation y_D that is the distance between the intersection of the blue and the red line and the optical axis z. The incident angle into the lens is in the object space for the blue ray by $\delta\theta$ greater than for the red ray. If the focus is changed the image plane shifts by Δb on the optical axis. Simultaneously, the distance of the intersection of the rays and the Gaussian image plane from the z axis is reduced by

$$\Delta y = \Delta b \cdot \theta' = \Delta b \cdot \frac{\theta}{M} \tag{10.308}$$

(θ': angle between ray and z axis in the image space).

With Eqs. (10.263) and (10.267) we get:

$$\Delta y = \Delta f \cdot \frac{b^2}{f^2} \cdot \frac{\theta}{M} = \Delta f \cdot \frac{M^2 \cdot g^2}{f^2} \cdot \frac{\theta}{M} = \Delta f \cdot M \cdot \theta \tag{10.309}$$

10.16 Delocalisation

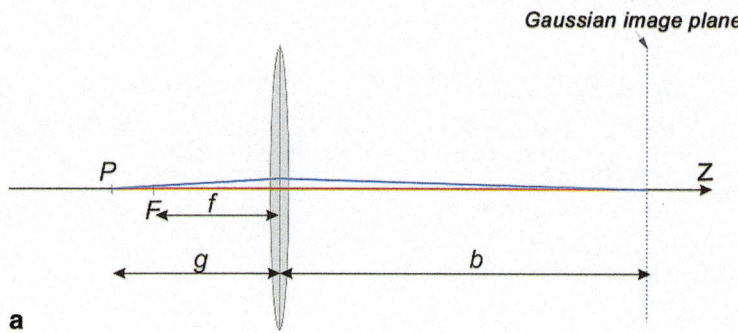

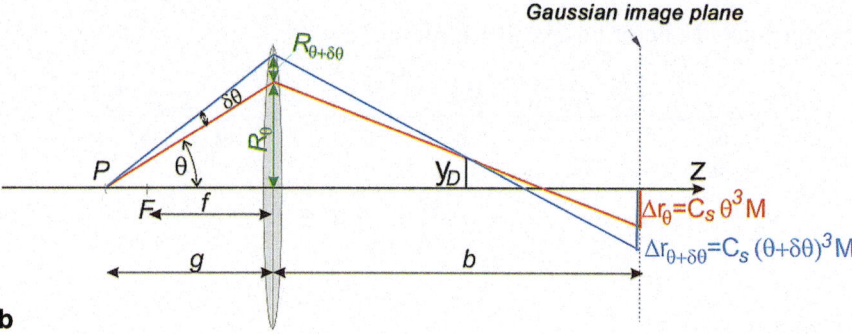

Fig. 10.37 Imaging with electron rays (waves, respectively) which are close to (**a**) and farther (**b**) from the optical axis z (*P*: object point, *F*: focal point, *M*: magnification)

The equations for the red and the blue line are:

$$y_{red} = R_\theta - \frac{\Delta r_\theta - \Delta y + R_\theta}{b} \cdot z$$
$$= g \cdot \theta - \frac{M \cdot C_s \cdot \theta^3 - \Delta f \cdot M \cdot \theta + g \cdot \theta}{b} \cdot z \quad (10.310)$$

and

$$y_{blue} = R_{\theta+\delta\theta} - \frac{\Delta r_{\theta+\delta\theta} - \Delta y + R_{\theta+\delta\theta}}{b} \cdot z$$
$$= g \cdot (\theta + \delta\theta) - \frac{M \cdot C_s \cdot (\theta+\delta\theta)^3 - \Delta f \cdot M \cdot \theta + g \cdot (\theta+\delta\theta)}{b} \cdot z \quad (10.311)$$

The z coordinate z_S of the intersection of both lines is:

$$g\cdot\theta - \frac{M\cdot C_S\cdot\theta^3 - \Delta f\cdot M\cdot\theta + g\cdot\theta}{b}\cdot z_S =$$

$$g\cdot(\theta+\delta\theta) - \frac{M\cdot C_S\cdot(\theta+\delta\theta)^3 - \Delta f\cdot M\cdot\theta + g\cdot(\theta+\delta\theta)}{b}\cdot z_S \qquad (10.312)$$

and after limitation to linear terms of $\delta\theta$:

$$z_S = \frac{g\cdot b}{3\cdot M\cdot C_S\cdot\theta^2 + g}. \qquad (10.313)$$

Insertion into the linear Eq. (10.310) leads to:

$$y_D = g\cdot\theta - \frac{M\cdot C_S\cdot\theta^3 - \Delta f\cdot M\cdot\theta + g\cdot\theta}{b}\cdot\frac{g\cdot b}{3\cdot M\cdot C_S\cdot\theta^2 + g}$$

$$y_D = g\cdot\theta\left[1 - \frac{M\cdot C_S\cdot\theta^2 - \Delta f\cdot M + g}{3\cdot M\cdot C_S\cdot\theta^2 + g}\right] \qquad (10.314)$$

and finally to

$$y_D = g\cdot\theta\cdot\frac{3\cdot M\cdot C_S\cdot\theta^2 + g - M\cdot C_S\cdot\theta^2 + \Delta f\cdot M - g}{3\cdot M\cdot C_S\cdot\theta^2 + g}$$

$$y_D = g\cdot\theta\cdot\frac{2\cdot C_S\cdot\theta^2 + \Delta f}{3\cdot C_S\cdot\theta^2 + g/M} \qquad (10.315)$$

We refer to the delocalisation on the object plane (i.e. division by magnification M), set $g=f$ (high magnification), and consider the relation $\theta=\lambda\cdot q$ between diffraction angle θ, electron wavelength λ, and space frequency q:

$$y_D = \frac{\lambda q\cdot(2\cdot C_S\cdot\lambda^2\cdot q^2 + \Delta f)}{3\cdot\lambda^2\cdot q^2\cdot M\cdot C_S/f + 1} \qquad (10.316)$$

The consequence can be seen in Fig. 10.38: With present spherical aberration the delocalisation can be avoided by defocussing only for a special space frequency. If the image contains various different lattice spacings some of them are delocalised anyway.

For comparison we cite the equation from the literature (see e.g. [5]) deduced using a wave-optical model which is more difficult to be understood:

$$y_D = \lambda q\cdot(C_S\cdot\lambda^2\cdot q^2 + \Delta f). \qquad (10.317)$$

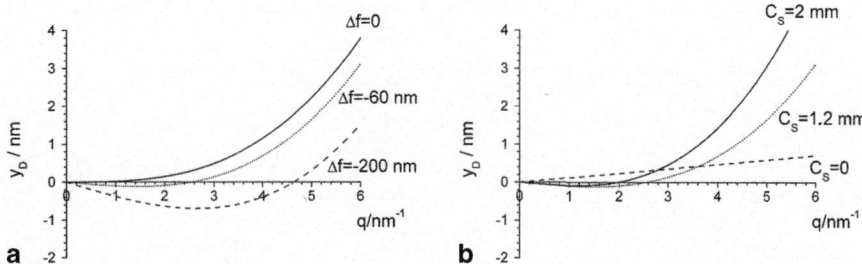

Fig. 10.38 Delocalisation in dependence on space frequency q (parameter: $E_0=300$ keV, i.e. $\lambda=1{,}97$ pm, magnification of the objective $M=100$, focal length $f=1$ mm). **a** Coefficient of spherical aberration $C_S=1.2$ mm, variation of the defocus Δf. **b** Defocus $\Delta f=-60$ nm, variation of C_S

Except for the factor 2 before the term containing the coefficient C_S of the spherical aberration the numerator of our Eq. (10.316) which is deduced using a simple geometric model agrees with Eq. (10.317). The deviation of the denominator from 1 remains less than 0.015 for parameters typical for a transmission electron microscope ($C_S \approx f$, $M \approx 100$, $E_0=300$ keV, $q<6$ nm^{-1}).

10.17 Potential in Electrostatic Multipoles

For stigmators and correctors of spherical and chromatic aberrations non-rotational symmetric electron-optical devices, so-called multipoles, are necessary. This has been well-known since Otto Scherzer's papers were published in 1936. On this fundament O. Rang realised a device for the correction of astigmatism named stigmator in 1949 [6]. An objective (with low resolution) only consisting of quadrupoles was constructed and tested by H.-D. Bauer in 1967 [7].

To understand the properties of multipoles we want to calculate and to illustrate the potential distribution of a quadrupole and an octupole. The method is known from Sect. 10.7: We have to solve Laplace's equation. In Cartesian coordinates it is:

$$\Delta U = \frac{\partial^2 U}{\partial x^2} + \frac{\partial^2 U}{\partial y^2} + \frac{\partial^2 U}{\partial z^2} = 0. \qquad (10.318)$$

Simplifying we assume an unlimitedly expanded multipole in z direction, i.e. the potential distribution is constant in z direction. Therewith Eq. (10.318) simplifies to

$$\frac{\partial^2 U}{\partial x^2} + \frac{\partial^2 U}{\partial y^2} = 0. \qquad (10.319)$$

To numerically solve it we transform this differential equation using the rules

$$\frac{\partial^2 U(x,z)}{\partial x^2} \rightarrow \frac{\Delta(\Delta U(x,y))}{(\Delta x)^2} = \frac{U(x+\Delta x, y) - 2 \cdot U(x,y) + U(x-\Delta x, y)}{(\Delta x)^2}$$

$$\frac{\partial^2 U(x,y)}{\partial y^2} \rightarrow \frac{\Delta(\Delta U(x,y))}{(\Delta y)^2} = \frac{U(x, y+\Delta y) - 2 \cdot U(x,y) + U(x, y-\Delta y)}{(\Delta y)^2} \quad (10.320)$$

into a difference equation. For equal differences Δx and Δy $(=\Delta)$ as well as

$$x = i \cdot \Delta \text{ and } y = j \cdot \Delta \quad (10.321)$$

we get

$$U(i+1, j) - 2 \cdot U(i, j) + U(i-1, j) + U(i, j+1) \\ - 2 \cdot U(i, j) + U(i, j-1) = 0 \quad (10.322)$$

and from this the recurrence formula

$$U(i, j) = \frac{1}{4}[U(i+1, j) + U(i-1, j) + U(i, j+1) + U(i, j-1)] \quad (10.323)$$

for the calculation of the potential distribution starting with the edge potentials. The edge potentials which are not directly given by the boundary conditions can be derived by the equations

$$U(i, 0) = \frac{1}{2}[U(i+1, 0) + U(i-1, 0)] \quad \text{for } i > 0$$

$$U(0, j) = \frac{1}{2}[U(0, j+1) + U(0, j-1)] \quad \text{for } j > 0 \quad (10.324)$$

For the definition of the boundary conditions we choose rods with a circular cross section as multipole elements arranged in the centre of the sides of a square and of a regular octagon, respectively (Fig. 10.39).

In principle the quadrupole is suitable to be used a as stigmator. To understand that let us have a look at Fig. 10.40. There the potential distribution is drawn in two different ways: in Fig. 10.40a colour-coded and in Fig. 10.40b as a potential surface. Figure 10.40b looks like a saddle. The point in which the electrons do not experience any deflecting force (potential equal to zero) is called "*saddle point*". We can also regard it as the optical image centre in the drawn multipole plane.

A bundle of electrons with a circular cross section is expanded in vertical direction and compressed in horizontal direction ("principal meridians"), i.e. it is elliptically distorted.

10.18 Electron Probe and Aberrations

Fig. 10.39 Schematic illustration of a quadrupole (**a**) and an octupole (**b**) with possible boundary potentials

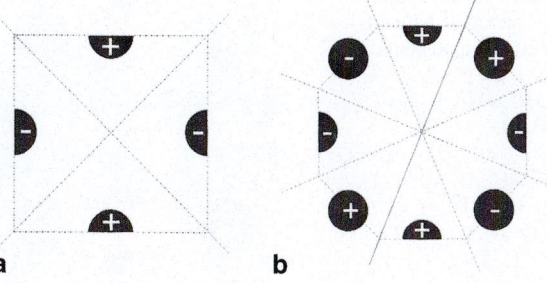

Fig. 10.40 Potential distribution in a quadrupole at pole potentials $\pm U_0$. **a** Colour-coded illustration (*red*: positive, *blue*: negative). **b** potential surface in perspective illustration

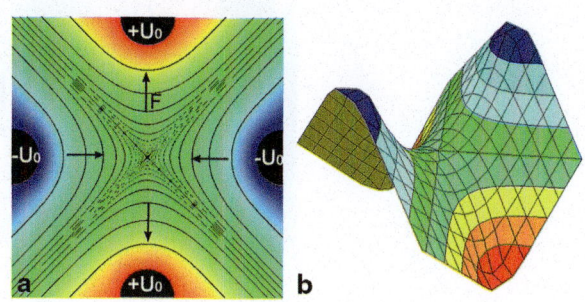

Simultaneously a difference of the focal lengths in both principal meridians is created. Just this is the indication of astigmatism. For optimal astigmatism correction the quadrupole has to be rotated against the astigmatic ellipse in a suitable way and its strength must be adjusted. To be able to do that without any mechanical adjustment an octupole is used instead of the "simple" quadrupole.

In Fig. 10.41 we see the potential distribution within an octupole with different pole potentials. With the aim of equipotential lines drawn in black the rotation of the potential field can be seen in the Figs. 10.41a–10.41c. In these figures the potential distribution is symmetric, i.e. opposite poles have equal potentials and the saddle point lies exactly in the centre. In Fig. 10.41d opposite poles have unequal potentials and the saddle point is shifted to the left side at the bottom. This demonstrates the possibility to align the octupole to the optical axis of another lens without any mechanical adjustment.

10.18 Electron Probe and Aberrations

In addition to Sect. 8.2 we would like to explain in more detail some of the facts and equations mentioned there. Contrary to the magnifying imaging an electron probe with very small cross section must be generated. The aberration disks directly limit the minimal probe size, their optical demagnification is impossible. Therefore, the equations written in this paragraph do not include the magnification.

Fig. 10.41 Potential field in an octupole with different pole potentials.
a $U_5=U_1$, $U_3=U_7=-U_1$, $U_2=U_4=U_6=U_8=0$
b $U_5=U_1$, $U_3=U_7=-U_1$, $U_2=U_6=-U_1$, $U_4=U_8=U_1/2$
c $U_5=U_1$, $U_3=U_7=-U_1$, $U_2=U_6=-U_1$, $U_4=U_8=U_1$
d $U_5=U_1/5$, $U_3=-U_1$, $U_7=U_3/5$, $U_2=U_4=U_6=U_8=0$

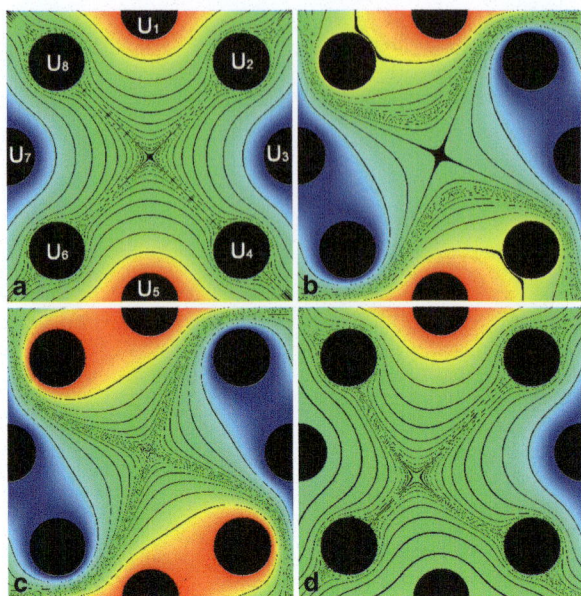

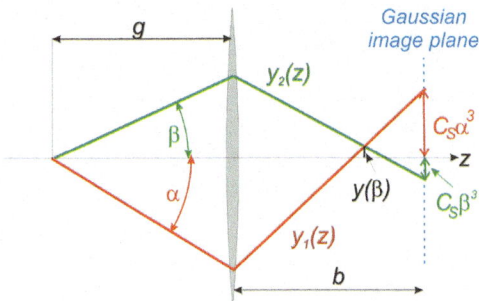

Fig. 10.42 For calculation of the probe radius in the plane of least confusion with spherical aberration. The lens centre is at $z=0$

Spherical aberration disk in the plane of least confusion The Gaussian image plane is not the optimal setting to reach the smallest probe diameter (Fig. 10.42).

We see that for an aperture angle α the radius $y(\beta)$ of the aberration disk is significantly smaller than $C_S \cdot \alpha^3$ in the Gaussian image plane. The plane with the smallest radius is called "*plane of least confusion*".

We would like to evaluate geometric-optically how small the minimal probe diameter (or radius, respectively) can be. For this purpose firstly we determine the linear equations $y_1(z)$ and $y_2(z)$:

$$y_1(z) = m_1 \cdot z + n_1, \quad y_1(0) = -g \cdot \alpha, \quad y_1(b) = C_S \cdot \alpha^3$$

$$y_1(z) = \frac{C_S \cdot \alpha^3 + g \cdot \alpha}{b} \cdot z - g \cdot \alpha \tag{10.325}$$

10.18 Electron Probe and Aberrations

as well as

$$y_2(z) = m_2 \cdot z + n_2, \; y_2(0) = g \cdot \beta, \; y_2(b) = -C_S \cdot \beta^3$$

$$y_2(z) = -\frac{C_S \cdot \beta^3 + g \cdot \beta}{b} \cdot z + g \cdot \beta. \tag{10.326}$$

Obviously, the radius y at the intersection of both lines depends on the angle β, in other words $y = y(\beta)$. The goal is to find this angle β which gives the maximum of y. This is the minimal probe radius.

At first, we have to determine the function $y(\beta)$. At the intersection of both lines (i.e. at z_S) it is:

$$y_1(z_s) = y_2(z_s)$$

$$\frac{C_S \cdot \alpha^3 + g \cdot \alpha}{b} \cdot z_s - g \cdot \alpha = -\frac{C_S \cdot \beta^3 + g \cdot \beta}{b} \cdot z_s + g \cdot \beta$$

$$\frac{z_s}{b}(C_S \cdot \alpha^3 + g \cdot \alpha + C_S \cdot \beta^3 + g \cdot \beta) = g \cdot (\alpha + \beta) \tag{10.327}$$

and finally

$$z_s = \frac{b \cdot g \cdot (\alpha + \beta)}{C_S \cdot (\alpha^3 + \beta^3) + g \cdot (\alpha + \beta)} \tag{10.328}$$

This is inserted into the linear equation $y_1(z)$:

$$y_1(z) = \frac{C_S \cdot \alpha^3 + g \cdot \alpha}{b} \cdot \frac{b \cdot g \cdot (\alpha + \beta)}{C_S \cdot (\alpha^3 + \beta^3) + g \cdot (\alpha + \beta)} - g \cdot \alpha \tag{10.329}$$

and after transformation we get the wanted function

$$y(\beta) = \frac{g \cdot C_S \cdot \alpha \cdot \beta (\alpha^2 - \beta^2)}{C_S \cdot (\alpha^3 + \beta^3) + g \cdot (\alpha + \beta)}. \tag{10.330}$$

From the object point a cone-shaped bundle with the half opening angle α starts, i.e. it is for β:

$$0 \le \beta \le \alpha$$

As mentioned above, the minimal probe size in the plane of least confusion is equal to the maximum of the function $y(\beta)$:

$$\frac{\partial y(\beta)}{\partial \beta}\bigg|_{Max} = 0 = g \cdot C_S \cdot \alpha \cdot \frac{(C_S \cdot \alpha^2 + g) \cdot (\alpha^3 - 3 \cdot \alpha \cdot \beta^2 - 2 \cdot \beta^3)}{C_S \cdot (\alpha^3 + \beta^3) + g \cdot (\alpha + \beta)^2}$$

$$0 = (C_S \cdot \alpha^2 + g) \cdot (\alpha^3 - 3 \cdot \alpha \cdot \beta^2 - 2 \cdot \beta^3)$$

$$\alpha^3 = 3\cdot\alpha\cdot\beta^2 + 2\cdot\beta^3. \tag{10.331}$$

Obviously this equation is fulfilled for $\beta=\alpha/2$. We insert this result into Eq. (10.330) and get for the radius of the minimal probe cross section:

$$r_{min} = \frac{g\cdot C_S \cdot \dfrac{\alpha^2}{2}\left(\alpha^2 - \dfrac{\alpha^2}{4}\right)}{C_S\cdot\left(\alpha^3 + \dfrac{\alpha^3}{8}\right) + g\cdot\dfrac{3}{2}\alpha} = \frac{3\cdot C_S\cdot\alpha^3}{9\dfrac{C_S}{g}\cdot\alpha^2 + 12}. \tag{10.332}$$

For $\alpha\ll 1$ and the condition $g>C_S$ (valid in case of demagnifying lenses) the denominator is:

$$9\frac{C_S}{g}\cdot\alpha^2 \ll 12 \tag{10.333}$$

and therewith

$$r_{min} = \frac{3\cdot C_S\cdot\alpha^3}{12} = \frac{1}{4}C_S\cdot\alpha^3, \tag{10.334}$$

i.e. the size of the spherical aberration disk in the plane of least confusion is only a quarter of that in the Gaussian image plane.

Chromatic aberration disk in the plane of least confusion We repeat this discussion for the chromatic aberration disk and have a look at Fig. 10.43.

Let the image point for electrons with the energy E, i.e. $\Delta E=0$, be in the Gaussian image plane. The relevant ray path is drawn green and labelled $y_2(z)$ in Fig. 10.43. Electrons with the energy $E-\Delta E$ ($y_1(z)$: red) are stronger refracted because of the chromatic aberration and they reach the Gaussian image plane in the distance $C_C\cdot(\Delta E/E)\cdot\alpha$ from the optical axis z (C_C: coefficient of chromatic aberration). For both linear equations one can read from Fig. 10.43:

$$y_1(z) = \left[C_C\cdot\left(\frac{\Delta E}{E}\right) + g\right]\cdot\frac{\alpha}{b}\cdot z - g\cdot\alpha$$

and

$$y_2(z) = -\frac{g\cdot\alpha}{b}\cdot z + g\cdot\alpha \tag{10.335}$$

10.18 Electron Probe and Aberrations

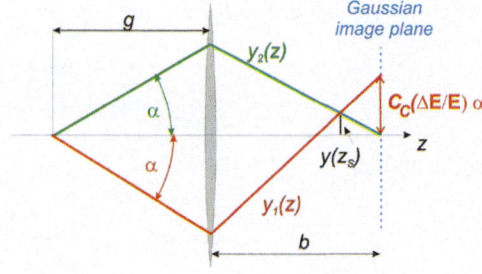

Fig. 10.43 For calculation of the probe radius in the plane of least confusion with chromatic aberration

At $z=z_s$ it is:

$$y_1(z_s) = y_2(z_s) \tag{10.336}$$

and

$$\left[C_c \cdot \left(\frac{\Delta E}{E}\right) + g\right] \cdot \frac{\alpha}{b} \cdot z_s - g \cdot \alpha = -\frac{g \cdot \alpha}{b} \cdot z_s + g \cdot \alpha$$

$$z_s = \frac{2 \cdot b}{\dfrac{C_c}{g} \cdot \left(\dfrac{\Delta E}{E}\right) + 2} \tag{10.337}$$

This is inserted into the linear equation for y_2:

$$y_2(z_s) = y(z_s) = -\frac{g \cdot \alpha}{b} \cdot \frac{2 \cdot b}{\dfrac{C_c}{g} \cdot \left(\dfrac{\Delta E}{E}\right) + 2} + g \cdot \alpha$$

$$= g \cdot \alpha \cdot \left[1 - \frac{2}{\dfrac{C_c}{g} \cdot \left(\dfrac{\Delta E}{E}\right) + 2}\right] \tag{10.338}$$

and we get for the minimal distance from the z axis:

$$r_{min} = \frac{C_c \cdot \left(\dfrac{\Delta E}{E}\right) \cdot \alpha}{\dfrac{C_c}{g} \cdot \left(\dfrac{\Delta E}{E}\right) + 2}. \tag{10.339}$$

As $g > C_C$ we can derive for normal experimental conditions:

$$\frac{C_C}{g} \cdot \left(\frac{\Delta E}{E}\right) \ll 2 \tag{10.340}$$

and with that

$$r_{min} = \frac{1}{2} \cdot C_C \cdot \left(\frac{\Delta E}{E}\right) \cdot \alpha, \tag{10.341}$$

i.e. the size of the chromatic aberration disk in the plane of least confusion is half of that in the Gaussian image plane.

Solution of the extremum problem for the determination of the optimal aperture: With consideration of astigmatism (astigmatic difference of the focal lengths Δf_A), spherical and chromatic aberration (coefficients C_S and C_C, respectively) the probe diameter d depends on the beam current I_P, the richtstrahlwert R, the relative energy spread $\Delta E/E$, and on the wavelength λ of the electrons (cf. Sect. 8.2, Eq. (8.10)):

$$d^2 = \frac{K_{IR\lambda}}{\alpha^2} + \frac{C_S^2}{4} \cdot \alpha^6 + K_{CA} \cdot \alpha^2 \tag{10.342}$$

with

$$K_{IR\lambda} = \frac{4 \cdot I_P}{\pi^2 \cdot R} + 1.5 \cdot \lambda^2 \text{ and } K_{CA} = C_C^2 \cdot \left(\frac{\Delta E}{E}\right)^2 + (\Delta f_A)^2. \tag{10.343}$$

To solve the extremum problem Eq. (10.342) has to be differentiated with respect to α and the result is set equal to zero:

$$\frac{\partial (d^2)}{\partial \alpha} = -\frac{2 \cdot K_{IR\lambda}}{\alpha^3} + \frac{3}{2} \cdot C_S^2 \cdot \alpha^5 + 2 \cdot K_{CA} \cdot \alpha$$

$$\frac{2 \cdot K_{IR\lambda}}{\alpha_{opt}^3} = \frac{3}{2} \cdot C_S^2 \cdot \alpha_{opt}^5 + 2 \cdot K_{CA} \cdot \alpha_{opt}$$

$$0 = \alpha_{opt}^8 + \frac{4}{3} \cdot \frac{K_{CA}}{C_S^2} \cdot \alpha_{opt}^4 - \frac{4}{3} \cdot \frac{K_{IR\lambda}}{C_S^2}. \tag{10.344}$$

We substitute

$$\alpha_{opt}^4 = x \tag{10.345}$$

10.18 Electron Probe and Aberrations

to get the quadratic equation

$$0 = x^2 + \frac{4}{3}\frac{K_{CA}}{C_S^2} \cdot x - \frac{4}{3}\frac{K_{IR\lambda}}{C_S^2}. \qquad (10.346)$$

Its solution is

$$0 = x^2 + \frac{4}{3}\frac{K_{CA}}{C_S^2} \cdot x - \frac{4}{3}\frac{K_{IR\lambda}}{C_S^2}$$

$$x_{1,2} = -\frac{2}{3}\frac{K_{CA}}{C_S^2} \pm \sqrt{\frac{4}{9}\left(\frac{K_{CA}}{C_S^2}\right)^2 + \frac{4}{3}\frac{K_{IR\lambda}}{C_S^2}} \qquad (10.347)$$

The solution must be positive, therefore only the positive sign in front of the radical term has to be used:

$$x = -\frac{2}{3}\frac{K_{CA}}{C_S^2} + \sqrt{\frac{4}{9}\left(\frac{K_{CA}}{C_S^2}\right)^2 + \frac{4}{3}\frac{K_{IR\lambda}}{C_S^2}} \qquad (10.348)$$

$$x = \frac{2}{3 \cdot C_S}\left(-\frac{K_{CA}}{C_S} + \sqrt{\frac{K_{CA}^2}{C_S^2} + 3 \cdot K_{IR\lambda}}\right). \qquad (10.349)$$

With consideration of Eq. (10.345) we get the optimal aperture

$$\alpha_{opt} = \sqrt[4]{\frac{2}{3 \cdot C_S}\left(-\frac{K_{CA}}{C_S} + \sqrt{\frac{K_{CA}^2}{C_S^2} + 3 \cdot K_{IR\lambda}}\right)}. \qquad (10.350)$$

Using a C_S probe corrector the spherical aberration can be corrected. In this case Eq. (10.342) with $C_S=0$ results after solution of the extremum problem in:

$$\alpha_{opt} = \sqrt[4]{\frac{K_{IR\lambda}}{K_{CA}}}. \qquad (10.351)$$

The consequences of these results on the resolution limit in the scanning transmission electron microscopic mode have been already discussed in Sect. 8.2.

Fig. 10.44 Momentum balances of the elastic (**a**) and inelastic (**b**) collision

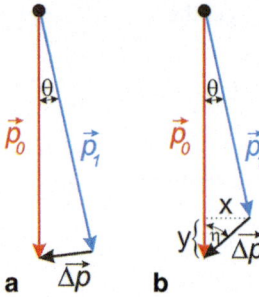

10.19 Classical Inelastic Collision

At inelastic interactions a part of the beam electron energy is used to induce changes of the atomic electron shell. Here, we want to discuss the consequences without consideration of the specific changes in the electron shell and use the model of the classical inelastic impact. In Fig. 10.44 the momentum balances of the elastic and inelastic collision in the case of small scattering angles ($\theta \ll 1$) are confronted (cf. also [8]).

At the elastic collision the momentum and the energy are conserved, i.e.

$$\theta = \frac{|\Delta \mathbf{p}|}{|\mathbf{p}_0|} = \frac{\Delta p}{p_0} = \frac{\Delta p}{p_1}. \tag{10.352}$$

In contrast, at an inelastic collision a part ΔE of the electron energy remains "in the atom" so that for the energy balance in non-relativistic approximation

$$\Delta E = \frac{p_0^2}{2 \cdot m} - \frac{p_1^2}{2 \cdot m} \tag{10.353}$$

(m: electron mass) and for momentum balance

$$\mathbf{p}_1 = \mathbf{p}_0 - \Delta \mathbf{p} \tag{10.354}$$

are given. After squaring and expansion of the scalar products it follows

$$(\mathbf{p}_1)^2 = (\mathbf{p}_0 - \Delta \mathbf{p})^2 \text{ and } p_1^2 = p_0^2 - 2 \cdot p_0 \cdot \Delta p \cdot \cos \eta + (\Delta p)^2 \tag{10.355}$$

and with consideration of Eq. (10.353) as well as $\Delta p \ll p_0$:

$$\Delta E = \frac{p_0 \cdot \Delta p}{m} \cdot \cos \eta. \tag{10.356}$$

10.20 Efficiency of Energy Dispersive X-ray Detectors

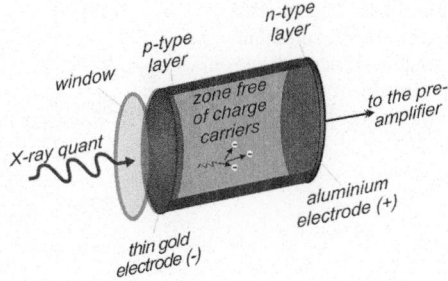

Fig. 10.45 Scheme of a semiconductor detector

From the small right-angled triangle of Fig. 10.44b it follows:

$$(\Delta p)^2 = x^2 + y^2 \approx p_0^2 \cdot \theta^2 + (\Delta p \cdot \cos \eta)^2. \tag{10.357}$$

According to Eq. (10.352) the first term $p_0 \cdot \theta$ characterises the change of the momentum direction for the elastic impact, in this sense the second term can be interpreted as a change of the direction for the inelastic impact. Let its characteristic angle be θ_{ie}. Therewith we get:

$$(\Delta p)^2 = p_0^2 \cdot (\theta^2 + \theta_{ie}^2) \tag{10.358}$$

as well as with consideration of Eq. (10.356) and

$$E_0 = \frac{p_0^2}{2 \cdot m} \tag{10.359}$$

$$\theta_{ie} = \frac{\Delta p \cdot \cos \eta}{p_0} = \frac{p_0 \cdot \Delta p \cdot \cos \eta}{p_0^2} = \frac{\Delta E \cdot m}{p_0^2} = \frac{\Delta E}{2 \cdot E_0}. \tag{10.360}$$

For a primary electron energy E_0 of 200 keV and an energy loss of 100 eV the characteristic deflection angle for inelastic scattering amounts to 0.25 mrad and is significantly smaller than that of the elastic scattering (cf. Sect. 10.11).

10.20 Efficiency of Energy Dispersive X-ray Detectors

The main part of an EDX detector is a cylindrical semiconductor crystal (mostly silicon) which is positioned within an electric field.

We assume both top surfaces of the cylinder as narrow p-type or n-type semiconductor zones, respectively. Between them a comparably wide zone free of charge carriers is located (Fig. 10.45). To calculate the efficiency of the detector we have to determine the probabilities for the X-ray quanta to overcome the barriers, as illustrated in Fig. 10.45.

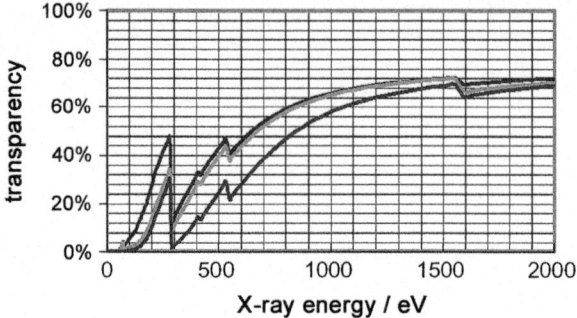

Fig. 10.46 Transparency T_W of different ultrathin windows in dependence on the energy of the X-radiation E_X. (Source: [9])

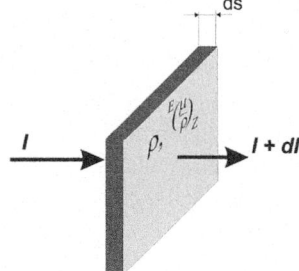

Fig. 10.47 Sketch for explanation of the attenuation law for X-radiation

Transparency of the window On the side of the microscopic column the detector tube is commonly closed by an ultrathin polymer window (thickness about 0.3 μm). This window represents the first barrier which has to be overcome by the X-radiation. The transparency of the window depends on the X-ray energy and is normally published by the manufacturer (Fig. 10.46).

The transparency $T_W(E_X)$ is equal to the probability of X-ray quant of the energy E_X to penetrate the window. The curves in Fig. 10.46 show four deviations from the monotonous trend at the energies 283 eV, 401 eV, 532 eV, and 1560 eV. These energies are necessary to transfer one electron of the K-shell of carbon, nitrogen, oxygen, and aluminium onto the first free energy state and are called *absorption edges*. The energies of the absorption edges are a bit larger than the energies of the X-ray K-radiation (cf. Sect. 9.1.2).

Up to now we have assumed that these energies come from the beam electrons, however, they can also arise from X-rays of sufficient energy (secondary fluorescence).

Obviously, the elements carbon, nitrogen and oxygen are the main components of the window material (not surprising for polymers). The absorption edge of aluminium originates from the aluminium support grid which is also responsible for the limitation of the transparency to about 80 % even at high X-ray energies.

10.20 Efficiency of Energy Dispersive X-ray Detectors

Table 10.4 Parameters for the evaluation of attenuation coefficients. [10]

		$E \geq E_K$	$E_K > E \geq E_L$	$E_L > E \geq E_M$
$3 \leq Z \leq 10$	C	$5.4 \cdot 10^{-3}$		
	α	2.92		
	β	3.07		
$11 \leq Z \leq 18$	C	$1.38 \cdot 10^{-2}$	$5.33 \cdot 10^{-4}$	
	α	2.79	2.74	
	β	2.73	3.03	
$19 \leq Z \leq 36$	C	$3.12 \cdot 10^{-2}$	$9.59 \cdot 10^{-4}$	$2.73 \cdot 10^{-5}$
	α	2.66	2.70	2.44
	β	2.47	2.90	3.47
$37 \leq Z \leq 54$	C		$1.03 \cdot 10^{-3}$	$2.73 \cdot 10^{-5}$
	α		2.70	2.44
	β		2.88	3.47
$55 \leq Z \leq 71$	C		$1.24 \cdot 10^{-3}$	$1.58 \cdot 10^{-4}$
	α		2.70	2.50
	β		2.83	2.98
$72 \leq Z \leq 86$	C		$1.03 \cdot 10^{-4}$	$9.39 \cdot 10^{-5}$
	α		2.50	2.55
	β		3.38	3.09
$87 \leq Z \leq 92$	C			$5.76 \cdot 10^{-7}$
	α			2.63
	β			4.26

Transparency of the gold electrode The thin gold electrode represents the second barrier. Generally, the attenuation of X-radiation in dependence on the material and the energy is described by *attenuation coefficients* (also: *absorption coefficients*).

While X-rays are going through a materials disc (thickness ds) characterised by a density ρ and an energy-dependent attenuation coefficient $^E(\mu/\rho)_Z$ their intensity is reduced by dI (Fig. 10.47):

$$dI = -I \cdot {}^E(\mu/\rho)_Z \cdot \rho \cdot ds. \tag{10.361}$$

From Eq. (10.361) we get for the intensity:

$$I = I_0 \cdot e^{-{}^E(\mu/\rho)_Z \cdot \rho \cdot s} \tag{10.362}$$

with I_0 as initial intensity and the total material's thickness s. The attenuation coefficient can be evaluated using the equation (source: [10])

$$\frac{{}^E(\mu/\rho)_Z}{cm^2 \cdot g^{-1}} = C \cdot \left(\frac{12.396}{E/keV}\right)^\alpha \cdot Z^\beta \tag{10.363}$$

(E: X-ray energy, Z: mean atomic number of the attenuating material) with the parameters C, α, and β from Table 10.4.

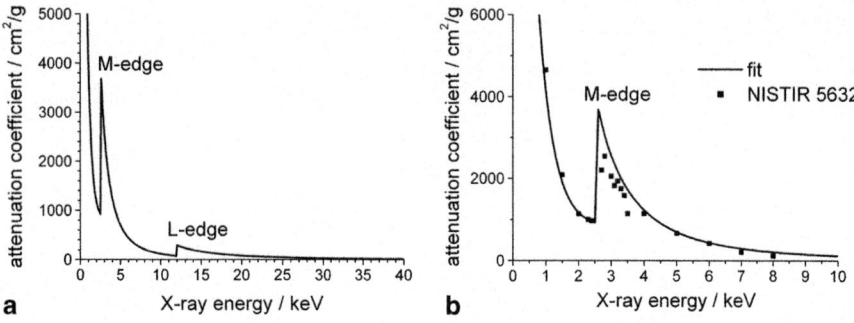

Fig. 10.48 Attenuation coefficient for gold in dependence on the X-ray energy. **a** Following Eqs. (10.363) and (10.364). **b** Comparison with measured values in NISTIR 5632 in the low energy range

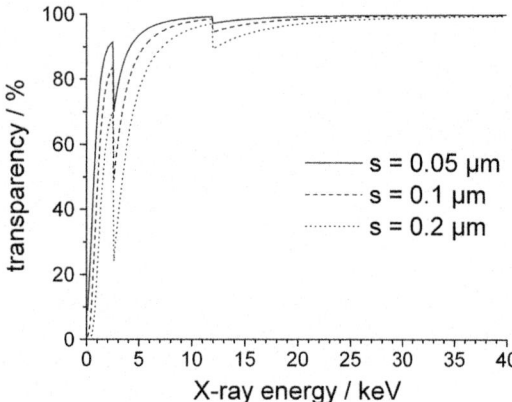

Fig. 10.49 Transparency T_{Au} of the gold electrode in dependence on the X-ray energy for three thicknesses s of the gold layer

Let us come back to the gold electrode. For gold (Z=79) the attenuation coefficients can be determined up to the M-edge (2.2 keV) using the values of Table 10.4. For lower energies it is possible to find values from the "National Institute of Standards and Technology (NIST)" of the USA in the internet [11]. In the energy range up to the M-edge these values of the attenuation coefficient K_{att} can be approximated by a function

$$\frac{K_{att}}{\text{cm}^2 \cdot \text{g}^{-1}} = 800 + 10000 \cdot e^{-2.2 \left(\frac{E}{\text{keV}} - 0.5\right)} \qquad (10.364)$$

(Fig. 10.48).

The energy-dependent transparency T_{Au} (Fig. 10.49) of the gold electrode with thickness s is given according to Eq. (10.362) by:

$$T_{Au} = \frac{I}{I_0} = e^{-E(\mu/\rho)_{79} \cdot \rho_{Au} \cdot s} \qquad (10.365)$$

(density of gold: $\rho_{Au} = 19.32$ g/cm^3).

10.20 Efficiency of Energy Dispersive X-ray Detectors

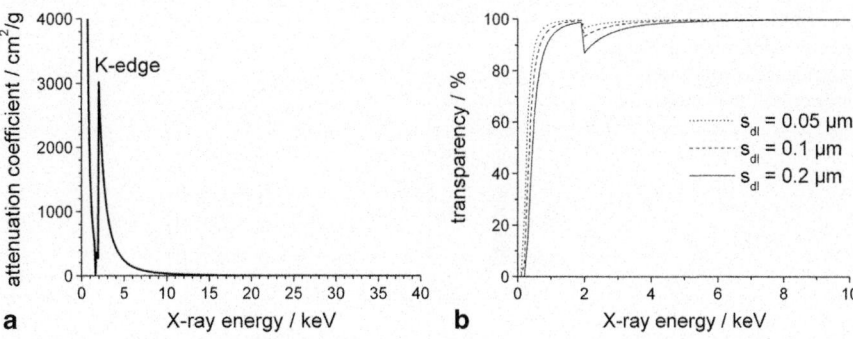

Fig. 10.50 Attenuation coefficients (**a**) and transparency (**b**) at different dead layer thicknesses s_{dl} of Si in dependence on the X-ray energy

Transparency of the p-type layer ("dead layer") The transparency T_{dl} of the dead layer of the detector crystal can be calculated in the same way like that of the gold electrode. We take the silicon crystal as starting point ($Z=14$, $\rho_{Si}=2.33$ g/cm³). For silicon Table 10.4 includes the complete energy range and we are able to evaluate the attenuation coefficients by use of Eq. (10.363) only (Fig. 10.50a).

Analogously to Eq. (10.365) we get

$$T_{dl} = \frac{I}{I_0} = e^{-E(\mu/\rho)_{14} \cdot \rho_{Si} \cdot s_{dl}} \tag{10.366}$$

(s_{dl}: thickness of the dead layer) for the transparency of the dead layer (Fig. 10.50b).

Absorption within the charge carrier free zone Finally, the X-radiation is absorbed within the charge carrier free zone of the silicon crystal and deposits its energy. The converted intensity is equal to the difference of the incident intensity I_0 and the intensity I_R leaving the crystal on its back surface:

$$I_A = I_0 - I_R = I_0 \cdot (1 - e^{-E(\mu/\rho)_{14} \cdot \rho_{Si} \cdot d}), \tag{10.367}$$

where d is the thickness of the crystal.

The probability for absorption is given by

$$w_A = \frac{I_A}{I_0} = 1 - e^{-E(\mu/\rho)_{14} \cdot \rho_{Si} \cdot d} \tag{10.368}$$

and plotted in Fig. 10.51. The results show that there is total absorption for X-ray energies up to ca. 20 keV.

The detector efficiency depends on additional (geometric) parameters: On the take off angle and on the acquired solid angle. However, generally the detector efficiency is given by a relative value based on a reference energy, e.g. on the Si Kα

Fig. 10.51 Probability w_A of X-ray absorption in Si in dependence on the energy for three crystal thicknesses d

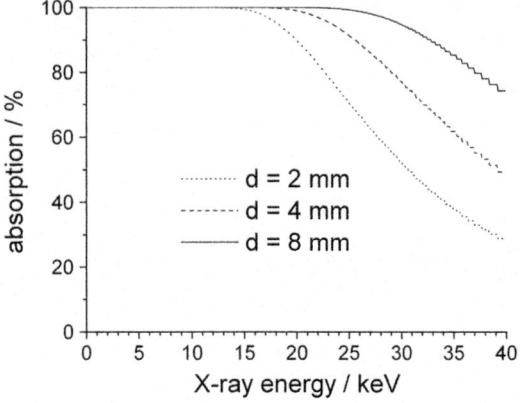

Fig. 10.52 Detector efficiency in dependence on the X-ray energy for two different parameter sets of the thicknesses of gold electrode (s), dead layer (s_{dl}), and crystal (d)

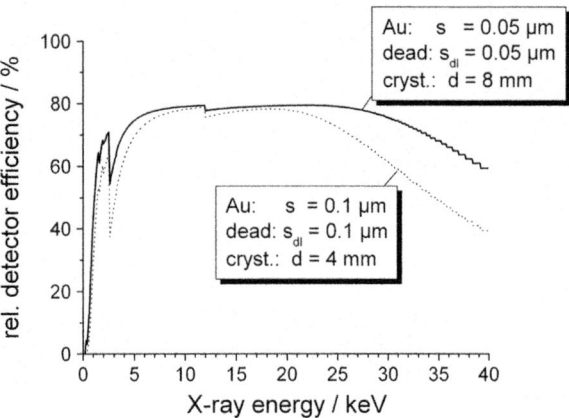

energy. This allows to avoid the inclusion of the experiment-depending geometric parameters. In this way only the energy dependence of the detector efficiency is of interest which is equal to the product of the probabilities calculated in this paragraph:

$$D_{eff} = T_W \cdot T_{Au} \cdot T_{dl} \cdot w_A \tag{10.369}$$

(Fig. 10.52).

In the low energy range we see the absorption edges of gold, i.e. the thickness of the gold electrode mainly determines the detector efficiency in this range. It seems possible to improve this by use of another electrode material. At high energies the size of the carrier free zone decides on the detector efficiency.

10.21 Calculation of Cliff-Lorimer k-factors

From Sect. 9.2.3 we know that the influence parameters on the sensitivity factors (*k-factors*) are given by the probability of X-ray emission after excitation by electrons and its detection: The ionisation of an atom occurs with the probability Q; the probability for the emission of an X-ray quant by this excited atom is ω (fluorescence yield) and finally, this X-ray quant can be detected with the probability D_{eff} (detector efficiency).

Let us assume we have a mixture of two elements A and B. The mentioned probabilities must be calculated for both elements. Additionally, the fraction a of the selected line (e.g. Kα) on the intensity of the total line series and, for values in mass percents, the atomic weight M plays a role. Therewith the ratio of the k-factors of the two elements A and B is:

$$\left(\frac{k_A}{k_B}\right)_M = \frac{Q_B \cdot \omega_B \cdot a_B \cdot D_{eff,B} \cdot M_B}{Q_A \cdot \omega_A \cdot a_A \cdot D_{eff,A} \cdot M_A}. \tag{10.370}$$

This is Eq. (9.34), expanded by the fraction a of the Kα-lines on the K-series. Commonly, for the L- and M-series the lines are close together and any distinctions within one series are impossible because of the limited energy resolution of the spectrometer. The fractions a are not applicable in this case. The detector efficiency has already been discussed in the previous Sect. 10.20, the three other influences are explained here, mainly by use of details described in the literature.

Cross section for ionisation Q Normally, for the calculation of the ionisation cross section $Q_{n,l}$ of an electron in the shell n with the orbital quantum number l an equation is used which was published by H. Bethe in 1930 [12]:

$$Q_{n,l} = \frac{2\pi \cdot e^4 \cdot Z_n \cdot b_{n,l}}{m \cdot v^2 \cdot E_{n,l}} \cdot \ln\left(\frac{2 \cdot m \cdot v^2}{E_{n,l}}\right) \tag{10.371}$$

(e: elementary electric charge, m: electron rest mass, v: velocity of the primary electrons, $E_{n,l}$: ionisation energy of the electron characterised by n and l, i.e. energy of the absorption edge, Z_n: number of electrons in shell n, $b_{n,l}$: quantum mechanical parameter which was calculated by Bethe to be equal to 0.2 … 0.6 using the electron eigenfunction of hydrogen. In non-relativistic approximation it is

$$\frac{m}{2} v^2 = e \cdot U_0 = E_0 \tag{10.372}$$

(U_0: acceleration voltage, E_0: energy of the primary electrons) and we can write

$$Q_{n,l} = \frac{\pi \cdot e^4 \cdot Z_n \cdot b_{n,l}}{E_0 \cdot E_{n,l}} \cdot \ln\left(\frac{4 \cdot E_0}{E_{n,l}}\right) \tag{10.373}$$

Table 10.5 Fit parameter c, b, and d for the calculation of ionisation cross sections

	Powell [14]	Schreiber and Wims [13] (Z: here atomic number)
K-shell ($n=1$, $Z_1=2$)	$c_K=0.94$ $b_K=0.64$ $d_K=1$	$c_K=1$ $b_K=8.874-8.158\times\ln Z+2.9055\times(\ln Z)^2-0.35778\times(\ln Z)^3$ for $Z\leq 30$ $b_K=0.661$ for $Z>30$ $d_K=1.0667-0.00476\times Z$
L-shell ($n=2$, $Z_2=8$)	$c_L=0.59$ $b_L=0.63$ $d_L=1$	$c_L=1$ $b_L=0.2704+0.00726\times(\ln Z)^3$ $d_L=1$
M-shell ($n=3$, $Z_3=18$)		$c_M=1$ $b_M=11.33-2.43\times\ln Z$ $c_M=1$

instead of Eq. (10.371). In practice the Eq. (10.373) is modified: The parameters $b_{n,l}$ and the ionisation energies are different only concerning the principal quantum number n, and the number 4 in the logarithm is substituted by an additional constant c_n:

$$Q_n = \frac{\pi \cdot e^4 \cdot Z_n \cdot b_n}{E_0 \cdot E_n} \cdot \ln\left(\frac{c_n \cdot E_0}{E_n}\right). \tag{10.374}$$

The constants b_n and c_n are determined by a fit of measurements. Using the *overvoltage ratio* $U_{OV}=E_0/E_n$ Eq. (10.374) can be written in the form

$$Q_n = \frac{\pi \cdot e^4 \cdot Z_n \cdot b_n}{E_n^2 \cdot U_{OV}^{d_n}} \cdot \ln(c_n \cdot U_{OV}). \tag{10.375}$$

The exponent d_n of the overvoltage ratio U_{OV} was added by Schreiber and Wims [13] to get a better agreement with measurements. The fit parameters by Schreiber and Wims [13] and Powell [14], respectively, are listed in Table 10.5.

For the further calculations we confine ourselves to the results by Schreiber and Wims [13]. But it has to be clear that the selection of the model influences the calculated k-factors and, therefore, this selection should be published in serious software descriptions by the manufacturer.

Commonly, the k-factors are referred to silicon, we would like to handle this ab initio and already refer the ionisation cross sections to those of the silicon K-shell (Fig. 10.53).

$$\frac{Q_{n,A}}{Q_{K,Si}} = \frac{Z_{n,A} \cdot b_{n,A} \cdot \ln(U_{OV,A})}{2 \cdot b_{K,Si} \cdot \ln(U_{OV,Si})} \cdot \frac{E_{K,Si}^{2-d_{KSi}}}{E_{n,A}^{2-d_n}}$$

$$= \frac{0.916 \cdot Z_{n,A} \cdot b_{n,A} \cdot \ln(U_{OV,A})}{(E_{n,A}/\text{keV})^{2-d_{nA}} \cdot \ln(0.5435 \cdot E_0/\text{keV})}. \tag{10.376}$$

10.21 Calculation of Cliff-Lorimer k-factors

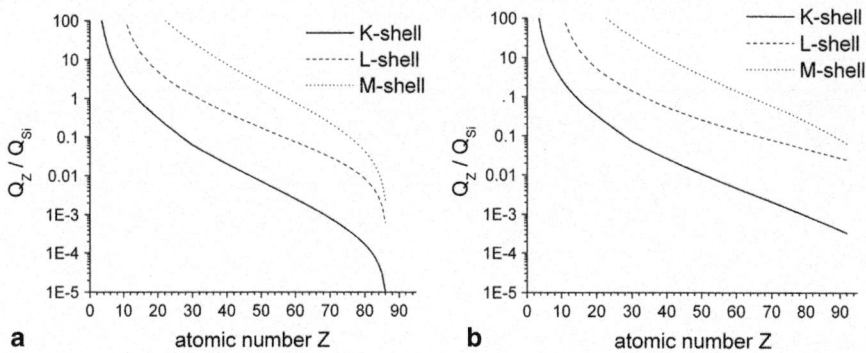

Fig. 10.53 Ionisation cross sections referred to Si-K, calculated using the parameters by Schreiber and Wims [13] for primary electron energies of 100 keV (**a**) and 300 keV (**b**)

We realise that the ionisation cross sections are approximately independent on the primary electron energy at low atomic numbers since the overvoltage ratio is large and its logarithmic change remains small at a variation of the primary electron energy. This changes if the energy of the absorption edge and that of the primary electrons come closer together (e.g. in low-voltage TEM).

Fluorescence yield ω The energy gain as a consequence of the refilling of an ionised core shell allows not only the emission of an X-ray quant but also the emission of an Auger[6] electron. The probability for the radiative transition is called *fluorescence yield*.

G. Wentzel [15] calculated quantum-mechanically the dependence of the fluorescence yield on the atomic number Z in the form

$$\omega = \frac{Z^4}{A+Z^4} \tag{10.377}$$

($A \approx 10^6$) in 1927, which is the basis of all further reflections regarding this topic. Similar to the ionisation cross section Eq. (10.377) is complemented by additional parameters to get a better agreement with measurements (compilation in [16]). A variant by E.H.S. Burhop [17] is based on the equation

$$\omega = \frac{(A+B \cdot Z+C \cdot Z^3)^4}{1+(A+B \cdot Z+C \cdot Z^3)^4} \tag{10.378}$$

The parameters A, B, and C are shell-specific and listed in Table 10.6.

We also refer the fluorescence yields to those of the Si-K shell and get the equation

$$\frac{\omega_{A,n}}{\omega_{Si,K}} = 28.05 \cdot \frac{(A_n + B_n \cdot Z_n + C_n \cdot Z_n^3)^4}{1+(A_n + B_n \cdot Z_n + C_n \cdot Z_n^3)^4} \tag{10.379}$$

(illustrated in Fig. 10.54).

---

[6] Pierre Victor Auger, French physicist, 1899–1993.

Table 10.6 Parameters A, B, and C for the calculation of fluorescence yields with Eq. (10.378)—see [17]

	A	B	C
K-shell	−0.03795	0.03426	−1.163·10⁻⁶
L-shell	−0.11107	0.01368	−2.177·10⁻⁷
M-shell	−0.00036	0.00368	2.010·10⁻⁷

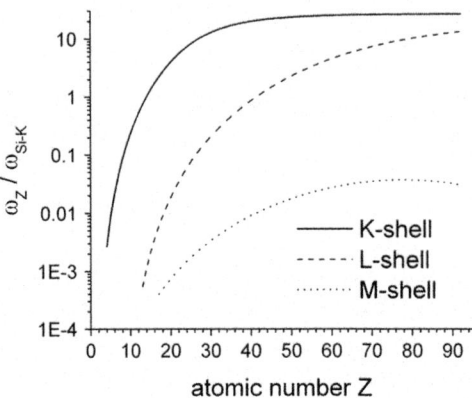

Fig. 10.54 Fluorescence yields, referred to Si-K and calculated according to the parameters in Table 10.6

Kα fraction a on the K-series: After ionisation of the K-shell it can be refilled by electrons from the L-shell (Kα- radiation) as well as (if existing) from the M-shell (Kβ-radiation). For silicon the Kα-line is at 1.74 keV, the Kβ-line at 1.83 keV. Both lines are more separated for elements with higher atomic numbers, i.e. they can be detected separately even with detectors with an energy resolution of about 130 eV. To separate the lines in the calculation we need the intensity ratio Kβ/Kα. A polynomial of the form

$$\frac{I_{K\beta}}{I_{K\alpha}} = A + B \cdot Z + C \cdot Z^2 + D \cdot Z^3 \tag{10.380}$$

(A, B, C, D see Table 10.7) fits to the values by Scofield [18] very well (Fig. 10.55).

Therewith the fraction a is:

$$a_A = \frac{I_{K\alpha,A}}{I_{K\alpha,A} + I_{K\beta,A}} = \frac{1}{1 + I_{K\beta,A}/I_{K\alpha,A}}. \tag{10.381}$$

For silicon the fraction is $a_{Si}=0.9736$. Therewith all preconditions are given for the calculation of the k-factors. We multiply the influence factors referred to silicon (results see Fig. 10.56):

10.21 Calculation of Cliff-Lorimer k-factors

Table 10.7 Parameters for calculation of the Kβ/Kα intensity ratio

Z	A	B	C	D
$11 \leq Z \leq 35$	−0.9834	0.1282	−0.00486	$6.134 \cdot 10^{-5}$
$Z > 35$	−0.065	0.0102	−1.12	$4.8 \cdot 10^{-7}$

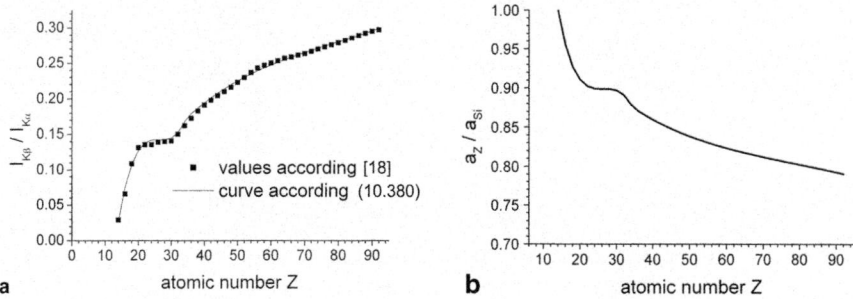

Fig. 10.55 Ratio of the Kβ- to the Kα-intensity in dependence on the atomic number. **a** Comparison between measured values and polynomial (10.380). **b** Fraction a referred to Si

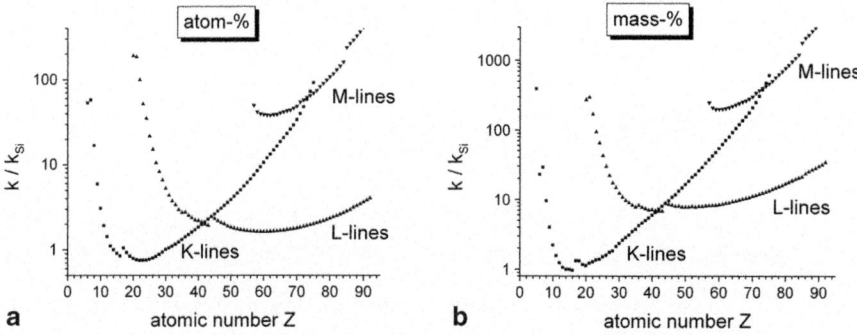

Fig. 10.56 Cliff-Lorimer k-factors referred to Si and 300 keV electrons in atom-% (**a**) and mass-% (**b**). The underlying models are explained in the text

$$\frac{k_A}{k_{Si}} = \frac{Q_{K,Si}}{Q_{n,A}} \cdot \frac{\omega_{Si,K}}{\omega_{A,n}} \cdot \frac{a_{Si}}{a_A} \cdot \frac{D_{eff,Si}}{D_{eff,A}} \text{ and } \left(\frac{k_A}{k_{Si}}\right)_M = \frac{k_A}{k_{Si}} \cdot \frac{M_A}{28.09}. \qquad (10.382)$$

Within the curve shapes some jumps attract attention. They are at X-ray energies for which the detector efficiency is non-monotonous because of the absorption edges of window and electrode materials.

Fig. 10.57 Sketch for explanation of the absorption of X-radiation within a thin sample of thickness t

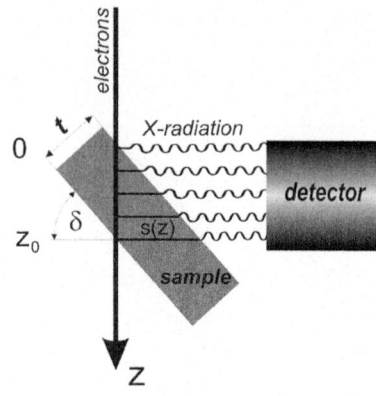

10.22 Correction of Absorption for EDXS

Although die samples for transmission electron microscopy are ultrathin especially the low-energetic X-radiation can be noticeably attenuated while it is penetrating the specimen leading to an underestimation of light elements at their quantification. In this paragraph we would like to evaluate among which circumstances this effect plays a role.

Figure 10.57 illustrates the problem. Along the path z of the electrons through the sample in the infinitesimal element dz X-radiation of the intensity dI_0 is generated. In total the intensity I_0 is generated on the complete path from $z=0$ to $z=z_0$. For a constant excitation along the complete path we can write:

$$dI_0 = \frac{I_0}{z_0} \cdot dz. \tag{10.383}$$

On the path to the detector X-rays have to travel the length $s(z)$ within the sample where they are partly absorbed. In the direction to the detector only the fraction dI penetrates the sample at the position z (cf. Sect. 10.20):

$$dI(z) = dI_0 \cdot e^{-{}^E(\mu/\rho)\cdot\rho\cdot s(z)} \tag{10.384}$$

($^E(\mu/\rho)$: attenuation coefficient of the sample material for X-quants of the energy E, ρ: density of the sample material). In Fig. 10.57 we read

$$s(z) = z \cdot \cot \delta \tag{10.385}$$

The detector is reached by the total intensity

$$I = \frac{I_0}{z_0} \cdot \int_0^{z_0} e^{-{}^E(\mu/\rho)\cdot\rho\cdot z \cdot \cot \delta} \cdot dz, \tag{10.386}$$

10.22 Correction of Absorption for EDXS

i.e. after solution of the integral:

$$I = \frac{I_0}{z_0} \cdot \left[\frac{-e^{-{}^E(\mu/\rho) \cdot \rho z \cdot \cot\delta}}{{}^E(\mu/\rho) \cdot \rho \cdot \cot\delta} \right]_0^{z_0} \qquad (10.387)$$

and after insertion of the bounds of integration as well as after consideration of

$$z_0 = \frac{t}{\cos\delta} = \frac{t}{\cot\delta \sin\delta} \qquad (10.388)$$

we get

$$I = \frac{I_0 \cdot \sin\delta}{{}^E(\mu/\rho) \cdot \rho t} \cdot 1 - e^{-{}^E(\mu/\rho) \cdot \rho t / \sin\delta}. \qquad (10.389)$$

We interpret the quotient of the X-ray intensity I which reaches the detector and the intensity I_0 which was originally generated within the sample as *absorption correction*

$$F_{AbsC} = \frac{I}{I_0} = \frac{\sin\delta}{{}^E(\mu/\rho) \cdot \rho t} \cdot 1 - e^{-{}^E(\mu/\rho) \cdot \rho t / \sin\delta}. \qquad (10.390)$$

For the quantification of the spectra the comparison of elements with different characteristic X-ray energies is important. However, the attenuation coefficients and the density depend on the composition of the sample and it is not sure whether these parameters are the same in thin layers and in bulk material. An approximation assumes that this the case and the attenuation coefficient as well the density are weighted linear combinations of the values of the elementary parts of the sample. This means for a sample consisting of two elements A and B:

$${}^E(\mu/\rho) = c_{A,M} \cdot {}^E(\mu/\rho)_A + c_{B,M} \cdot {}^E(\mu/\rho)_B, \qquad (10.391)$$

respectively,

$$\rho = c_{A,M} \cdot \rho_A + c_{B,M} \cdot \rho_B. \qquad (10.392)$$

The ratio of the absorption corrections for the elements A and B with the X-ray energies EA and EB is given by:

$$\frac{F_{AbsC,A}}{F_{AbsC,B}} = \frac{{}^{EB}(\mu/\rho)}{{}^{EA}(\mu/\rho)} \cdot \frac{1 - e^{-EA(\mu/\rho) \cdot \rho t / \sin\delta}}{1 - e^{-EB(\mu/\rho) \cdot \rho t / \sin\delta}}. \qquad (10.393)$$

Since the layer thickness is in the order of 0.1 µm we transform the exponential functions in series and terminate them after the quadratic term:

$$\frac{F_{AbsC,A}}{F_{AbsC,B}} = \frac{{}^{EB}(\mu/\rho) \cdot {}^{EA}(\mu/\rho) \cdot \rho t / \sin\delta - \frac{1}{2}{}^{EA}(\mu/\rho)^2 \cdot \rho^2 \cdot t^2 / \sin^2\delta}{{}^{EA}(\mu/\rho) \cdot {}^{EB}(\mu/\rho) \cdot \rho t / \sin\delta - \frac{1}{2}{}^{EB}(\mu/\rho)^2 \cdot \rho^2 \cdot t^2 / \sin^2\delta}$$

$$= \frac{1 - \frac{1}{2}{}^{EA}(\mu/\rho) \cdot \rho t / \sin\delta}{1 - \frac{1}{2}{}^{EB}(\mu/\rho) \cdot \rho t / \sin\delta}. \qquad (10.394)$$

This function is transformed in a series again and is terminated after the second term:

$$\frac{F_{AbsC,A}}{F_{AbsC,B}} = 1 + \frac{1}{2}({}^{EB}(\mu/\rho) - {}^{EA}(\mu/\rho)) \cdot \frac{\rho t}{\sin\delta}. \qquad (10.395)$$

If the absorption error has to be less than, for example, 10% the condition

$$\frac{1}{2} \cdot \frac{\rho t}{\sin\delta} \left| ({}^{EB}(\mu/\rho) - {}^{EA}(\mu/\rho)) \right| < 0.1. \qquad (10.396)$$

must be fulfilled. Except for the factor 1/sinδ this formula agrees with the "*thin film criterion*" by Goldstein et al. [19]. The *take-off angle* δ consists of two parts: The tilt angle of the sample around the specimen holder axis and a contribution which exists if the detector axis is located at a higher or inclined position relative to the tilt axis of the specimen.

10.23 Prisms for Electrons

To generate electron energy loss spectra we need a dispersion element similar to a glass prism for light. The glass prism differentiates white light in its colours, i.e. waves with different wavelengths exit the prism at different angles. We know that electrons can be deflected in electric and magnetic fields and want to think about the angles at which electrons of different velocity (energy) exit such fields.

We assume a magnetic field (*magnetic prism*) whose field lines are perpendicular to the drawing plane (Fig. 10.58a). The magnetic induction **B** has the same direction. Electrons with the initial energy, i.e. such ones without energy loss, incide vertically with the velocity $\mathbf{v}_0$ into this magnetic field and experience the Lorentz force

$$\mathbf{F}_L = -e \cdot \mathbf{v}_0 \times \mathbf{B}. \qquad (10.397)$$

The velocity vector lies in the drawing plane, the magnetic induction is perpendicular to it, i.e. velocity and induction are at a right angle. Thus, the modulus of the

10.23 Prisms for Electrons

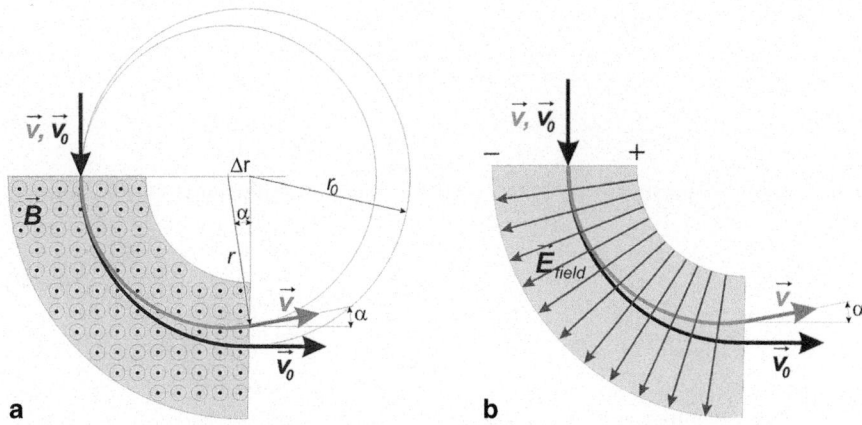

Fig. 10.58 Magnetic (**a**) and electrostatic (**b**) prism for electrons (explanations in the text)

Lorentz force is equal to the product of the elementary electric charge e as well as the moduli of velocity v_0 and magnetic induction B:

$$F_L = e \cdot v_0 \cdot B. \tag{10.398}$$

According to Eq. (10.397) this force is perpendicular to the plane which is spanned by $\mathbf{v_0}$ and $\mathbf{B}$, i.e. always perpendicular to the electron path. This is the typical feature of a radial force

$$F_r = m \cdot \frac{v_0^2}{r_0} \tag{10.399}$$

at a circular motion with the radius r_0 (m: mass of the electron):

$$m \cdot \frac{v_0^2}{r_0} = e \cdot v_0 \cdot B, \tag{10.400}$$

i.e. the electron with the velocity v_0 runs within the magnetic field on a circular orbit with the radius

$$r_0 = \frac{m \cdot v_0}{e \cdot B}. \tag{10.401}$$

An electron which experiences an energy loss ΔE during its penetration of the sample has a lower velocity v and runs on a circular orbit with the radius

$$r = \frac{m \cdot v}{e \cdot B}. \tag{10.402}$$

In Fig. 10.58a we read the angle α:

$$\sin \alpha = \frac{\Delta r}{r} = \frac{(r_0 - r)}{r} = \frac{r_0}{r} - 1 = \frac{v_0}{v} - 1. \tag{10.403}$$

The relation between velocity v and energy E is in non-relativistic approximation:

$$v = \sqrt{\frac{2}{m} \cdot E}, \tag{10.404}$$

i.e.

$$\sin \alpha = \sqrt{\frac{E_0}{E}} - 1 = \sqrt{\frac{E_0}{E_0 - \Delta E}} - 1. \tag{10.405}$$

For $\Delta E \ll E_0$ we can approximate $\sin\alpha \approx \alpha$ and after a series expansion of Eq. (10.405) with termination after the linear term it follows

$$\alpha = \frac{\Delta E}{2 \cdot E_0}, \tag{10.406}$$

i.e. a linear relation between deflection angle and energy loss exists.

Now, let us consider a second possibility, the *electrostatic prism* as sketched in Fig. 10.58b. We assume two circularly curved plates in a distance d and a voltage U between them. With the condition that d is much smaller than the mean radius of curvature the electric field strength between the plates is approximately given by

$$E_{field} = \frac{U}{d}. \tag{10.407}$$

Furthermore, we suggest that the electrostatic force constrains electrons with the velocity v_0 on an orbit with the radius r_M by a suitable choice of the plate voltage U_M. There is a balance between electrostatic and centrifugal force in the coordinate system which is moving together with the electrons:

$$m \cdot \frac{v_0^2}{r_M} = e \cdot \frac{U_M}{d}. \tag{10.408}$$

An electron with the velocity v experiences an inwards directed force F which is equal to the difference of the electrostatic and the centrifugal force:

$$F = e \cdot \frac{U_M}{d} - m \cdot \frac{v^2}{r}. \tag{10.409}$$

10.23 Prisms for Electrons

For the inwards directed acceleration a we get:

$$a = \ddot{s} = \frac{F}{m} = \frac{e}{m} \cdot \frac{U_M}{d} - \frac{v^2}{r} \tag{10.410}$$

and, after double integration over the time t, for the path:

$$s = \left(\frac{e}{m} \cdot \frac{U_M}{d} - \frac{v^2}{r}\right) \frac{t^2}{2}. \tag{10.411}$$

On the other hand, in the time t the electron travels the path l between the curved plates:

$$l = v \cdot t \tag{10.412}$$

From Eqs. (10.411) and (10.412) it follows for the trajectory:

$$s(l) = \frac{1}{2}\left(\frac{e}{m} \cdot \frac{U_M}{d \cdot v^2} - \frac{1}{r}\right) \cdot l^2. \tag{10.413}$$

With $r \approx r_M$ the angle deviation from the orbit with the radius r_M (this is the first derivation s'(l)) and with consideration of Eq. (10.408) we get:

$$s'(l) = \left(\frac{v_0^2}{r_M \cdot v^2} - \frac{1}{r_M}\right) \cdot l. \tag{10.414}$$

At the position $l = \pi \cdot r_M/2$ (exit from the prism) s' is:

$$s'(\tfrac{1}{2} \pi \cdot r_M) = \tan \alpha = \frac{\pi}{2}\left(\frac{v_0^2}{v^2} - 1\right) \tag{10.415}$$

and, respectively,

$$\tan \alpha = \frac{\pi}{2}\left(\frac{E_0}{E} - 1\right) = \frac{\pi}{2} \cdot \frac{\Delta E}{E}. \tag{10.416}$$

For $\Delta E \ll E_0$ and small angles we get for the electrostatic prism:

$$\alpha = \frac{\pi}{2} \cdot \frac{\Delta E}{E_0} \tag{10.417}$$

which is a linear relation, too.

324   10 Basics Explained in More Detail (with a Bit More Mathematics)

In practice magnetic prisms as deflection units are preferred for a technical implementation. On conditions typical for transmission electron microscopy magnetic fields of 5 mT reach the same deflection radii as electric fields of 10 kV/cm and are easier controllable.

10.24 Convolution of Functions

The convolution of functions is a procedure which allows the calculation of overlays of different influences leading to a "spreading" of measured results. Examples are: The overlay of aberration disks to evaluate the optical resolution limit as well as the widening of peaks in energy dispersive X-ray spectra or in electron energy loss spectra due to the limited energy resolution.

We would like to explain the principle of the convolution of functions using a simple example: We assume that during a production process plugs are pressed into a tube end, followed by a small widening of the tube diameter. Let the widening be equal to the difference of the plug diameter and the inside diameter of the tube. The thickness of the tube wall is constant, so that the frequency distributions $r(d)$ of the inside and the outside diameters of the tubes before pressing-in the plugs are equal, too. This frequency distribution and that of the plug diameter $s(d)$ should be known. We want to calculate the frequency distribution $R(d)$ of the tube end diameters after pressing-in of the plugs. For the measurement of the frequency distributions the diameters are separated in intervals of the uniform width Δd. Therewith each diameter can be considered as a product of an integer number i and the interval width Δd:

$$d_i = i \cdot \Delta d. \tag{10.418}$$

Let the tube diameter before pressing-in vary from d_{t1} to d_{t2}, the plug diameter from d_{p1} to d_{p2}. The corresponding integer numbers should be i_{t1}, i_{t2}, i_{p1} and i_{p2}.

By normalisation following the rules

$$^t w_i = \frac{r(d_i)}{A_t} = \frac{r_i}{A_t} \quad \text{with} \quad A_t = \sum_{i=i_{t1}}^{i_{t2}} r_i \tag{10.419}$$

and

$$^p w_i = \frac{s(d_i)}{A_p} = \frac{s_i}{A_p} \quad \text{with} \quad A_p = \sum_{i=i_{p1}}^{i_{p2}} s_i \tag{10.420}$$

we get the probabilities $^t w_i$ and $^p w_i$.

The smallest diameter R_{min} after pressing-in arises when the smallest tube diameter $r_{min} = i_{t1} \cdot \Delta d$ is combined with the smallest plug diameter $s_{min} = i_{p1} \cdot \Delta d$:

$$R_{min} = r_{min} + s_{min} = (i_{r1} + i_{s1}) \cdot \Delta d. \tag{10.421}$$

10.24 Convolution of Functions

The probability $^R w_{min}$ of the existence of R_{min} is equal to the product of the probabilities of the existence of r_{min} and s_{min}:

$$^R w_{min} = {^t w_{it1}} \cdot {^P w_{ip1}}. \qquad (10.422)$$

The counting of the diameters can be generalised and it can be started at zero in principle. Absent diameter values have the probability zero:

$$R_0 = r_0 + s_0 = (0+0) \cdot \Delta d \qquad (10.423)$$

with

$$^R w_0 = {^t w_0} \cdot {^P w_0}. \qquad (10.424)$$

The diameter value next in size can be reached by two combinations:

$$R_1 = r_1 + s_0 = (1+0) \cdot \Delta d = r_0 + s_1 = (0+1) \cdot \Delta d \qquad (10.425)$$

with

$$^R w_1 = {^t w_1} \cdot {^P w_0} + {^t w_0} \cdot {^P w_1}. \qquad (10.426)$$

This can be continued up to the largest possible value:

$$R_{max} = r_{max} + s_{max} = (i_{t2} + i_{p2}) \cdot \Delta d \qquad (10.427)$$

with

$$^R w_{max} = {^t w_{it2}} \cdot {^P w_{ip2}}. \qquad (10.428)$$

Obviously, the probability of the diameter value R_i is given by

$$^R w_i = \sum_{k=0}^{i} {^t w_k} \cdot {^P w_{i-k}}. \qquad (10.429)$$

After insertion of two functions $f(x)$ and $g(x)$ instead of the probabilities w and the transition to infinitesimally small interval widths dx we get the result $F(\xi)$

$$F(\xi) = \int_{-\infty}^{\infty} f(x) \cdot g(\xi - x) \cdot dx \qquad (10.430)$$

which is the well-known formula for convolution operations.

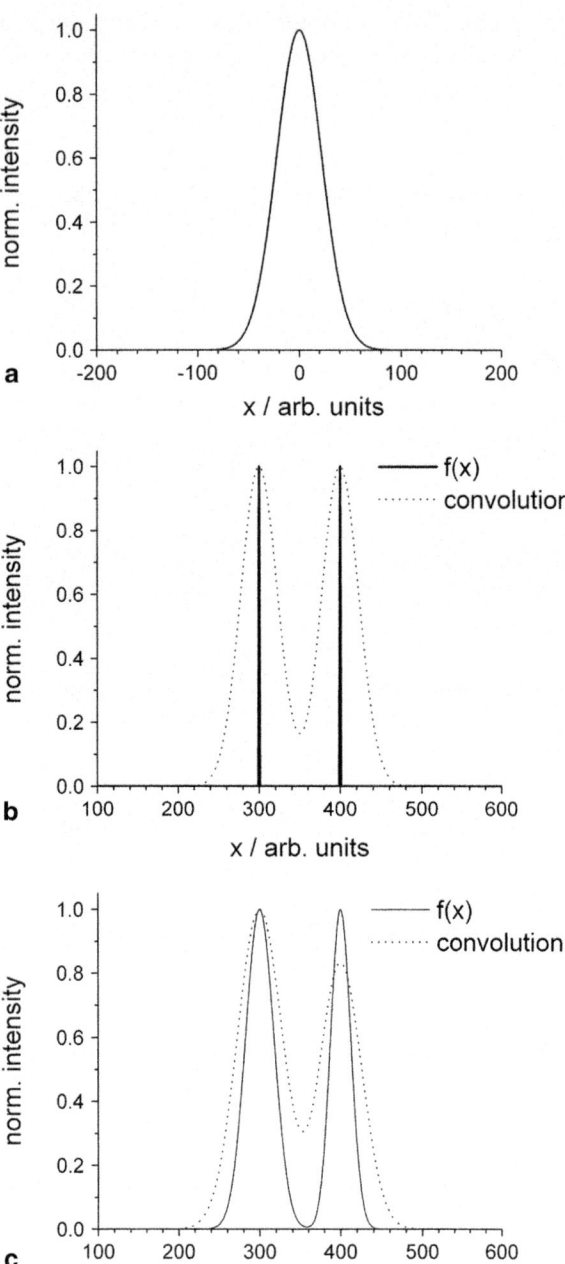

Fig. 10.59 Convolution with a Gaussian curve. **a** Gaussian curve which is used as basis for the convolution. **b** Convolution of two sharp δ-functions with the Gaussian curve of (**a**). **c** Convolution of two Gaussian curves of equal maximal height but different full widths half maximum with the Gaussian curve of (**a**)

We would like to demonstrate two consequences of such convolutions with the help of Fig. 10.59. In part 10.59b it can be seen how two sharp lines are widened by convolution with the Gaussian shown in Fig. 10.59a. The convoluted function does not reach the abscissa (ordinate value 0) between both lines. In Fig. 10.59c the initial function is given by two Gaussians with different full widths half maximum whose maxima are shifted against each other. As a result of the convolution besides the widening a height difference of both maxima also arises.

The inversion, i.e. the deconvolution of measured curves, is much more difficult. Theoretically, the numerical deconvolution seems to be possible but it fails in practice since the calculation has to start with the values at the edge and the next values are based on the previous ones. Unfortunately, the small edge values are, not least due to the noise of measured results, extremely unsure.

Nevertheless, in principle there are two possibilities to try the deconvolution: The assumption of functions with subsequent convolution and comparison with the measurement ("trial and error" method) or the calculation in the Fourier space. The convolution of functions corresponds to the multiplication of their Fourier transforms and subsequent inverse transformation back into the position space. Consequently, the deconvolution corresponds to the division of the Fourier transforms. Often, software packages for mathematical operations include such possibilities for deconvolutions. However, the noise often pretends frequencies and, connected with it, Fourier coefficients which are not part of the measured result. As a consequence the deconvoluted measuring curve oscillates and includes a pretended fine structure.

The convolution of functions is of principal importance for the interpretation of measurements. It is the mathematical basis for the consideration of the overlay of several influences on the measured result. The limited resolution of the spectrometer and the inelastic multiple scattering at EELS as shown in Sect. 9.6.3 are examples for such influences.

References

1. Glaser, W.: Grundlagen der Elektronenoptik, p. 113. Springer-Verlag, Wien (1952)
2. Glaser, W.: ibidem, p. 297
3. Debye, P., Bueche, A.M.: Scattering by an inhomogenous solid. J. Appl. Phys. **20**, 518–525 (1949)
4. Reimer, L.: Transmission Electron Microscopy, Physics of Image Formation and Microanalysis, p. 71. Springer-Verlag, Berlin (1997). (et sqq)
5. Williams, D.B., Carter, C.B.: Transmission Electron Microscopy, p. 498. Springer-Verlag, New York (2009)
6. Rang, O: Der elektronenoptische Stigmator, ein Korrektiv für astigmatische Elektronenlinsen. Optik **5**, 518–530 (1949)
7. Bauer, H.-D.: Elektronenoptische Eigenschaften einer elektrostatischen Objektivlinse mit Vierpolsymmetrie. Optik **23**, 596–609 (1965/1966)
8. Reimer, L.: Transmission Electron Microscopy, Physics of Image Formation and Microanalysis, p. 160. Springer-Verlag, Berlin (1997)
9. http://www.moxtek.com/x-ray-windows/ap3-ultra-thin-polymer-windows.html

10. Theisen, R., Vollath, D.: Tabellen der Massenschwächungskoeffizienten von Röntgenstrahlen, p. 11. Verlag Stahleisen mbH, Düsseldorf (1967). (et seqq)
11. National Institute of Standards and Technology: NISTIR 5632: http://physics.nist.gov/PhysRefData/XrayMassCoef/ElemTab/z79.html. Accessed 20 March 2012
12. Bethe, H.: Zur Theorie des Durchgangs schneller Korpuskularstrahlung durch Materie. Annalen der Physik **5**, 325–400 (1930)
13. Schreiber, T.P., Wims, A.M.: A quantitative X-ray microanalysis thin film method using K-L- and M-lines. Ultramicroscopy **6**, 323–334 (1981)
14. Powell, C.J.: Cross sections for ionization of inner shell electrons by electrons. Rev. Mod. Phys. **48**, 33–47 (1976)
15. Wentzel, G.: Über strahlungslose Quantensprünge. Zeitschrift für Physik, **43**, 524–530 (1927)
16. Klaar, H.J., Schwaab, P.: Röntgenmikroanalyse im Elektronenmikroskop, part 1: Grundlagen und Messtechnik. Prakt. Metallogr. **27**, 319–331, part 2: Quantitative Analyse, 373–384, part 3: Vergleich mit anderen Verfahren, 429–438 (1990)
17. Burhop, E.H.S.: Le rendement de fluorescence. J. Phys. Radium **16**, 625–629 (1955). (cited in [16])
18. Scofield, J.H.: Radiative Transitions in: Atomic inner shell processes, vol. I, pp. 265–292 (Ed. B. Crasemann). Academic Press, New York (1975)
19. Goldstein, J.I., Costley, J.L., Lorimer, G.W., Reed, S.J.B.: Quantitative X-ray analysis in the electron microscope. In: Johari, O. (ed.) Scanning Electron Microscopy, vol. 1, pp. 315–324. ITTRI, Chicago (1977), (cited in [16])

Summary and Outlook

In the footnotes of the ten chapters in this book the names of 70 well-known scientists and engineers are listed whose work contributed to fundament and develop the transmission electron microscopy. Twenty one of them are Nobel laureates: *"The electron microscopy is based on the shoulders of giants."*

Hans-Dietrich Bauer who dealt with the experimental problems using quadrupole lenses in a transmission electron microscope at Alfred Recknagel's institute at the Technical University Dresden already in the 1960s began his lecture on transmission electron microscopy with two (not really serious) fundamental theorems:

1st It is impossible to get no image!
2nd It is impossible to get a sharp image!

The background of the "second theorem" is the fact that in rotational symmetric fields which are free of space charges and temporarily constant the spherical aberration cannot be avoided. In principle the alternative was evident: The use of multipoles instead of rotational symmetric lenses. Nevertheless, it lasted until about 2000 to implement such multipole units for the correction of the spherical aberration into commercial instruments. On the one hand it needed the ideas by Harald Rose, Max Haider, and Knut Urban and on the other hand fast computers were necessary. The latter is because one has to be able to measure and correct the spherical aberration quicker than the alignments drift away. Highly optimised image processing and correction coil adjustment algorithms are absolutely necessary.

After correction of the spherical aberration the point resolution reaches the information limit. Now the correction of the chromatic aberration comes into the centre of interest. This is also possible by the use of multipole units in which electrostatic and magnetic dipoles are combined. Therewith the information limit with electrons of different energy (EFTEM) is improved.

The information limit, or more precisely: the highest transferred space frequencies, is not only determined by the chromatic aberration. Mechanical vibrations, temperature fluctuations, lens current instabilities, and external alternating magnetic fields deteriorate it, too. With the improvement of electron lenses the attention is more and more directed to these ambient influences: Modern high end microscopes are located within a completely closed box. Houses on special basements far away from busy city centres are used as laboratories. Such houses must be built only for the electron microscope.

A consequence of this development is that not only the resolution limit, i.e. the ability to separate smallest distances, is considered but also the measurement accuracy of such smallest distances. At this point one reaches the picometres range and it is possible to determine the deviation of single atomic columns within the crystal lattice from their regular position.

In the analytical transmission electron microscopy the high spatial resolution is only one issue. It has to be combined with X-ray and electron energy loss spectroscopy. For the measurement of bonding states a high energy resolution of the energy loss spectrometer in the order of 0.1 eV is needed. At structure sizes in the sub-nanometre range the measuring time is limited by (minimal) mechanical drift. To get a sufficient signal-to-noise ratio it is important to improve the efficiency of the detectors and to increase the beam current, especially in the scanning transmission electron microscopic mode. To get a high beam current into smallest probes an electron gun with high richtstrahlwert (exclusively field emission cathodes) and a corrector for the spherical aberration of the condenser system are needed. In terms of the X-ray spectroscopy the detector efficiency can be improved by an increase of the acquired solid angle, e.g. by use of multiple detectors.

It is very expensive to create ideal conditions for the laboratory, to say nothing about the costs of a high end instrument. In biology and materials research it is often unnecessary to use such a high end machine. Electron diffraction methods for phase analysis, diffraction contrast investigations for the determination of the microstructure, and the imaging of cell structures are examples for which a "normal" transmission electron microscope is good enough. More important than an excellent performance is in these cases that the microscope is quickly accessible, i.e. in immediate proximity to the other laboratories. At last we should not forget in this context that ultrathin specimens are needed for transmission electron microscopic investigations: The better the performance of the microscope the higher are the requirements on the quality of the specimens.

There are also "mixed" cases, e.g. in the semiconductor industry. At semiconductor devices the thickness of interlayers can only be measured with the sufficient accuracy using a transmission electron microscope. It should be equipped with a corrector for the spherical aberration to avoid uncertainties by delocalisation. However, complete investigations including specimen preparation need to be performed as rapidly as somehow possible.

As a consequence there is a request to the manufacturers of transmission electron microscopes: Reduction of the sensitivity against ambient influences.

When the reader after studying this book is convinced that the work with a transmission electron microscope and the profound interpretation of its results need more than the knowledge "to tilt on a special knob for reaching a special purpose" the authors have achieved their main goal.

We wish all the current and prospective electron microscopists "the right touch" and, please, be careful and think always while interpreting electron microscopic images and other analytical electron microscopic results:

> Believe only what you see after you have understood why you see it!

Physical Constants

Values of physical constants mentioned within the text of the book:

Avogadro's constant	N_A	$= 6.022 \cdot 10^{23}$ /mol
Boltzmann's constant	k	$= 1.381 \cdot 10^{-23}$ J/K
Elementary charge	e	$= 1.602 \cdot 10^{-19}$ A·s
Gas constant	R	$= 8.315$ J/(mol·K)
Permittivity of vacuum	ε_0	$= 8.854 \cdot 10^{-12}$ A·s/(V·m)
Velocity of light in vacuum	c	$= 2.998 \cdot 10^8$ m/s
Planck's constant	h	$= 6.626 \cdot 10^{-34}$ J·s
Richardson's constant	A	$= 120$ A·cm^{-2}·K^{-2}
Rest mass of the electron	m_0	$= 9.109 \cdot 10^{-31}$ kg
Specific charge of the electron	$-e/m_0$	$= -1.759 \cdot 10^{11}$ A·s/kg

Combinations of often needed constants:

$\dfrac{h \cdot c}{e} = 1.24 \cdot 10^{-6}$ V·m

$\dfrac{m_0 \cdot c^2}{e} = 511059$ V

$\dfrac{h}{m_0 \cdot c} = 2.4263 \cdot 10^{-12}$ m $= 2.4263$ pm

$\sqrt{1 - \left(\dfrac{v}{c}\right)^2} = \dfrac{511.06 \text{ kV}}{U_0 + 511.06 \text{ kV}}$ (U_0: accelerating voltage)

$\dfrac{1}{4\pi \cdot \varepsilon_0} = 8.99 \cdot 10^9$ V·m/(A·s)

Conversions:

1 N = 1 kg·m/s^2
1 J = 1 N·m = 1 V·A·s = 1 W·s = $6.242 \cdot 10^{18}$ eV
1 Pa = 1 N/m^2 1 Torr = 133 Pa   1 bar = 10^5 Pa
1 mbar = 100 Pa
$R = k \cdot N_A$

Hints for Further Reading

(Analytical) transmission electron microscopy:

1. Hirsch, P. B.: Electron Microscopy of Thin Crystals. Krieger Publishing Company, Malabar (1977)
2. Lechene, C. P.: Microbeam Analysis in Biology. Academic Press, New York (1979)
3. Mulvey, T.: Advances in Imaging and Electron Physics—The Growth of Electron Microscopy, vol. 96. Academic Press, San Diego (1996)
4. Hirsch, P. B.: Topics in Electron Diffraction and Microscopy of Materials. IOP Publishing Ltd., London (1999)
5. Ernst, F., Rühle, M. (eds.): High-Resolution Imaging and Spectrometry of Materials. Springer, Berlin (2003)
6. De Graaf, M.: Introduction to Conventional Transmission Electron Microscopy. Cambridge University Press, Cambridge (2003)
7. Sigle, W.: Analytical transmission electron microscopy. Annual Rev. Mater. Res. **35**, 239–314 (2005)
8. Hawkes, P., Spence, J.C.H.: Science of Microscopy, vol. I. Springer, New York (2007)
9. Reimer, L., Kohl, H.: Transmission Electron Microscopy: Physics of Image Formation. Springer, Berlin (2008)
10. Williams, D. B., Carter, C. B.: Transmission Electron Microscopy – A Textbook for Materials Science, Springer Science + Business Media, LLC, New York (2009)
11. Zewail, A. H., Thomas, J. M.: 4D Electron Microscopy. Imperial College Press, London (2010)
12. Fultz, B., Howe, J.: Transmission Electron Microscopy and Diffractometry of Materials Science. Springer, Heidelberg (2013)

Electron optics:

13. Glaser, W.: Grundlagen der Elektronenoptik. Springer, Wien (1952)
14. Hawkes, P.W., Kasper, E.: Principles of Electron Optics. Academic Press, London (1989)
15. Rose, H.: Geometrical Charged-Particle Optics. Springer, Berlin (2009)

Electron diffraction:

16. Hahn, T. (ed.): International Tables for Crystallography, Vol. **A**, Space-Group Symmetry. Kluwer Academic Publishers, Dordrecht (1996)
17. Champness, P.E.: Electron Diffraction in the Transmission Electron Microscope. BIOS Scientific Publishers Ltd., Oxford (2001)
18. Spence, J.C.H., Zuo, J.M.: Electron Microdiffraction. Plenum Press, New York (1992)
19. Morniroli, J.-P.: Large-Angle Convergent-Beam Electron Diffraction (LACBED) – Applications to crystal defects, Monograph of the French Society of Microscopies, Paris (2004)

High resolution transmission electron microscopy:

20. Buseck, P.R., Cowley, J. M., Eyring, L.: High-Resolution Transmission Electron Microscopy and Associated Techniques. Oxford University Press, Oxford (1988)
21. Spence, J.C.H.: High-Resolution Electron Microscopy. Oxford University Press, Oxford (2008)

Scanning transmission electron microscopy:

22. Pennycook, S.J., Nellist, P.D. (eds.): Scanning Transmission Electron Microscopy. Springer, New York (2011)

X-ray and electron energy loss spectroscopy:

23. Reimer, L. (ed.): Energy-Filtering Transmission Electron Microscopy. Springer, Berlin (1995)
24. Egerton, R. F.: Electron-Energy Loss Spectroscopy in the Electron Microscope, Plenum Press, 2nd Edition, New York, London (1996)
25. Garret-Reed, A. J., Bell, D. C.: Energy Dispersive X-Ray Analysis in the Electron Microscope, Taylor & Francis Ldt, Oxford UK (2003)
26. Ahn, C.C. (ed.): Transmission Electron ELS in Materials Science and the EELS Atlas (2005), 2nd edn. Wiley-VHC, New York (2004)

Electron microscopic preparation:

27. Goodhew, P.J.: Specimen preparation for transmission electron microscopy of materials. Oxford University Press, Oxford (1984)
28. Petzow, G.: Metallographic Etching: Techniques for Metallography, Ceramography, Plastography. ASM International, Ohio (1999)
29. Ayache, J., Beaunier, L., Boumendil, J., Ehret, G., Laub, D.: Sample Preparation Handbook for Transmission Electron Microscopy – Methodology. Springer, New York, Dordrecht, Heidelberg, London (2010)
30. Ayache, J., Beaunier, L., Boumendil, J., Ehret, G., Laub, D.: Sample Preparation Handbook for Transmission Electron Microscopy – Techniques. Springer, New York, Dordrecht, Heidelberg, London (2010)

Index

A
Abbe, Ernst 3
aberration 15, 16, 18, 122
    chromatic 17, 18, 145, 146, 158, 160, 166-168, 177, 224, 288, 297, 302-304
    correction 168
    diffraction 5, 19
    disk 16, 17, 19, 20, 167, 224, 243, 244, 288, 299, 300, 302, 304, 324
    -free 4, 15, 16, 149, 282
    spherical 16-19, 28, 31, 87, 88, 135, 142-145, 152, 158, 160, 161, 164, 167-169, 243, 282, 283, 287, 288, 290, 292, 294, 296, 297, 300, 302, 305
absorption
    coefficient 199, 200, 309, 310, 311, 318, 319. *See also* attenuation coefficient
    correction (EDXS) 199-201, 319
    edge 193, 197, 209, 210, 308, 312, 313, 315, 317
air conditioner 39
alignment 57, 58, 61, 62, 88, 121, 224
    apertures, 77
    beam 38
    condenser aperture 62, 63, 65
    electron gun 63, 64, 165
    eucentric height 65, 66
    pivot points 66, 67
    rotation centre 67, 68, 69
    state 57, 61, 165
    spectrometer 206
aluminium (Al) 32, 33, 51, 109, 128, 189, 217, 308
amorphous 73, 110, 112, 152, 153, 155-157, 270
    film 111, 150, 154, 155, 157, 160, 170
    layer 55, 142, 152

material 137, 219, 268
sample 48, 60, 110-112, 152, 153, 170
analytics 166
    nano 168, 169
angle
    acceptance 116-118, 172, 173, 215, 222, 225, 277, 279
    azimuthal 14, 15, 155, 160
    covergence 89, 90, 106, 108, 220
    diffraction 4, 78, 84, 86, 87, 91, 93, 98, 108, 143, 144, 149, 249-251, 259, 294, 296
    solid 24, 25, 115, 165, 198, 220, 269, 275, 277, 311
    visual 1, 2, 12
anisotropy 111, 152
anode 7, 23, 24, 34, 39, 46, 204, 206, 247
aperture 18, 19, 26, 29, 35, 38, 57, 58, 88, 116, 151, 165, 166-168, 206, 243, 282, 284, 300
    condenser (or C2) 28, 29, 38, 62-65, 89, 107, 171, 174
    contrast (or objective) 38, 117-123, 125, 127, 129, 132, 138, 150, 151, 160, 220, 221, 224
    entrance 171, 173, 220, 225
    holder 88
    illumination 28, 29, 89, 106, 174
    numeric 5
    optimal 19, 168, 244, 304, 305
    ring-shaped 151
    selected area 38, 88, 161, 221
apparatus constant 88, 91, 92, 94, 96
Ardenne, Manfred von 6, 163
astigmatism 16, 61, 155, 156, 158, 166, 167, 299, 304
    axial 18
    condenser 63, 65

astigmatism (*continued*)
   correction 154-156, 158, 160, 168, 170, 297
   two-fold 18, 158, 161, 170
atomic
   arrangement 98, 110, 111, 126, 131, 149, 150, 161, 216, 251, 262, 263, 268
   columns 41, 55, 137, 138, 141, 143, 150, 151, 169, 174
   density 111, 116, 152
   number 43, 116, 170, 171, 172, 174, 178, 179, 183, 185, 190, 191, 193, 198-200, 203, 210, 215, 219, 221, 225, 272, 277, 309, 314-317
   position 79-81, 83, 94, 97, 152, 174
   weight 195, 198, 311
attenuation coefficient 199, 309-311, 318, 319
Avogadro, Amedeo 34
   constant 34, 116, 170, 277
azimuthal 14, 15, 121, 155, 158, 160, 173

B

back-focal
   plane 29, 30, 86, 117, 127, 149, 150, 165, 171, 221
   point 3, 29
bending contours 123-125
Bethe, Hans 220, 313
   surface 220
binding 183. (*See* also bonding)
   kind of 183
   model 185
   specific 177
binning 33
Bloch, Felix 134
   wall 134, 135
blooming 33
Boersch, Hans 22, 88
   diffraction 88
   effect 22
Bohr, Niels 177
   model 177, 179
Boltzmann, Ludwig 21
   constant 21, 34
bond(ing) 184, 185, 208, 210, 225
   analysis 216
   chemical 48, 210
   covalent 183
   hybrid orbitals 184
   ionic 183
   kind 184, 217
   metallic 183
   specific 221
   type 205

Borries, Bodo von 6, 24
Bragg, William Henry and Lawrence 78
   angle 89, 90, 106, 108, 123, 129, 130, 186
   condition 123
   direction 90
   law 78, 86, 90, 93, 99, 106, 118, 128, 130, 144, 149, 249, 250, 261, 285
   orientation 132
   position 108, 118, 119, 123, 124, 128- 131
bremsstrahlung 178, 182, 187, 190, 191, 196, 197, 201, 208, 211
brightness 24, 28, 33, 51, 64, 86, 91, 92, 99, 110, 117, 119-121, 141, 143, 144, 147-150, 153, 155, 157, 161, 173, 174, 206, 213, 221, 222, 223, 280, 281, 286
brightfield image 120, 121, 123, 124, 128, 131, 163, 217, 221-223
Broglie, Louis de 6
   formula 6, 180, 232, 233
Brown, Robert 35
bulk metallic glass 73, 111
Burgers, Johannes Martinius 125
   circulation 125, 126
   vector 125-127
Busch, Hans 6

C

camera length 87, 92, 127, 171, 173, 174
carbon (C) 45, 59, 60, 73, 171-173, 187, 203, 210, 308
   film 43, 45, 69, 70, 91, 117-119, 171, 194, 217
   graphitic 209, 210, 217
   nanotubes 44, 202
cathode 7, 17, 21-26, 34, 39, 46, 204, 206, 208, 218, 244-247, 288,
   chamber 23, 35
   cold field emission 22, 25, 218
   field emission 23, 24, 168,
   LaB_6 21-23, 25, 37, 166, 288, 289
   Schottky field emission 22, 25, 37, 166, 168, 218, 224, 244, 288, 289
   tungsten hairpin 21-23, 25, 166
   type 288
cathodoluminescence 32
CBED. *See* diffraction, convergent beam electron
CCD 9
   array 32, 33
   camera 9, 32-34, 91, 140, 155, 206
   element 33, 34
charge carrier 33, 190, 192-194, 307, 311
chemical shift 217

Index   337

chromium (Cr) 45, 182, 191-193, 199-201,
    203, 204, 217
  oxide 192
classical approximation 7, 219. *See also* non-
    relativistic approximation
cleavage 47
Cliff-Lorimer
  equation 195
  factor 195, 199, 313, 317. *See also*
    k-factors
cobalt (Co) 1, 135, 136
coefficient 146, 147, 238
  aberration 18, 161
  chromatic aberration (C_C) 17, 18, 145, 167,
    224, 290, 302, 304
  Fourier 147, 327
  mass absorption (or attenuation) 199, 200,
    309-311, 318, 319
  spherical aberration (C_S) 16, 18, 143, 145,
    152, 167, 169, 243, 282, 285, 290, 297,
    304
  thermal expansion 31
  transmission cross 147
cold trap 38, 59, 60, 71
collision 34-36, 115, 220
  elastic 306
  inelastic 184, 306
concentration 71, 195, 196, 198, 199, 219, 223
conical darkfield 122, 123, 160
contamination 38, 52, 59, 70-72, 88, 192
contrast 43, 44, 66, 69, 115, 117-120, 124,
    127, 128, 136, 138, 141, 144, 147, 151,
    153, 153, 156, 157, 163, 287, 290, 292
  amorphous material 152-154
  aperture. *See* aperture, contrast
  bending contours 123-125
  coffee-bean-like 127
  diffraction 116, 118, 119, 123, 125, 127,
    138, 221
  dislocation 124, 125, 128
  magnetic domains 133, 135, 136
  mass thickness 116, 118, 119, 124, 138,
    150, 152, 212, 221, 294
  orientation-dependent 132
  phase 137, 141, 143, 145, 150, 282, 287,
    290, 294
  STEM 164, 170, 171, 173, 174
  semicoherent particles, *138*
  stacking faults 131, 132
  thickness contours 129, 130
  transfer 142, 144, 145, 147, 152, 289
  twins 132
contrast transfer function 145, 146, 151, 153-
    157, 160, 161, 282, 287, 289-294
  damped 288

Convergent Beam Electron Diffraction
    (CBED) *See* diffraction, convergent
    beam electron
convolution 218, 324, 326, 327
  formula 325
coordinates
  Cartesian 235, 236, 247, 248, 254, 297
  cylindrical 14, 234-236, 247
  orthogonal 99, 253, 256, 258-260
  polar 91, 99
copper (Cu) 1, 38, 43, 44, 52, 59, 60, 71, 109,
    128-130, 135, 136, 142, 143, 193, 209,
    210, 276, 279
Coulomb, Charles Augustin 41
  force 41, 178, 179, 183, 244, 272
  interaction 115, 177
counterforce 159
Crewe, Albert V. 163
cross over 22, 24, 28, 142, 165, 169
cross section 24, 55, 63, 166, 178, 189, 277,
    298, 299, 302
  differential 116, 275, 276
  ionisation 36, 198, 209, 219, 220, 313- 315
  scattering 93, 112, 115, 116, 268, 275, 276
cross-section preparation 48, 50-52, 135, 136,
    203
cryo-cycle 38
crystal 44, 45, 59, 77, 78, 88, 98, 99, 106- 109,
    123-125, 127, 130, 131, 133, 137, 139,
    141, 142, 186, 187, 189, 190, 192-194,
    196, 212, 213, 232, 259, 264, 265, 267,
    279, 311
  angles 78, 79
  axis 79, 108, 260
  lattice 31, 73, 76, 78, 79, 88, 90, 91, 106,
    109-111, 118, 128, 129, 131, 137-140,
    142, 183, 252, 259, 261, 285, 286, 294
  orientation 60, 90, 105, 132,
  planes 76, 78, 152,
  potential 109, 139, 141, 183,
  semiconductor 187, 188, 190, 307,
  size 98,
  structure 53, 75, 79, 83, 90, 103, 138, 259,
  surface, 57
  thickness 131, 139, 312,
crystal system 80, 83, 84, 99, 103-105, 131,
    257, 259
  cubic 79, 80, 83-85, 92-95, 99, 100, 103,
    104, 121, 125, 137, 249, 258
  hexagonal 80, 83-85, 97, 100, 217, 258,
  monoclinic 80, 85, 102, 258,
  orthorhombic 80, 85, 100, 127
  rhombohedral 80, 85, 100
  tetragonal 80, 83-85, 100, 258

crystal system (*continued*)
  triclinic 79, 80, 85, 102, 253, 254, 259
  trigonal 80
CTF. *See* contrast transfer function

D

damage 33, 73, 91, 202, 206
damping 112, 146, 157, 158, 288, 289
darkfield image 120-123, 126, 127, 163, 173, 174
Davisson, Clint 75
Debye, Peter 268
  scattering 268
deconvolution 152, 211, 218, 220, 271, 327
decoration 45
defocus 143, 145, 152, 154-156, 161, 287, 290, 293, 297
  optimal 292, 293
  Scherzer 290, 293, 294
    defocussing 142, 282-284, 292, 296,
delocalisation 142, 294, 296, 297
demagnification 165, 166, 169, 299
density 43, 116, 138, 170, 172, 199, 200, 203, 215, 275, 309, 310, 318, 319
  atomic 93, 111, 116, 131, 152
  charge 185, 233
  current 21, 24, 25, 27, 28, 63, 165
  electron 110, 134, 135
  fluctuations 152
  particle 34, 35, 44, 70
  gold 310
  occupation 93, 131
  line 280
  (of) states (DOS) 208, 209, 216,
  radial 112, 270,
density functional theory (DFT) 217
detector 59, 163, 164, 189, 190, 193, 198, 201, 202, 307, 308, 316, 318-320
  crystal 59, 178, 189, 192-194, 196, 311
  EDX 187, 225, 318
  efficiency 178, 190, 193, 194, 198, 307, 311-313, 317
  STEM 165, 171-174
  window 189, 192, 198
Dewar, Sir James 60, 71, 189
diffraction 2, 3, 5, 19, 53, 60, 75, 76, 77, 86, 87, 90, 92, 93, 96, 111, 116, 118, 119, 123, 127, 128, 130, 132, 133, 138, 152, 157, 165, 171, 220, 225, 226, 251, 262
  angle 4, 78, 84, 86-88, 91, 93, 98, 108, 143, 144, 149, 249, 250, 251, 259, 294, 296
  base equation 78
  convergent beam electron 84, 89, 90, 108, 110, 164

diagram 87, 91, 96, 103, 105, 110, 111, 122, 123, 219
disk 5, 19, 89, 90, 110, 167, 170, 243,
double 133, 157, 158
electron 44, 75, 78, 79, 87-91, 98, 248, 252, 259
excitation error 89, 123, 237
focus(sing) 86, 106
indexing of reflections 103, 105, 259
intensity of reflections 98, 263
lens 86
nano beam electron (NBED) 89
pattern 75, 79, 86-89, 92, 96, 98, 99, 103, 108, 111, 120, 121, 133, 142, 149, 150, 177, 221, 252, 262, 267
plane 29
precessing electron (PED) 110
reflections 75, 78, 89, 93, 99, 103, 108 -110, 122, 133, 149, 249, 252, 257, 261, 267, 268
selected area electron (SAED) 86, 88-90, 104, 110, 161, 221
X-ray 75, 78
diffraction contrast.. *See* contrast, diffraction
diffraction mode 87, 88, 127, 142, 152, 165, 171, 220
Dirac, Paul 217
dirty darkfield 122
dislocation 41, 123-128
dispersion 205, 206, 218, 320
distance
  image 3, 11, 12, 18, 29, 65, 282
  next neighbour 219
  object 3, 12, 18, 29, 31, 65, 282
  optimal viewing 1
distribution 45, 111, 112, 120, 185, 187, 204, 208, 222, 269-271, 324
  intensity 5, 110, 227, 228, 262, 265
  map 221-223
  particles 43, 45
  periodic 147, 280
  potential 247-250, 297-299
  radial brightness 91, 92, 110, 111
  radial density 268
  thickness 222
velocity 35

E

EDXS 53, 60, 185, 187, 193, 204, 210, 212, 221, 224, 225
  absorption correction 199, 319
  absorption correction by Horita 200, 201
  background 179, 187, 190, 191, 196, 197, 201, 225
  Cliff-Lorimer 195, 199, 313, 317

Index   339

detector efficiency 178, 190, 193, 194, 198, 307, 311-313, 317
escape peak 192, 193
line scan 202
peak intensities 196, 197
peak overlap 192, 194, 197
proportionality factors 195-197
quantification of spectra 196-200, 220, 318, 319
random errors 201
shading artifact 201
spatial resolution 203, 225
spectrometer 186, 187, 192-194, 197, 225, 313
spectrum 188, 201,
standardless quantification 197, 198,
sum peak 193
EELS 204-206, 208, 212, 215, 216, 221, 225, 327
background 207-214, 220-225
bonding analysis 216
characteristic edges 208
core-loss 206, 208, 209, 211
delayed edge 209, 210, 225
edge fine structure 208, 209, 216-219, 221
element analysis 209, 210
energy window 202, 206, 220-225
low-loss 206, 208, 210, 211, 213, 215, 218, 220, 222
natural line width 218
onset 208-210, 213, 216, 217
pre-peak 209, 210, 217
quantification of spectra 217, 220, 225
spectrometer 205, 206, 208-210, 215, 218, 220, 225, 327
spectrum 206-214, 218-223
thickness measurement 213-216
white line 209, 210, 225
zero-loss 206, 213, 214, 218, 222, 223
EFTEM 221-224
background 221-224
jump-ratio method 222
post-edge window 222, 223
pre-edge window 222, 223
slit 221, 223, 224
spatial resolution 224, 225
three-windows method 222
two-windows method 222
electron energy loss 17, 41, 55, 177, 182, 221
Electron Energy Loss Spectroscopy. *See* EELS
electron gas 184, 185, 206
electron gun 8, 20, 23, 24, 27, 28, 34, 35, 37, 38, 59, 61, 165, 218, 223, 224, 247, 250
alignment 63, 64
electron probe 26, 51, 71, 163, 165, 169, 170, 193, 202-204, 221,

aberrations 165, 299-305
broadening 203
elemental mapping 202, 204
elementary electric charge 7, 13, 116, 134, 139, 170, 179, 184, 224, 232-234, 245, 272, 277, 313, 321
ELNES. *See* Energy Loss Near Edge Fine Structure
emission 22, 165, 187
field 22, 23, 25, 35, 37
radiation 178, 181, 182, 185
thermal 20-22, 25
X-ray 177, 178, 182, 185, 187, 193, 194, 198, 201, 204, 220, 313, 315
energy
electron 17, 33, 70, 118, 145, 170, 179, 181, 182, 206, 224, 272, 276, 277, 279, 306, 307, 315,
gap 186, 187
kinetic 6, 179, 180, 183, 211
level 20, 21, 72, 73, 180-185, 187, 191, 206, 209-211, 219, 245
potential 180, 183, 231, 245, 246
primary electron 116, 118, 145, 167, 170, 203, 205, 215, 220, 224, 288, 307, 313, 315
spread 17, 22, 25, 167, 168, 206, 208, 218, 223, 224, 288, 304
states 181
width 22, 24, 145
energy band 183, 184, 186, 206
conduction 186, 187, 211
valence 186, 211
Energy Dispersive X-ray Spectroscopy. *See* EDXS
Energy Filtered Transmission Electron Microscopy. *See* EFTEM
Energy Loss Near Edge Fine Structure (ELNES) 216, 217
eucentric height 65, 66, 88, 225
eucentric focus 65
Euler, Leonhard 94
equation 94, 266
number 214, 289
evaporation 45, 46,
Ewald, Paul Peter 251
construction 99, 109, 249, 252, 264
sphere 108, 109, 252
Extended Energy Loss Fine Structure (EXELFS) 216, 219
excitation error 89, 123, 237
extinction distance 129, 130
extraction replica 45
extractor 24, 25, 247,
extractor hood 46
eye 8, 9, 12, 31, 65, 194

F
face-to-face preparation 50, 51
Fermi, Enrico 21
  energy 21, 245
FIB 51, 54, 55. *See also* focussed ion beam
  cutting out, welding 54, 55
  cutting out, placing on grid 53, 54
  H-bar 52, 53,
FIB lamella 52-55, 161, 202
field
  electric 6, 7, 22-25, 159, 160, 178, 184, 187, 189, 205, 228, 244-246, 307, 320, 322, 324
  magnetic 6, 13-15, 17, 36, 38, 60, 61, 135, 158-160, 164, 178, 228, 238, 242, 320, 323, 324
  magnetic, homogenous 14, 158
  magnetic, inhomogeneous 14
  rotational-symmetric 12, 16, 27, 134, 227, 234
  strength 7, 13, 17, 22-25, 242, 244, 322
fixed beam 163, 166, 170, 174
fluorescence 308
  yield 198, 313, 315, 316
focal length 3, 11, 12, 16, 18, 28, 29, 31, 62-69, 86, 87, 89, 127, 135, 142, 143, 145, 151, 155, 158, 163, 164, 167, 169, 170, 189, 224, 282, 284, 297, 299, 304
focal point 29, 31, 117, 164, 242, 282, 295
focal-series reconstruction 161
focussed ion beam (FIB) 51
focussing 22, 24, 56, 65-67, 69, 143, 145, 167, 169, 170, 206, 223, 249, 282,
forbidden reflections 93, 262
  rules for 93, 96, 97, 263
force 7, 13-15, 22, 72, 183, 228, 245, 249, 272, 298, 321, 322
  adhesive 53
  central 272, 274
  centrifugal 322
  Coulomb 41, 178, 179, 183, 244, 272
  driving 71
  electric 179,
  electrostatic 180, 322,
  field 111, 178, 183, 244, 272, 273
  interatomic 111, 131
  Lorentz 13-15, 17, 134, 158, 159, 234, 236, 241, 320, 321
  magnetic 61
  press-on 49
  radial 179, 321
  restoring 184
Fourier, Joseph 147
  analysis 147, 148
  coefficients 147, 327
  filtered 153

  integral 271
  space 150, 327
  transform 147-150, 153-157, 160, 327
Fresnel, Augustin Jean 228
  fringes 228
Friedrich, Walter 75

G
gallium (Ga) 52, 55,
  amorphisation 52
  implantation 52
  ion beam 52
gas 35, 36, 38, 52, 59, 189
  constant 34
  hydrocarbon 52, 54
  ideal 34
  injection 52, 54
  law 34
  mass 34
  molecules 35-37, 70
  noble 183
  particle density 34, 35, 70
  pressure 23, 34, 37
  residual 23
  volume 34
Gauss, Carl Friedrich 16
  curve 187, 197, 326
  function 63, 187, 197, 327
  image plane 16, 17, 142, 143, 145, 167, 282, 294, 300, 302, 304
  profile 63
generalised oscillator strength (GOS) 220
Germer, Lester 75
Glaser, Walter 239, 242
  bell-shaped curve 239, 240, 242,
gluon 179
gold (Au) 44, 91, 92, 94, 118, 157, 171-173, 191, 203, 263, 276, 279, 310, 312
  atoms 45, 94
  electrode 189, 309-312
  film 96, 171
  islands 91, 117-119
  layer 45, 310
  lattice constant 94
  -palladium 91
  particles 45, 69
    gun. *See also* electron gun
  lens 24, 247
  shift 65
  tilt 64

H
harmonic oscillation 185, 229, 230
H-bar 52, 53

Index 341

Heisenberg, Werner 218
  uncertainty principle 218
hexagonal system 84. *See also* crystal system, hexagonal
  four-fold indexing 84
hexapole 158
High Resolution Transmission Electron Microscopy (HRTEM) 55, 137, 138, 143, 147, 152, 161
(hkl) indices 83, 92, 93, 95, 99, 106, 107, 251, 263,
holography 162
HOLZ 106, 108, 109
HRTEM. *See* High Resolution Transmission Electron Microscopy
Huygens, Christiaan 2
  principle 2, 3, 227, 228
hydrocarbon 38, 52, 54, 59, 70, 71

I

illumination 6, 33, 62, 65, 71, 72, 224
  aperture 28, 29, 89, 106, 174
  coherent 170
  conditions 28
  convergent 28, 108, 142, 163, 170, 220
  parallel 28, 86, 88, 89, 106, 110, 142
  system 27, 37, 63, 163, 165
image 2-5, 8, 9, 14, 29, 30, 32, 34, 44, 57, 63, 65-69, 88, 91, 118, 119, 123, 131, 133, 140, 141, 143, 147, 149-151, 156, 157, 160, 174, 202, 204, 222, 286, 296, 298
  brightfield 120, 121, 123, 124, 128, 131, 163, 217, 221-223
  brigtness 28, 33, 117, 120, 141, 143, 161
  contrast 68, 69, 115, 116, 117, 132, 136, 144, 145, 170
  darkfield 120-123, 126, 127, 163, 173, 174
  defocussed 69, 70, 153
  distance 3, 11, 18, 29, 65, 282
  final 12, 27, 29, 32, 171, 221
  focussed 68, 69
  HRTEM 137, 138
  intermediate 12, 29, 86, 127, 158, 221
  interpretation 145, 151, 161, 287, 289
  overfocussed 136
  plane. *See* Gauss, image plane
  point 3, 15, 16, 242, 282, 302
  processing 91, 153
  ratio 12, 65
  record 31, 33, 37
  rotation 69, 127
  STEM 171, 212
  STEM-HAADF 174, 202, 212

  simulation 142, 143, 147
  sharp 2, 65, 69, 135, 170
  space 142, 164, 294
  underfocussed 136
imaging 3, 4, 11, 17, 25, 30, 52, 67, 88, 127, 128, 131, 137, 143, 144, 147, 150, 151, 161, 174, 220, 223, 224, 294, 295
  aberration-free 4, 15, 16, 149, 282
  brigthfield 120, 120
  carbon structures 70
  crystal lattices 137-139, 142
  darkfield 120-123, 160, 173
  energy filtered 221-223
  fixed beam 170, 174
  high resolution 44, 157, 169
  magnetic specimen 15
  multistage 11, 29, 57, 86
  optical 6, 15, 20, 51, 65, 142
  plane 16
  position 32
  STEM 165, 166, 171
  system 29, 31, 37, 57, 152
imaging energy filter 221
improvement of FIB lamellas 55
information limit 146, 147, 151, 157, 158, 289, 290
interaction 68, 70, 73, 109, 115, 127, 138, 141, 170, 182, 211, 261
  Coulomb 115, 177
  elastic 115, 177, 272, 273
  inelastic 17, 177, 179, 181, 184, 208, 214, 306
interference 3-5, 36, 69, 76, 128, 133, 140, 141, 143, 144, 150, 157, 161, 216, 227, 228, 288, 294
  constructive 4, 76, 77, 141, 144, 219
  destructive 76, 143, 144, 179, 219
International Tables of Crystallography 82
ionisation 36, 72, 198, 206, 208, 219, 313, 316
  cross section 198, 209, 219, 220, 313-315
  edge 216
  energy 186, 209, 313, 314
ion milling. *See* thinning, ion milling
irradiation 57, 59

J

JEMS by P. Stadelmann 142
jump-ratio method 222

K

Kα fraction 313, 316, 317
k-factors 198, 199, 313, 314, 316, 318. *See also* Cliff-Lorimer factors

Kikuchi, Seishi 90
  lines 90, 106
  pattern 106-108, 142
Knipping, Paul 75
Knoll, Max 6
Köhler, August 28
  illumination 28
Kramers, Hendrik Anthony 178
  equation 178

L
lamella 52-55, 161, 193, 202
Laplace, Pierre-Simon 231
  equation 236-238, 247, 248, 297
  operator 231
lattice. *See also* crystal, lattice
  distances 78, 130
  factor 98, 261, 262, 264, 265, 267
  plane 76-78, 83-86, 90, 92, 93, 95, 98, 99, 103, 106-108, 118, 121, 123, 124, 126, 127-133, 138, 149, 152, 249-251, 255, 259, 262, 263
  reciprocal 94-96, 99, 104, 106, 108, 110, 249, 251-253, 256-259, 261, 262, 264, 267
  rotated 280, 281
  spacing 90, 91, 93-96, 99, 104, 133, 144, 149, 152, 157, 249-251, 257, 279, 296
  vacancies 73
Laue, Max von 75, 76
  equations 249, 251
  zones 108
lens
  aberration 11, 15, 16
  condenser 11, 12, 26-29, 61, 62, 64, 86, 88, 89, 163, 164, 168, 169, 224
  electron 12-16, 27, 134, 158
  equation 3, 4, 11, 12, 29, 65, 284
  field 14, 15, 17, 18, 31, 57, 158, 165, 167, 189, 239, 243
  Lorentz 135
  magnetic 14-17, 19, 30, 51, 65, 68, 127, 235, 239, 242, 243
  objective 12, 15, 17, 29, 31, 39, 61, 65-69, 86-90, 117, 125, 127, 135, 142, 145, 149-153, 155, 164, 165, 189, 221, 282, 288, 294, 297
  power 17
  projective 12, 27, 29, 30, 61, 66, 86, 87, 127, 135, 165, 221
  saturated pole piece 242
life time 21
  of states 219
line scan 202, 203, 221, 225

lithium (Li) 188
long-range order 111, 152
Lorentz, Hendrik Antoon 13
  force. See force, Lorentz
  lens 135
  microscopy 133, 135, 136

M
magnetic induction 13-15, 17, 134, 236-241, 243, 320, 321
magnification 1, 2, 6, 8, 9, 12, 16-19, 26, 29, 31, 59, 64, 66, 69, 71, 87, 88, 91, 127, 137, 156, 282, 283, 288, 295-297, 299
  high 18, 38, 66, 115, 152, 296
  low 61, 64, 135, 189
  moderate 62
  STEM 169, 170
  total 11
  useful 8, 9, 11, 30, 57, 170
manganese (Mn) 187, 211, 212, 217
mass 219, 233, 235
  absorption coefficient 199, 200, 309-311, 318, 319
  atomic 195
  electron 115, 139, 179, 185, 272, 306, 321
  gas 34
  percent 195, 198, 313, 317
  proton 115
  relativistic 233
  rest 6, 7, 233, 313
  sample 73
mass thickness 220
  contrast. *See* contrast, mass thickness
material 21, 42-44, 47, 48, 52, 58, 78, 90, 170-173, 197, 199, 200, 203, 215, 309, 312, 318, 319
  amorphous 110, 137, 152, 219, 268
  bulk 48, 319
  crystalline 31, 183
  ductile 48
  lens 4
  luminescent 32
  magnetic 15, 18
  phase 78, 122
  photo 64
  pole piece 18, 242
  polycrystalline 60
  sputtered 50
  window 308, 317
materials science 1, 42, 47, 57, 58, 75, 78, 90
Maxwell, James Clerk 35
  equations 178, 184, 236
  velocity distribution 35

Index 343

mean free path 34
  elastic scattering 116, 170, 172, 277-279
  inelastic scattering 214-216, 218
microscope
  electron 6, 8, 11, 12, 18, 23, 24, 27, 30, 32-39, 43, 57, 59, 61, 64, 65, 66, 70, 127, 135, 150, 151, 153, 158, 187, 189, 218, 221, 234
  light optical 3, 5, 6, 8
  scanning electron 11, 46, 51, 52, 71, 163, 190
  scanning ion 51, 52
  scanning transmission electron 163. *See also* STEM
  transmission electron. *See* transmission electron microscope
Miller, William Hallowes 83
  indices 83, 98,
Miller's indices. *See also* (hkl) indices
model
  dynamical 109, 110
  geometric 130, 261, 294, 297
  kinematical 98, 109, 261
Moiré 132
  distances 279
  fringes 133, 134
  pattern 132, 133, 279
  stripes 132, 279-282
molecular weight 172
momentum 6, 139, 219, 231, 272, 273, 306, 307
  angular 180, 273
  balance 306
  conservation 273
  orbital angular 180, 184
  quantum number 181
monochromator 218
Monte-Carlo simulation 171, 203
multilayer 1, 203, 204
multipole 156, 158, 168, 205, 297, 298

N
nanoprobe 164, 165
nanostructures 1, 2, 6, 89, 169
NBED. *See* diffraction, nano beam electron
neutron 179
nickel (Ni) 44, 75, 122, 119, 123, 207, 209-213, 218,
NIST 310
nitrogen (N) 35, 36, 43, 73, 217, 308
  gas 59
  liquid 38, 42, 59, 60, 187, 189, 190
non-relativistic approximation 139, 180, 219, 235, 272, 306, 313, 322. See also classical approximation

O
octupole 158, 160, 297, 299, 300
offset 208
onset 208-210, 213, 216, 217
optical axis 3, 4, 13-16, 24, 29-31, 65-68, 86, 88-90, 117, 120, 122, 143, 152, 155, 161, 164, 168, 238, 282, 294, 295, 299, 302
optics
  electron 6, 11, 24, 163, 218
  geometric 3, 14
optimal viewing distance 1
overvoltage ratio 314, 315
oxygen (O) 35, 36, 43, 59, 60, 73, 183, 191, 192, 209, 210, 221-223, 225
  K-edge 209, 210
  K-peak 192

P
path
  difference 4, 5, 76, 77, 143, 219, 261, 284, 285,
  electron beam 234, 242, 243
  geometric 3, 4
  optical 3, 4, 142
Pauli, Wolfgang 20
  principle 20, 180, 181
Pearson, Frederic Treadwell 79, 80
  symbol 80
Peltier, Jean 33, 188, 190
Penning, Frans Michel 37
permittivity of vacuum 116, 170, 185, 245, 272, 277
phase
  difference 141, 142, 285,
  modulation 140-142, 286, 287
  shift 3, 76, 115, 131, 139- 144, 229, 261, 282, 284-286, 288
phase contrast transfer function 145, 287, 291. *See also* contrast transfer function
phonon 185, 208
photoelectric effect 186
Pirani, Marcello 37
pivot point 66, 67, 121, 164
pixel 9, 33, 34, 92, 94, 95, 96, 149, 169-172, 206, 212, 221-223
  brightness 91, 149
  size 9, 33, 34, 51, 92, 96, 206
Planck, Max 6
  constant 6, 139, 178, 180, 185, 219, 231, 233
plane
  angle-selective 29, 30, 117, 221
  back-focal 29, 30, 86, 117, 127, 149, 150, 165, 171, 221

plane (*continued*)
   energy-selective 206, 208, 221, 224
   final image 27, 29, 32, 171, 221
   intermediate image 29, 127, 158
   least confusion 167, 300-304
   object 11, 16, 17, 19, 26, 29, 31, 66, 67, 135, 157, 161, 169, 296
   position-selective 221
   specimen 31, 163
plan-view 48, 50
plasmacleaner 59, 60, 71
plasmon 185, 206, 210
platinum (Pt) 52
point diagram 89, 106
pole piece 13, 18, 31, 32, 61, 164, 234, 239, 242, 243
   bore 18, 242
   gap 14, 189, 235, 242, 243
post-edge 222
post-field (objective lens) 164, 165, 171
potential 20-24, 72, 138-140, 180, 183, 237, 245-247, 298, 299
   crystal 109, 139, 141, 183, 231, 232
   distribution 247-249, 297-299
   electric 116, 240, 249
   gun 248
   magnetic 236, 238-240, 242, 243
   multipoles 299, 300
   projected 143
   wall 20, 22, 245, 246
power spectrum 149, 155, 156, 160, 170
power supply 18
pre-egde 222
pre-field (objective lens) 164, 165, 169, 170,
prevacuum 36
prism for electrons 205, 206, 221, 227, 320, 323
   electrostatic 321-323
   magnetic 320, 321
product 3, 6, 87, 88, 98, 233, 253, 262, 312, 321, 324, 325
   cross (vector) 13, 14, 126, 252, 259, 260
   scalar 252, 253, 257, 259, 264, 269
pulse processor 189, 190, 193
pump. *See* vacuum pump

Q

quadrupole 158, 160, 297, 298, 299
quantum number 20, 180, 181, 183
   magnetic 180
   orbital 180, 313
   principal 180, 314
   spin 180
   total momentum 181

R

radial brightness distribution 91, 92, 110
radial density function 112
radiation 178, 181, 193, 194
   damage 73
   electromagnetic 32, 178, 179
   emission 182, 185
   heat 73
   ionising 38
   low-energetic 225
   X-ray 177, 178, 182, 185-187, 190, 193, 194, 198-201, 204, 210, 225, 308, 309, 311, 316, 318
Rayleigh, Lord John 5
   criterion 5
reciprocal 92, 96, 198
   distance 96, 104
   lattice 94-96, 99, 104, 106, 108, 110, 249, 251-253, 256-259, 261, 262, 264, 267
   lattice point 251, 252
   lattice spacing 94-96, 104
   lattice vector 106, 251, 256, 257, 259, 262
   length 96, 144, 149
   ratio 173, 219
   space 150, 267
   value 83, 145, 146, 157, 251, 252, 280, 290
recurrence formula 248, 249, 298
refractive index 3-5, 19, 243
relativistic 7, 8, 233, 234
   correction factor 215, 234
resolution 2, 5, 34, 70, 157, 170, 177, 243, 297
   atomic 151, 169
   energy 187, 190, 192, 194, 197, 206, 208, 210, 218, 223, 225, 313, 316, 324
   high 18, 23, 44, 55, 60, 137, 138, 142, 143, 151-153, 161, 162, 169, 174
   high, interpretation 161, 162
   human eye 8, 31
   lateral 33, 55
   light optical microscope 3, 5, 8, 43, 54,
   limit 3, 5, 6, 8, 9, 19, 20, 57, 69, 115, 135, 137, 145, 147, 156, 157, 163, 166, 243, 287, 289, 290, 324, 327
   measurement 156, 157
   point 145, 146, 290
   spatial 163, 203, 224, 225
   STEM 163, 165, 170, 305
Richardson, Owen Williams 21
   constant 21
   equation 21, 25
richtstrahlwert 24-26, 28, 165, 166, 224, 288, 304

Index 345

right-hand rule 13
ring diagram 88-90, 92, 98, 110
Ronchi, Vacco 170
Ronchigram 170
Röntgen, Wilhelm Conrad 75
Rose criterion 194
rotation centre 67- 69
Ruska, Ernst 6, 24, 43, 70
Ruska, Helmut 43
Rutherford, Ernest 115
   -Bohr's model 177, 179
   differential cross section 115
   scattering formula 276

S
SAED. *See* diffraction, selected area electron
sample 8, 11, 27, 28, 38, 41-55, 58, 61, 66, 70, 71, 88, 108, 115-120, 123-125, 127, 128, 138, 142, 161, 163, 164, 172, 193, 198-200, 203, 204, 213, 222, 224, 318, 320. *See also* specimen
   amorphous 48, 60, 110-112, 152, 153, 170
   bent 124
   composition 200
   crystalline 45, 84, 86, 91, 120, 149, 152
   damaging 70, 72, 73
   heat-sensitive 73
   holder. *See* specimen holder
   magnetic 61, 134-136
   plane 67, 121, 122, 165. *See also* plane, object
   preparation 41-55
   requirements 210
   stage 65, 119
   thickness 41, 46, 109, 116, 118, 129, 132, 190, 191, 200, 201, 261, 318
saddle point 298, 299
scanning transmission electron microscope or microscopy. *See* STEM
scattering 76, 109, 110, 115, 116, 118, 164, 171, 172, 219, 261, 262, 268-270, 274
   angle 110, 112, 116, 170, 173, 220, 306
   capability 93, 116, 129, 130, 131, 138, 170, 171
   cross section 93, 115, 116, 268, 275, 276
   curve 110-112, 152, 268, 270, 271
   Debye 268
   elastic 115, 116, 170, 172, 177, 203, 206, 225, 250, 276-279
   factor 94, 98, 99, 112, 268, 271
   inelastic 177, 214-216, 218, 220, 225, 307
   multiple inelastic 210, 211, 213, 218
   vector 91, 98, 112, 250, 251, 261, 262, 264, 268

Scherzer, Otto 16, 297
   focus 145, 146, 151, 153-155, 290, 293, 297
Schottky, Walter 22
   effect 22, 25, 244
   field emission cathode 22, 25, 37, 166, 218, 224, 288, 289
   reduction of work function 246
Schrödinger, Erwin 109
   equation 109, 138, 183, 217, 220, 231, 232,
SDD. *See* silicon drift detector
secondary electron 51, 183, 211
secondary fluorescence 308
segregation 45, 127
semicoherent particles 123
semiconductor 186-188, 190, 307
   n-type 307
   p-type 307
SF_6. *See* sulphur hexafluoride
sharpness 69, 98, 169, 170
shell
   core 181, 187, 198, 206, 208, 219, 315
   electron 72, 116, 177, 179, 182, 183, 186, 204, 272, 276, 306
   K 180-183, 192, 193, 308, 314, 316
   L 180-183, 187, 209, 314, 316
   M 180, 183, 187, 314, 316
shielding 38, 116, 272, 276
short-range order 48, 60, 111, 152
short-wave limit 178
silicon (Si) 33, 43, 47, 48, 99, 103-105, 107, 124, 125, 135, 137, 138, 153, 186, 188, 192-194, 198-201, 307, 311, 316,
   as reference 314,
silicon drift detector (SDD) 190
slit 4, 5, 43, 221, 223, 224
Sommerfeld, Arnold 75
soundproof housing 38
space frequency 144-149, 151-153, 157, 280, 281, 287-290, 292, 293, 296, 297
space group 79- 82, 92, 97, 137
specimen 11, 12, 15, 17, 20, 27-31, 34, 38, 41-44, 47-51, 53, 54, 56-61, 66- 68, 70-73, 75, 87-89, 91, 110, 115, 118, 119, 123, 127, 132, 135, 142, 150, 155-157, 161, 163, 164, 169-171, 182, 183, 190, 193, 197, 199-203, 206, 221, 225, 242, 287, 318, 320. *See also* sample
   chamber 38, 39
   density 199
   drift 71, 158, 169, 222
   geometry 47
   magnetic 15, 134, 242

plane 31, 163. *See also* plane, object
position 211
preparation 41, 43, 47, 56, 142
stage 27, 30-37, 65, 66
thickness 41, 53, 89, 138, 140, 141, 158, 161, 172, 174, 178, 192, 199, 200, 203, 204, 211-214, 216, 218, 222, 224, 225
specimen holder 31, 58, 59, 201, 320
    cooling 60, 73
    double-tilt 60, 142
    heating 45
    low-background 60, 193
    side-entry 31, 32, 59, 60, 119
    single-tilt 60, 242
    special 60
    top-entry 32
spectrometer 177, 190, 194, 197, 205, 206, 208, 209, 218, 225
    acceptance angle 215, 220
    dispersion 205, 206, 218
    EDXS 186, 187, 192-194, 210, 313
    EELS 205, 210, 327
    efficiency 192, 197
    energy resolution 187, 197, 206, 210, 218
    software 190, 194,197, 225
spin 180,
stability 17, 27, 61, 169
    electric 17, 18, 39, 160, 218
    mechanical 27, 31, 32, 36, 38, 57, 161
    thermal 31, 39
stacking faults 128, 131, 132
STEM 163, 164, 165, 170, 174, 204, 212, 213, 216, 220, 221, 225
    brightfield image 163, 171
    contrast 170, 171, 173
    darkfield image 163, 174
    electron-optically 163-165
    focussing 169
    resolution 165, 170
STEM detector 165,
    brigthfield 171
    darkfield 171, 173
    HAADF 173, 174
stigmator 57, 156, 297, 298
    condenser 63, 65
    objective 155
structure factor 93, 94, 98, 130, 261-263
sulphur hexafluoride (SF_6) 59
super cell 161
superconductor 42
support grid 43-47, 54, 193, 201, 202, 308
    bringing thin films on 44
    gluing on 44
    heat-resistant 45
suppressor 24, 247

T

take-off angle 198-200, 320
tantalum (Ta) 50, 202, 203, 221-223
texture 89, 98, 99, 120, 121
thermal expansion 31, 72
thickness contour 128-131
thin film criterion 320
thinning 50, 54, 55
    dimpling 49-51
    electrolytic 46, 47, 58,
    FIB. *See* FIB lamella
    ion milling 47, 49, 50, 51, 58, 124
titanium (Ti) 36, 37, 217, 225
trajectory 158-160, 272, 274, 323
transformation matrix 257
transition 110, 136, 143, 187, 204, 206, 208, 217
    between energy levels 181, 211, 219
    inter-band 211
    Kα 187, 191
    radiative 181, 182, 185, 191, 315
transmission electron microscope or microscopy (TEM) 8, 9, 11, 20, 27, 41-44, 51-54, 56- 61, 65, 66, 69-72, 75, 78, 84, 87, 88, 110, 115, 119, 123, 124, 126-128, 133, 137, 138, 140, 147, 149, 152, 156, 161-164, 164, 166, 170, 171, 177, 186, 190, 193, 194, 197, 200, 202, 205, 206, 215, 220, 221, 224, 227, 242, 243, 249, 250, 252, 259, 268, 276, 297, 305, 318, 324
transparency
    dead layer 311
    electron 44, 52
    gold electrode 309, 310
    window 308
trial and error 327
tripod 49
tungsten (W) 21-23, 25, 48, 52, 166
tunnelling 22, 25, 32, 60
twins 128, 131, 132

U

ultramicrotome 47, 58
ultrasonic
    bath 44
    core hole drill 48
unit cell 78, 79, 83, 84, 93, 94, 98, 99, 104, 137, 143, 174, 262-265, 267, 268
    body-centred 97
    diamond lattice 97, 137, 138
    face-centred 81, 93, 95
    hexagonal close-packing 97
    packing 79, 92

primitive 92, 93, 97
volume 80, 81, 98, 130, 253, 256, 265

V

vacuum 19, 32, 36, 38, 44, 60, 61, 69, 70, 189, 211
   energy level 21, 22, 245
   leakproffness 32
   system 27, 34, 37, 38
   tweezers 58
   ultrahigh 23, 35
vacuum gauge 37
   Penning 37
   Pirani 37
vacuum pump 35-38, 59, 60, 189
   cryo 71
   current 36
   diaphragm 35, 58
   getter 36, 37
   oil diffusion 36
   prevacuum 36
   roots 35
   turbomolecular 35, 38, 59
vector 13, 14, 79, 233, 235, 249, 251-253, 260, 262, 264, 265, 272, 281, 320
   Burgers. *See* Burgers, vector
   product 13, 14, 126, 252, 259, 260
   reciprocal lattice 106, 251, 256, 257, 259, 262
   scattering 91, 98, 99, 112, 250, 251, 261, 262, 264, 268
   wave number 77, 112, 147, 250, 252, 268
velocity 6, 13-15, 17, 179, 219, 229, 233, 235, 272, 320-322
   distribution 35
   electron 17, 25, 115, 134, 158-160, 178, 234, 272, 313, 320
   light 7, 178, 233
   phase 229
viewing distance 1
voltage 23-25, 36, 39, 46, 218, 314, 315, 322
   accelerating 7, 8, 17, 25, 33, 39, 69, 88, 92, 118, 139, 140, 204, 206, 224, 233, 244, 292, 313
   centre 69
   proof 34

W

wave 2-4, 6, 15, 22, 69, 70, 75-77, 87, 90, 93, 109, 131, 138, 141-145, 147, 150, 152, 178-180, 227, 228, 231, 229, 248, 250, 268, 270, 271, 285, 288, 294, 320
   amplitude 140, 261

diffracted 4, 109, 128, 129, 131, 141-144, 228, 261, 282, 285
electromagnetic 178
electron 6, 19, 76, 86, 98, 109, 111, 115, 128-131, 133, 138, 141, 216, 219, 259, 261, 268, 294
elementary 3, 4, 227
exit 140, 141, 161, 286
front 3, 227, 284
function for electrons 180, 228, 229- 233, 269, 286
initial 4, 86, 109, 112, 129, 131, 133, 219, 261, 268, 285
light 2, 3, 4, 115
-optical 145, 146, 290, 296
phase 78, 141, 230
plane 140, 141, 228, 285, 286
propagation 3, 76, 109, 227
reference 161, 162
wavelength 4- 6, 17, 32, 76, 77, 78, 87, 92, 141, 160, 178, 182, 186, 219, 228, 229, 250, 261, 268, 287, 288, 320
   electron 6-8, 17, 23, 69, 78, 93, 109, 130, 138, 139, 143, 145, 152, 158, 159, 167, 177, 179, 232-234, 252, 284, 296, 304
   light 5, 32
   X-ray 186
WDXS 186
Wehnelt, Arthur 23
   electrode 23, 24, 247
Weiss, Pierre-Ernest 134
   domains 134
white line 209, 210, 225
Wien, Wilhelm 159
   filter 159, 160, 218
window 43, 53, 154, 190, 192, 213, 308, 317
   detector 192, 198
   energy 202, 206, 211-213, 220-225
   lead glass 32, 38
   polymer 189, 190, 192, 308
   transparency 308
wobbling 66-68
work 6, 21
   function 21, 22, 25, 245, 246
working distance 59, 163, 164
Wyckoff, Ralph Walter Graystone 80
   symbol 80, 83,

X

X-radiation 177, 178, 182, 185-187, 190, 193, 194, 198-200, 204, 210, 308, 309, 311, 318
   characteristic 182

X-ray 34, 38, 75, 78, 110, 178, 182, 187, 189-193, 199-202, 220, 225, 308, 309, 318, 319
  absorption 190, 312
  characteristic energies 177
  detector 59, 225, 227, 307
  emission 178, 182, 194, 198, 201, 313
  energy 178, 185-187, 190-193, 197, 198, 308-312, 317, 319
  peak 187, 190, 191, 195, 197, 201, 202, 225
  quant 34, 182, 183, 186, 187, 189, 190, 192-194, 198, 307, 308, 313, 315
  spectrum 178, 186, 187, 188, 190-194, 196, 197, 209, 210, 225, 324
  spectrometer 186, 192, 210
  spectroscopy 53, 60, 185, 187, 192, 194, 221, 225

Y
YAG 33, 34
Young, Thomas 157
  fringes 157, 158
  interference pattern 157

Z
Zeiss, Carl 3
Zemlin tableau 160
zone axis 103, 105-109, 142, 259, 260
  low-indexed 108, 142

The manufacturer's authorised representative in the EU is Springer Nature Customer Service Centre GmbH, Europaplatz 3, 69115 Heidelberg, Germany. If you have any concerns regarding our products, please contact ProductSafety@springernature.com

Printed and bound by CPI Group (UK) Ltd, Croydon, CR0 4YY
25/03/2026
02078169-0007